T0228059

Symbolic and Numeric Computation Series

Edited by Richard Fateman

Dean Duffy, *Transform Methods for Solving Partial Differential Equations*

Victor G. Ganzha and Evgenii V. Vorozhtsov, *Numerical Solutions for Partial Differential Equations: Problem Solving Using Mathematica*

Prem K. Kythe, *Introduction to Boundary Element Methods*

H. T. Lau, *A Numerical Library in C for Scientists and Engineers*

Patrick J. Roache, *Elliptic Marching Methods and Domain Decomposition*

William E. Schiesser, *Computational Mathematics in Engineering and Applied Science*

Mikhail Shashkov, *Conservative Finite-Difference Methods on General Grids*

Conservative Finite-Difference Methods on General Grids

Mikhail Shashkov

Edited by
Stanly Steinberg

CRC Press
Taylor & Francis Group
Boca Raton London New York

CRC Press is an imprint of the
Taylor & Francis Group, an **informa** business

The software mentioned in this book is now available for download on our Web site at: http://www.crcpress.com/e_products/downloads/default.asp

CRC Press
Taylor & Francis Group
6000 Broken Sound Parkway NW, Suite 300
Boca Raton, FL 33487-2742

© 1996 by Taylor & Francis Group, LLC
CRC Press is an imprint of Taylor & Francis Group, an Informa business

First issued in paperback 2019

No claim to original U.S. Government works

ISBN-13: 978-0-367-44874-5 (pbk)
ISBN-13: 978-0-8493-7375-6 (hbk)

This book contains information obtained from authentic and highly regarded sources. Reasonable efforts have been made to publish reliable data and information, but the author and publisher cannot assume responsibility for the validity of all materials or the consequences of their use. The authors and publishers have attempted to trace the copyright holders of all material reproduced in this publication and apologize to copyright holders if permission to publish in this form has not been obtained. If any copyright material has not been acknowledged please write and let us know so we may rectify in any future reprint.

Except as permitted under U.S. Copyright Law, no part of this book may be reprinted, reproduced, transmitted, or utilized in any form by any electronic, mechanical, or other means, now known or hereafter invented, including photocopying, microfilming, and recording, or in any information storage or retrieval system, without written permission from the publishers.

For permission to photocopy or use material electronically from this work, please access www. copyright.com (http://www.copyright.com/) or contact the Copyright Clearance Center, Inc. (CCC), 222 Rosewood Drive, Danvers, MA 01923, 978-750-8400. CCC is a not-for-profit organiza-tion that provides licenses and registration for a variety of users. For organizations that have been granted a photocopy license by the CCC, a separate system of payment has been arranged.

Trademark Notice: Product or corporate names may be trademarks or registered trademarks, and are used only for identification and explanation without intent to infringe.

Visit the Taylor & Francis Web site at
http://www.taylorandfrancis.com

and the CRC Press Web site at
http://www.crcpress.com

Library of Congress Cataloging-in-Publication Data

Catalog record is available from the Library of Congress

PREFACE

The desire to solve new challenging problems imposes additional requirements for the quality and robustness of numerical algorithms used to simulate such problems. A number of factors in real problems make designing algorithms difficult, including non-linearity, discontinuity of coefficients or solutions, complexity of the physical processes, and irregular geometry. Experience has shown that the best results usually are obtained by using discrete models that reproduce fundamental properties of the original continuum model of the underlying physical problem. Some of these important properties are conservation, symmetries of the solution, etc.

Modeling becomes increasing difficult as the number of physical processes increases, the shape of the physical domain becomes more complex, and boundary conditions are made more realistic. The development of the discrete algorithms that capture all the important characteristics of such physical problems becomes more difficult with this increasing complexity. Thus, it is important to have a discretisation method that is sufficiently general that it applies to a wide range of physical systems.

Many important simulation algorithms necessarily involve the numerical solution of partial differential equations (PDEs). Most of these PDEs can be formulated using invariant first-order differential operators, such as the divergence of vectors and tensors, the gradient of scalars and vectors, and the curl of vectors. Therefore, it is possible to construct "good" finite-difference schemes (FDSs) for approximating a wide class of problems, if it is possible to construct discrete analogs of these invariant operators so that the discrete operators also have "good" properties that are analogous to those of the continuum operators.

First-order differential operators are the main objects in the vector and tensor analysis. These operators satisfy certain integral identities, usually associated with the names of Green, Gauss, and Stokes, that are closely related to the conservation laws of the continuum models. Therefore, to obtain high-quality FDSs, it is important to construct discrete analogs of the continuum divergence, gradient, and curl, which will satisfy discrete analogs of these integral identities.

In this book, finite-difference schemes are considered to be a discrete model of the continuum physical system. It is emphasized that, when computing numerical solutions to partial differential equations (PDEs), the discrete difference operators should mimic the crucial properties of the continuum differential operators, and that such discretisations are usually more accurate than those that do not mimic these properties. Properties such as symmetry, conservation, stability, and the integral identities between the gradient, curl, and divergence are all important. The *support-operators* method for constructing finite-difference schemes that mimic these properties is emphasized in this book. The terminology "support operators" comes from the translation of papers in Russian. Unfortunately, this is not a good translation. More appropriate would be fundamental, basic, or reference operators. As many papers have been translated from Russian using this terminology, we have elected to not change.

Many of the standard finite difference methods are special cases of the support-operators method. Also, it is important to realize that the popular finite-volume methods are special cases of the support-operators method. In this case the finite-volume approach gives an approximation of the divergence. Unlike elementary finite difference methods, the support-operators method can be used to construct finite-difference schemes on grids of arbitrary structure. In addition, because invariant operators are used, the method can be easily used in any coordinate system. This gives some idea of the universality of the support-operators method.

This book deals with the construction of finite-difference (FD) algorithms, in Cartesian coordinates, for three main types of equations: elliptic or steady-state equations, parabolic or diffusion equations, and gas dynamics equations in Lagrangian form. The methods apply to domains of arbitrary shape. Therefore this book gives scientists and engineers tools for solving real practical problems. The presentation of the material in this book is constructive; all of the details of algorithms needed for its implementation are given. The theoretical material is covered briefly and only when this contributes to a better understanding of the applications and properties of the algorithms. Many illustrations are used to help the reader understand the material. In addition, the results of solving many test problems are also presented. To increase the utility of the book, computer codes implementing the basic algorithms are provided on the floppy disk that is included with this book. This book is unique because it is the first book not in Russian to present the support-operators ideas.

The background needed to understand this book is matrix theory, vector calculus, the elementary theory of partial differential equations, introductory finite-difference techniques, and elementary numerical programming techniques.

Chapter Summaries

In the *Introduction*, we consider statements of problems for partial differential equations used in this book, and the main notions and ideas of theory of finite-difference schemes. The main notations related to grids, grid functions, and finite-difference operators are also described. Uniform, tensor-product, and generally logically rectangular grids are introduced, and notions of smooth and non-smooth grids in 1-D and 2-D are also given. Nodal and cell-valued types of discretisation are considered. The notion of finite-difference operators is introduced. Then we consider simplest finite-difference operators which correspond to derivatives and introduce the notion of truncation error. We emphasize that the value of truncation error strongly depends on the definition of the projection operators, which project functions from continuous space to discrete space, and on the norms chosen in the discrete spaces. We also consider two different approaches to the investigation of truncation error, which are based on assumptions about properties of the grid. Next we consider the notion of FDS for partial differential equations, and the notions of accuracy and stability. The correspondence between FDS and related matrix problems is examined. Some questions of implementation of finite-difference schemes are considered too, e.g., linear and non-linear systems of algebraic equations related to FDS, and methods for their solution.

The general description of the support-operators method is given in *Chapter 2*. The main stages for constructing finite-difference schemes by this method are considered for examples of elliptic equations and the equations of gas dynamics. Attention is focused on determining the connection between conservation laws and properties of first-order operators. This analysis gives an idea of what properties of differential operators have to be mimicked to obtain conservative FDSs. To define system of discrete operators we introduce spaces of discrete scalar and vector functions and inner products in these spaces. For 1-D and 2-D cases, systems of consistent finite-difference analogs of the operators div and grad are constructed. These finite-difference operators are used in the next chapters to construct finite-difference schemes for PDEs.

The construction of finite-difference schemes for general *elliptic* equations with non-diagonal matrices of coefficients and Dirichlet and Robin boundary conditions is considered in *Chapter 3*. In the introduction, we considered properties of the continuum elliptic problem, and analyzed the relationship between the properties of first-order operators and properties of the boundary value problem. These considerations, in continuous case, illuminate the properties of the first order operators that must be preserved when the finite-difference schemes are constructed. It is important to note here that the case of Robin boundary conditions handled naturally in a framework of the support-operators method by appropriate definition of operators on the boundary and the use of inner products. The cases of nodal and cell-centered discretisation of unknown functions in 1-D and 2-D are considered, and for both cases truncation error and accuracy of the finite-difference scheme are investigated. The results of solving test problems for the case of general matrix of coefficients in 1-D and 2-D are presented. By numerical experiment, we demonstrate the dependence of the error on quality of the grid. Results of computations on very distorted grids confirm robustness of our algorithms. Discrete operators constructed in this chapter are used in the next two chapters for approximation of the heat equation and equations of gas dynamics in Lagrangian form.

The construction of finite-difference schemes for the *heat equation* are considered in *Chapter 4*. Analysis of conservation law for the heat equation and other desired properties of FDSs shows that the finite-difference analogs of operators div and grad constructed in the previous chapter for elliptic equations can be used for approximating spatial differential operators in the heat equation. Therefore the main subject of this chapter is: what new features are brought in by the presence of the time derivative? Next, we consider ideas related to discretisation in time. The notion of explicit and implicit finite-difference schemes are introduced and stability of finite-difference schemes and the related restriction on the time step are described. Unstable behavior of discrete solution is demonstrated on numerical examples. Some methods of investigating stability are briefly described. The cases of nodal and cell-centered discretisation of unknown functions in 1-D and 2-D are considered. The results of solving test problems are presented. We also consider the problem of violation of conservation laws when solving systems of linear equations by iterative methods, and a new approach which preserves the conservation during the iteration. The problem of symmetry preserving finite-difference schemes and properties of iteration methods is also considered.

In *Chapter 5* questions related to the construction of finite-difference schemes for the *equation of gas dynamics in Lagrangian form* are presented. In the first part of this chapter some physical and mathematical background for gas dynamics are given. This material is presented here to make this chapter self-consistent, so the reader will not need to refer to other books while reading this chapter. The background material includes the notion of the mathematical model of gas dynamics (approach of continuum media, the main features of continuum media, and so on). Both the integral and differential forms of Lagrangian gas dynamics equations are presented. We consider the acoustic equations as a simplest linear model of gas dynamics equations, and introduce such notions as characteristics for this example. For the example of 1-D equations, we describe the characteristic form of gas dynamics equations, the notions of Riemann's invariants, domain of dependence, and range of influence. A special section deals with the description of discontinuous solutions of gas dynamics equations and Hugoniot relations on the discontinuity surface. We also present short descriptions of shocks in ideal gas, structure of shock in viscous media, equations of state, and boundary and initial conditions.

In a special section we analyze the connection between conservation laws for the gas dynamics equations and properties of first order differential operators div and grad. It is shown that the natural approximation of div has to be consistent with the formula for change of elementary volume with time. The detailed description of the application of the support-operators method is presented for 1-D and 2-D. A method of introducing artificial viscosity needed for solving problems with shocks is considered. Explicit and implicit schemes are considered. The methods of Newton and parallel chords for solving the non-linear algebraic equations related to implicit FDS are given. Results of the solutions of test problems are presented.

In *Appendix A* we describe the contents of the floppy diskette attached to the book. The programs for elliptic equations include programs for 1-D and 2-D cases for Dirichlet and Robin boundary conditions. The programs for heat equations include programs for explicit and implicit FDSs for 1-D and 2-D cases for Dirichlet and Robin boundary conditions. The programs for gas dynamics equations include explicit and implicit FDSs for test problems in 1-D and 2-D that are described in Chapter 5.

In the Bibliography section we collected numerous references which can be useful for readers who are interested in more advanced applications and in examples of using algorithms described in this book for solving real physical problems. This section can also be useful for readers because many references to papers by Russian authors, which are translated to English, are given.

ACKNOWLEDGMENTS

The material in the book arose from the author's original research on constructing finite-difference methods of the support-operators type.

The author belongs to the scientific school of Academician A.A. Samarskii and the plan of this book follows the ideas developed by this school at the Keldysh Institute of Applied Mathematics and Institute of Mathematical Modeling of Russian Academy of Sciences, which is directed by Academician A.A. Samarskii. The main concept of this book is close to the books of Academician A.A. Samarskii, e.g. [108, 109, 110, 111, 107]. In particular I am using very useful interpretation of FDSs as operator equations in abstract linear Euclidean spaces. Thanks to my teachers Academician A.A. Samarskii and Prof. A.P. Favorskii and to Prof. V.F. Tishkin, who was co-author of many papers related to method of support-operators.

Thanks to Dr. B. Swartz, Prof. S. Steinberg, Dr. J.M. Hyman, Dr. L. Margolin, Prof. J. Brackbill, Dr. J. Dukowicz, Dr. B. Wendroff, and Dr. J. Castillo for many helpful discussions of the material in this book.

ABOUT THE AUTHOR

Mikhail Ju. Shashkov is a staff member of Theoretical Division at Los Alamos National Laboratory. He received his degree of Candidate of Science (Ph.D.) in computational mathematics from Keldysh Institute of Applied Mathematics in Moscow, Russia, in 1979, and his degree of Doctor of Science in computational mathematics from Moscow State University, Russia, in 1990. From 1979 through 1994 he worked as research mathematician in Keldysh Institute of Applied Mathematics and Institute of Mathematical Modeling in Moscow. From 1985 through 1991 he lectured in Moscow State Pedagogical Institute in information science and computational mathematics.

He has published more than 100 papers (more than 30 of them in English) in the numerical solution of partial differential equations, computational fluid dynamics, and symbolic computations.

His recent work has focused on theory of finite-difference schemes for diffusion equation with rough coefficients, construction of high-order mimetic finite-difference schemes and developing a general discrete vector and tensor analysis.

Contents

1 **Introduction**   1

   1   Governing Equations 1

      1.1   Elliptic Equations 1

      1.2   Heat Equations 3

      1.3   Equations of Gas Dynamics in Lagrangian Form . . 4

   2   The Main Ideas of Finite-Difference Algorithms 5

      2.1   1-D Case 6

         2.1.1   Grid Functions 7

         2.1.2   Smooth and Non-Smooth Grids 8

         2.1.3   Different Types of Grid Functions 9

         2.1.4   Finite-Difference Operators,
                Truncation Error 13

         2.1.5   Two Approaches for the Investigation
                of Truncation Errors 16

         2.1.6   Approximation of Second-Order
                Derivatives 18

         2.1.7   Finite-Difference Schemes for the
                1-D Poison Equation 19

         2.1.8   Finite-Difference Operators and the
                Related Matrix 23

         2.1.9   Finite-Difference Schemes and Related
                Matrix Problems 24

      2.2   2-D case 26

         2.2.1   Grid in 2-D 26

         2.2.2   Grid Functions and Difference
                Operators in 2-D 32

      2.3   Methods for the Solution of the System of
         Linear Algebraic Equations 37

         2.3.1   Application of the Gauss-Seidel Method
                for a Model Problem 40

      2.4   Methods for the Solution of the System of
         Non-Linear Equations 41

2 Method of Support-Operators   47

   1   Main Stages 48

   2   The Elliptic Equations 49

   3   Gas Dynamic Equations 51

   4   System of Consistent Difference Operators in 1-D 53

       4.1   Inner Product in Spaces of Discrete Functions
           and Properties of Difference Operators 56

   5   System of Consistent Discrete Operators in 2-D 57

3 The Elliptic Equations   65

   1   Introduction 65

       1.1   Continuum Elliptic Problems. Dirichlet
           Boundary Conditions   65

           1.1.1   The Properties of the Differential
                 Operator 66

           1.1.2   The Properties of First-Order Operators
                 and the Operator of the BVP 67

       1.2   Continuum Elliptic Problems. Robin Boundary
           Conditions   68

           1.2.1   The Properties of the Differential
                 Operator 68

           1.2.2   The Properties of First-Order Operators
                 and the Operator of the BVP 69

   2   One-Dimensional Support-Operators Algorithms 70

       2.1   Nodal Discretisation of Scalar Functions and
           Cell-Centered Discretisation of Vector Functions . . 71

           2.1.1   Spaces of Discrete Functions 72

           2.1.2   The Prime Operator 73

           2.1.3   The Derived Operator 73

           2.1.4   Multiplication by a Matrix K 74

           2.1.5   The Finite-Difference Scheme for the
                 Dirichlet Boundary Value Problem 75

           2.1.6   The Matrix Problem 77

           2.1.7   The Finite-Difference Scheme for the
                 Robin Boundary Problem 77

           2.1.8   The Matrix Problem for Robin
                 Boundary Conditions   78

           2.1.9   Approximation Properties 79

           2.1.10   The Numerical Solution of Test
                 Problems 84

       2.2   Cell-Valued Discretisation of Scalar Functions
           and Nodal Discretisation of Vector Functions 90

           2.2.1   Spaces of Discrete Functions 94

           2.2.2   The Prime Operator 95

		2.2.3	The Derived Operator	96
		2.2.4	Multiplication by a Matrix and the Operator $\mathcal{D}$	96
		2.2.5	The Finite-Difference Scheme for Dirichlet Boundary Conditions	97
		2.2.6	The Matrix Problem	98
		2.2.7	The Finite-Difference Scheme for Robin Boundary Conditions	98
		2.2.8	The Matrix Problem	99
		2.2.9	Approximation Properties	99
	2.3		Numerical Solution for Test Problems	103
3		Two-Dimensional Support-Operators Algorithms	108
	3.1		Nodal Discretisation of Scalar Functions and Cell-Valued Discretisation of Vector Functions . . .	108
		3.1.1	The Finite-Difference Scheme for the Dirichlet Boundary Problem	121
		3.1.2	The Finite-Difference Scheme for the Robin Boundary Problem	123
	3.2		Cell-Valued Discretisation of Scalar Functions and Nodal Discretisation of Vector Functions	127
		3.2.1	The Finite-Difference Scheme for the Dirichlet Boundary Problem	134
		3.2.2	The Finite-Difference Scheme for the Robin Boundary Problem	136
	3.3		The Numerical Solution of Test Problems	140
		3.3.1	The Dirichlet Boundary Value Problem . .	140
		3.3.2	The Robin Boundary Value Problem	146
4		Conclusion	147

4	**The Heat Equation**			**149**
1	Introduction			149
	1.1	The Conservation Law		149
	1.2	Time Discretisation		150
	1.3	Explicit and Implicit Finite-Difference Schemes . . .		151
	1.4	Stability of the Finite-Difference Scheme for the Heat Equation		154
	1.5	The Positivity Preserving Methods		159
	1.6	The Method of Lines		160
2	Finite-Difference Schemes for the Heat Equation in 1-D . .			161
	2.1	Introduction		161
	2.2	Nodal Discretisation for Scalar Functions and Cell-Valued Discretisation for Vector Functions . . .		162
		2.2.1	Stability of the Finite-Difference Scheme and Results of the Solution for Test Problems .	165

	2.3	Cell-Valued Discretisation of Scalar Functions and Nodal Discretisation of Vector Functions	173
	2.4	Conservation Laws and the Iteration Process	176
3		Finite-Difference Schemes for the Heat Equation in 2-D	180
	3.1	Stability Conditions in 2-D	180
	3.2	Symmetry Preserving Finite-Difference Schemes and Iteration Methods	181
	3.3	Numerical Examples	186

5 Lagrangian Gas Dynamics — **193**

1		Fundamentals of Gas Dynamics	195
	1.1	The Integral Form of Gas Dynamics Equations	201
	1.2	Integral Equations for the One-Dimensional Case	202
	1.3	Differential Equations of Gas Dynamics in Lagrangian Form	206
	1.4	The Differential Equations in 1-D Lagrangian Mass Coordinates	207
	1.5	Statements for Gas Dynamics Problems in Lagrange Variables	209
	1.6	Different Forms of Energy Equations	210
	1.7	Acoustic Equations	212
	1.8	Reference Information	214
	1.9	The Characteristic Form of Gas Dynamics Equations	216
	1.10	Riemann's Invariants	217
	1.11	Discontinuous Solutions	220
		1.11.1 Contact Discontinuity	223
		1.11.2 Shock Waves. The Hugoniot Adiabat	224
		1.11.3 Shock Waves in an Ideal Gas	224
		1.11.4 Zemplen Theorem	227
		1.11.5 Approximation of the "Strong Wave".	228
		1.11.6 Structure of the Shock Front	229
2		Conservation Laws and Properties of Operators	233
3		Finite-Difference Algorithms in 1-D	237
	3.1	Discretisation in 1-D	237
	3.2	Discrete Operators in 1-D	238
		3.2.1 The Prime Operator	238
		3.2.2 The Derived Operator	239
		3.2.3 Boundary Conditions and Discretisations	239
	3.3	Semi-Discrete Finite-Difference Schemes in 1-D	240
	3.4	Fully Discrete, Explicit, Computational Algorithms	242

	3.5	Fully Discrete, Implicit, Computational Algorithm	248
	3.6	Stability Conditions	254
		3.6.1 General Remarks	254
		3.6.2 One-Dimensional Transport Equations	255
		3.6.3 Finite-Difference Schemes for the One-Dimensional Transport Equation	255
		3.6.4 Stability Conditions for 1-D Acoustic Equations	259
	3.7	Homogeneous Finite-Difference Schemes. Artificial Viscosity	260
	3.8	Artificial Viscosity in 1-D	263
	3.9	The Numerical Example	265
4	**The Finite-Difference Algorithm in 2-D**		**268**
	4.1	Discretisation in 2-D	268
	4.2	Discrete Operators in 2-D	269
		4.2.1 The Prime Operator	269
		4.2.2 The Derived Operator	271
		4.2.3 Boundary Conditions and Discretisations	272
	4.3	The Semi-Discrete Finite-Difference Scheme in 2-D	272
	4.4	The Finite-Difference Algorithm in 2-D	276
		4.4.1 The Fully Discrete, Explicit, Computational Algorithm	276
	4.5	Implicit Finite-Difference Scheme	284
		4.5.1 The Method of Parallel Chords	292
		4.5.2 Boundary Conditions	293
		4.5.3 Properties of the System of Linear Equations	298
		4.5.4 Some Properties of Algorithm	301
	4.6	Artificial Viscosity in 2-D	302
	4.7	The Numerical Example	304
6	**Conclusion**		**311**
A	**Fortran Code Directory**		**313**
	1	The Structure of Directories on the Disk	313
	2	Programs for Elliptic Equations	315
	2.1	Programs for 1-D Equations	316
		2.1.1 Dirichlet Boundary Conditions	316
		2.1.2 Robin Boundary Conditions	318
	2.2	Programs for 2-D Equations	319
		2.2.1 Dirichlet Boundary Conditions	319

	2.2.2	Robin Boundary Conditions	322
3	Programs for Heat Equations	325
	3.1	Programs for 1-D Equations	325
		3.1.1	Dirichlet Boundary Conditions	325
		3.1.2	Robin Boundary Conditions	327
	3.2	Programs for 2-D Equations	330
		3.2.1	Dirichlet Boundary Conditions	330
		3.2.2	Robin Boundary Conditions	332
4	Programs for Gas Dynamics Equations	336
	4.1	Programs for 1-D Equations	336
		4.1.1	Explicit Finite-Difference Scheme	336
		4.1.2	Implicit Finite-Difference Scheme	336
	4.2	Programs for 2-D Equations	337
		4.2.1	Explicit Finite-Difference Scheme.	337
		4.2.2	Implicit Finite-Difference Scheme	338

Bibliography — 341

Index — 355

NOTATIONS

Symbol	Explanation
x, y	Cartesian coordinates in 2-D physical space
ξ, η	Cartesian coordinates in 2-D logical space
$X(\xi),$ $X(\xi, \eta), Y(\xi, \eta)$	Functions which determine mapping between logical and physical space
V, Ω	Two-dimensional domain
$\partial V, S$	Boundary of 2-D domain
$\vec{n}$	Vector of unit outward normal
t	Time variable in heat and gas dynamics equations
L	Differential operator
L^*	Differential operator which is adjoint to L
div	Differential divergence operator
grad	Differential gradient operator
curl	Differential curl operator
H	Space of continuous scalar functions
$\overset{0}{H}$	Subspace of space H, $u(x, y) \in H$ and $u(x, y) = 0$ for $(x, y) \in \partial V$
$(u, v)_H$	Inner product of u and v in H
$\mathcal{H}$	Space of continuous vector functions
$\vec{A}$	Vector function
AX, AY	Cartesian components of vector $\vec{A}$
$\left(\vec{A}, \vec{B} \right)$	Dot, or scalar product of two vectors
$\left(\vec{A}, \vec{B} \right)_{\mathcal{H}}$	Inner product of vector functions in $\mathcal{H}$
$KXX, KXY,$ KYY	Elements of matrix K in elliptic operator div K grad
$\|u\|_C$	Maximum norm
$\|u\|_{L_2}$	L_2 norm
h, hx, hy	Steps of spatial grids
$\tau_n, \Delta t$	Steps of time discretisation
$t_n = n \Delta t$	n-th time level
$x_i \, ; x_{i,j}, y_{i,j}$	Coordinates of grid nodes in 1-D and 2-D in physical space
$\xi_i \, ; \xi_{i,j}, \eta_{i,j}$	Coordinates of grid nodes in 1-D and 2-D in logical space

Symbol	Explanation
Ω_{ij} , $V_{i,j}$	Grid cell in 2-D
$l\xi_{ij}$, $l\eta_{i,j}$	Sides of cell in 2-D
φ_{ij}^{kl}	Angle in cell ij with vertex kl
VC_{ij} , VN_{ij}	Volumes related to cell
	and node ij correspondingly
x_i^* , $x_{i,j}^*$, $y_{i,j}^*$	Coordinates of center of the cell
C_1 , C_2 ,	Different constants in inequalities
C_{min} , C_{max} $\cdots$,	
u^h , u_i^h , u_{ij}^h	Discrete scalar function in context,
	where it is necessary to distinguish
	it from continuous function u
u_i	Discrete scalar function in context,
u_{ij}	where it does not lead to misunderstanding
u_i^n , u_{ij}^n	Values of discrete functions, which
	depend on time
u^σ	$(1-\sigma)\,u^n + \sigma\,u^{n+1}$
ψ^h , ψ	Truncation error
ψ_{DIV}	Truncation error for divergence
ψ_{GRAD}	Truncation error for gradient
z^h	Error
p_h , $\mathcal{P}_h$	Projection operators
	from continuous to discrete spaces
L_h	Discrete operator
l_h	Discrete operator
D_x , $\mathcal{D}_x$, D_y , $\mathcal{D}_y$	Discrete analogs of first derivatives
D_{xx} , $\mathcal{D}_{xx}$	Discrete analogs of second derivatives
HC , HN	Spaces of discrete scalar functions,
	nodal, and cell-valued discretisations correspondingly
$\mathcal{HC}$, $\mathcal{HN}$	Spaces of discrete vector functions,
	nodal, and cell-valued discretisations correspondingly
C-N discretisation	Cell-valued discretisation for scalar functions
	and nodal discretisation vector functions
N-C discretisation	Nodal discretisation for scalar functions
	and cell-valued discretisation vector functions
DIV , $\mathcal{DIV}$	Discrete analogs of differential operator divergence
GRAD , $\mathcal{GRAD}$	Discrete analogs of differential operator gradient
$St(i)$, $St(i,j)$	Different stencils for discrete operators

Notations

Symbol	Explanation
$I, \vec{I}$	Discrete scalar and vector functions, which are equal to 1 in all nodes, or cells
$\vec{r}$	Radius-vector
p	Pressure
ρ	Density
η	Specific volume
ε	Specific internal energy
T	Temperature
S	Entropy
R	Gas constant
γ	Ratio of heat capacities for given pressure and given volume
$\vec{W}$	Velocity
$V(t)$	Fluid volume
Δ	Jacobian of transformation from Eulerian to Lagrangian coordinates
s	Lagrange mass coordinate, Chapter 5
c	Speed of sound
$D, \mathcal{D}$	Speed of shock
$F_{i,j}(p, \rho, \varepsilon)$	Equation of state for cell (i, j)
ω	Artificial viscosity
$m_i^C, m_{i,j}^C$	Mass of the cell
$m_i^N, m_{i,j}^N$	Mass of the node

Chapter 1

Introduction

The main goal of this book is to construct the finite-difference algorithms for solutions to differential equations of mathematical physics. To emphasize the relationship between these algorithms and the usual university course for equations of mathematical physics, we describe the typical equation and statements of boundary problems for it. We do not have room for a detailed description of physical problems which lead to the described equations in addition to the theory of boundary value problems so we refer readers to the following books related to these topics: [11], [26], [52], [133], [149].

1 Governing Equations

There are two main types of the physical process: non-stationary (time dependent) and stationary (which is not depend on time). The examples of non-stationary processes in this book are heat equations (equations of parabolic type) and equations of gas dynamics in Lagrangian form (the system of equations of hyperbolic type). The stationary processes are presented by the stationary heat equation (equations of elliptic type). Following is a more detailed description of these equations.

1.1 Elliptic Equations

The simplest example of equations of elliptic type is the Laplace equation

$$\Delta u = \sum_{\alpha=1}^{3} \frac{\partial^2 u}{\partial x_\alpha^2} = 0, \tag{1.1}$$

where $u = u(x)$ is the unknown function of x, $x = (x_1,\, x_2,\, x_3)$ is a point in 3-D Cartesian space.

The non-homogeneous equation

$$\Delta\,u = -f(x),$$

is called the Poisson equation. The problems associated with stationary heat transfer, the stationary diffusion process, and with electrostatics lead to the Laplace and Poisson equations.

Stationary distribution of temperature $u = u(x_1, x_2, x_3)$ in homogeneous media can be described by the Poisson equation

$$\Delta u = -f/k,$$

where $f(x)$ is the density of heat sources, and $k = \text{const} > 0$ is a coefficient of heat conductivity.

For non-homogeneous media the coefficient of heat conductivity depends on the position $x, k = k(x)$ and instead of the Poisson equation we get

$$\operatorname{div}(k\operatorname{grad}u) = \sum_{\alpha=1}^{3} \frac{\partial}{\partial x_\alpha}\left(k(x)\frac{\partial u}{\partial x_\alpha}\right) = -f(x),$$

where $k(x) > 0$.

For anisotropic media temperature $u(x)$ satisfies the following equation:

$$L\,u = \sum_{\alpha,\beta=1}^{3} \frac{\partial}{\partial x_\alpha}\left(K_{\alpha\beta}\frac{\partial u}{\partial x_\beta}\right) = -f(x).$$

This equation is the equation of elliptic type if the quadratic form is positive definite (for all $x = (x_1, x_2, x_3)$)

$$\sum_{\alpha,\beta=1}^{3} K_{\alpha\beta}\,\xi_\alpha\,\xi_\beta > 0,$$

where ξ_1, ξ_2, ξ_3 are the coordinates of an arbitrary nonzero vector.

For setting up the problem for equations of elliptic type, we must add the boundary conditions. In this section we describe the statement of boundary value problems for two-dimensional Poisson equations.

Let G be the bounded region on the plane $x = (x_1, x_2)$ and Γ the boundary of G. The statement of boundary value problems usually looks like this: To find the continuity solution of equation

$$\Delta\,u = -f(x), \quad x \in G,$$

in closed domain $G + \Gamma$, which satisfies one of the following boundary conditions:

- Dirichlet

$$u = \mu_1(x), \quad x \in \Gamma,$$

means that temperature is given on the boundary.

- Neumann

$$\frac{\partial u}{\partial n} = \mu_2(x), \quad x \in \Gamma,$$

where $\frac{\partial u}{\partial n}$ is the directional derivative in direction $\vec{n}$, and $\vec{n}$ is the outward unit normal to Γ, means that heat flux is given on the boundary.

- Robin

$$\frac{\partial u}{\partial n} + \sigma \left(u - \mu_3(x) \right) = 0, \quad x \in \Gamma.$$

From the physical viewpoint, this condition means that there is heat exchange, by Newton's law, with external media that has temperature $\mu_3(x)$.

In these conditions $\mu_i(x)$, $i = 1, 2, 3$ and $\sigma \geq 0$ are given functions.

The detailed description of boundary value problems for elliptic equations of general form will be given in Chapter 3.

We will not formulate the conditions which stipulate the uniqueness and existence of the solution for boundary value problems. In this book it is assumed that the described problems of mathematical physics have a unique solution and all functions have all derivatives that are needed. Readers who are interested in the detailed theory of elliptic equations can find more information in the following books: [133], [73], [11], [26], [45].

1.2   Heat Equations

The typical equation of parabolic type is the non-stationary heat equation, which describes the non-stationary process of heat transfer. For isotropic media this equation has the following form:

$$c(x, t) \frac{\partial u}{\partial t} = L u + f(x, t), \quad x \in G, t > 0, \tag{1.2}$$

where

$$L u = \text{div} \left(k \, \text{grad} \, u \right) = \sum_{\alpha=1}^{2} \frac{\partial}{\partial x_\alpha} \left(k(x, t) \frac{\partial u}{\partial x_\alpha} \right),$$

where $k = k(x, t) > 0$ is the coefficient of heat conductivity, and $c = c(x, t) > 0$ is heat capacity of unit volume. The temperature is a function of x and time t, $u = u(x, t)$.

In the initial moment of time, when $t = 0$, the value of temperature is given

$$u(x, 0) = u_0(x), \quad x \in G + \Gamma.$$

The statement of boundary conditions is the same as for elliptic equations, that is we will consider Dirichlet, Neumann, and Robin boundary conditions, the only difference is that all functions in these conditions also depend on time. All assumptions about operator L are the same as in the previous section.

Readers who are interested in the detailed theory of parabolic equations can find more information in the following books: [133], [74], [11], [26].

1.3 Equations of Gas Dynamics in Lagrangian Form

The construction of finite-difference algorithms for equations of hyperbolic types is described by the example of equations of gas dynamics in Lagrangian form.

When using the Lagrange approach for the description of motion of continuum media, we observe a particular fluid particle and trace how the parameters of the particle change with time. In this case, the role of the independent variable, except time, has some identifying marks, which then gives us the ability to distinguish one particle from another (so-called the Lagrangian variable). Therefore, the velocity in the sense of Lagrange is the velocity of particular fluid particle, as it is in the classical particle dynamics. The density, pressure, and so on, are the parameters which describe the state of a particular particle, which moves with media.

It is common for the role of Lagrange variables to have the coordinates of a particular particle in some initial moment of time. In this case the motion of each particle can be written as follows:

$$x_i = g_i(x_1^0,\ x_2^0, t),\quad i = 1, 2\,,$$

where x_i are the current coordinates of the particle and x_i^0 are the coordinates of the particle at the initial moment.

Let

$$J = \frac{D(x_1, x_2)}{D(x_1^0, x_2^0)} = \frac{\partial x_1}{\partial x_1^0}\frac{\partial x_2}{\partial x_2^0} - \frac{\partial x_1}{\partial x_2^0}\frac{\partial x_2}{\partial x_1^0}$$

be the Jacobian for the transformation from variables x_i to variables x_i^0, then the system of equations of gas dynamics in Lagrangian form has the following form:

$$\frac{d}{dt}(\rho J) = 0\,,$$

$$\rho\frac{d\vec{W}}{dt} = -\operatorname{grad} p + \vec{F}\,, \tag{1.3}$$

$$\rho\frac{d}{dt}(\varepsilon + \frac{\vec{W}^2}{2}) = -\operatorname{div}(p\vec{W}) + Q + \vec{F}\cdot\vec{W} - \operatorname{div}(\vec{H})\,,$$

where ρ is density, $\vec{W}$ is the velocity vector, p is pressure, $\vec{F}$ is density of external body force, Q is the power of body sources of energy, and $\vec{H}$ is the vector of density of heat flux. The time derivative in 1.3 is so-called *substantive derivative* or the *derivative following the motion* because dG/dt, for some G, is the rate of change of G measured by an observer moving with a fluid particle.

The equations of state, for example, in form

$$p = p(\rho, T), \quad \varepsilon = \varepsilon(\rho, T), \quad T - \text{temperature},$$

complete this system of equations.

The boundary condition must be given on the boundary. If pressure is given on the boundary as a function of time and position of points, then this type of boundary is called a *free boundary*. Another example of a boundary condition is a condition where the projection of velocity to a direction of normal to boundary is given. We will consider a different type of boundary condition in detail in a corresponding chapter.

Readers who are interested in a detailed theory of fluid dynamics equations can find more information in the following books: [96], [10], [103].

2   The Main Ideas of Finite-Difference Algorithms

A significant number of physical and engineering problems lead to differential equations with partial derivatives (equations of mathematical physics). Explicit solutions to equations of mathematical physics are obtainable only in special cases. Therefore these problems are generally solved approximately by using some numerical method.

One of the most universal and effective methods in wide use today for approximately solving equations of mathematical physics is the method of *finite differences*.

The finite-difference solution for mathematical-physics differential equations is carried out in two stages:

- The writing of the finite-difference scheme (a difference approximation to the differential equation on a grid).

- The computer solution for difference equations, which is written in the form of a high-order system of linear or non-linear algebraic equations.

The essence of the *method of finite difference* is as follows:

1. The continuous domain (for example, an interval, a rectangle, or a domain of arbitrary shape) is replaced by a discrete set of points (*nodes*), called the *grid*.

2. Instead of a function of continuous arguments, a function of discrete arguments is considered. The value of this function is defined at the

nodes of the grid or at other elements of the grid (for example, in some cells that have a vertex in nodes) and is called the *grid function*.

3. The derivatives entering into the differential equations and the boundary conditions are approximated by the difference expression, thus the differential problem is transformed into a system of linear or nonlinear algebraic equations (*grid or difference equations*). Such a system is often called the *finite-difference scheme*.

2.1 1-D Case

The unknown functions in differential equations are the functions of *continuous arguments*. For example, the heat equation function $u(x, y, z, t)$ gives us the value of temperature in each point of space in time t. This means that at time moment t to each point with coordinates (x, y, z) we can associate the value of temperature $u(x, y, z, t)$. In the simplest stationary, 1-D, u is the function of one spatial argument $x = x_1$, $u(x)$ and $x \in [a, b]$. Function $u(x)$ of the continuous argument $x \in [a, b]$ is the element of some functional space H, and some norm is usually used for comparison of the two functions $u(x)$, $v(x) \in H$. For example,

$$\| u \|_C = \max_{a \le x \le b} | u(x) | \tag{2.1}$$

or

$$\| u \|_{L_2} = \left(\int_a^b u^2(x) dx \right)^{1/2}. \tag{2.2}$$

The choice of norm for the comparison of these functions depends on the physical meaning of function u.

To understand what the *function of discrete arguments* is, let us consider the process of measurement of a point in space by an instrument. As a result of these measurements, we get a bounded set of numbers, each of them corresponding to a measurement of a point. These values can be enumerated in some way, for example, by using a number of instruments that give these values. The values can then be denoted as u_i. In some physical applications the instruments can give values averaged over a small area or volume. Therefore in this case, the values do not correspond to a specific point. It is clear that only index i does not give us information about the location of instrument number i. Therefore the notion of the function of discrete arguments includes information about the location of points, volumes, or other geometrical figures and values related to these geometrical elements. The examples of grids and grid functions are given below.

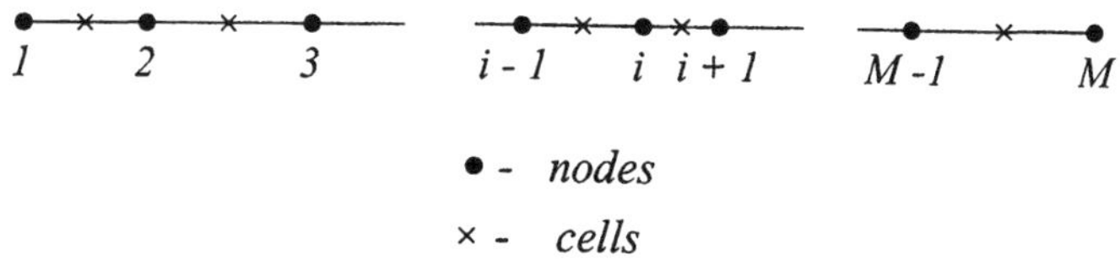

$\bullet$ - *nodes*

$\times$ - *cells*

Figure 2.1: *1-D Non-uniform grid.*

2.1.1 Grid Functions

A simple example of a grid is the grid in a 1-D case. Let the domain of variable x be the interval $0 \leq x \leq 1$. We split this interval by M nodes $\{x_i, \ i = 1, \dots, M\}$

$$0 = x_1 \leq x_2 < \cdots < x_i < x_{i+1} < \cdots < x_{M-1} \leq x_M = 1.$$

These nodes split the interval $[0, 1]$ into $M - 1$ subintervals called *cells*

$$\{[x_i, x_{i+1}], \ i = 1, \dots, M - 1\}.$$

This is a *non-uniform* grid in a 1-D case (see Figure 2.1)
    For the identification of nodes, we can use index i. If

$$x_i = h * (i - 1), \quad h = 1/(N - 1),$$

then the grid is a *uniform grid.*
    In the case of *uniform grids,* density of nodes distribution can be characterized by the parameter h, which tends to go to zero when the number of nodes increases.
    For *non-uniform grids* it is usually assumed that there are some constants C_{max} and C_{min}, which are not dependent on h, that

$$C_{min} \, h \leq x_{i+1} - x_i \leq C_{max} \, h. \tag{2.3}$$

If this assumption called *regularity* is valid, then if h tends to go to zero then the length of each cell of the grid also tends to go to zero. Therefore, for the non-uniform case, grids are characterized by parameter h and constants C_{max} and C_{min}.
    The notion of regularity can be explained by the following example. Let us consider the following family of grids on segment $[0, 1]$:

$$x_1 = 0,$$
$$x_i = (i - 1) \, h + R_i \, \eta \, h, \quad i = 2, \dots, M - 1,$$
$$x_M = 1,$$

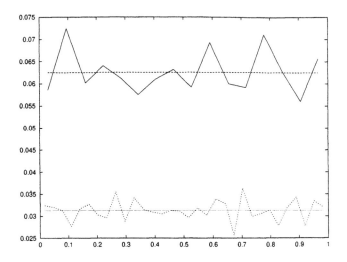

Figure 2.2: *Mesh size for* $M = 17$, $M = 33$, *and* $\eta = 0.2$.

where $h = 1/(M - 1)$ is the mesh size for a uniform grid, $R_j \in (-0.5, 0.5)$ is a random number, $\eta < 1$ is the parameter determining the relative size of the perturbations of the uniform grid. To demonstrate how a random grid looks we will present how mesh size depends on x for $M = 17$ and $M = 33$ for $\eta = 0.2$ On the top part of Figure 2.2 we present how mesh size depends on x for $M = 17$ (broken line), the straight line presents value $h = 1/(M - 1) = 0.0625$, which is mesh size for a uniform grid. On the bottom part of the Figure 2.2 we present mesh size for $M = 33$. For a given η we have the following equation for mesh size:

$$x_{i+1} - x_i = h + \eta\, h\, (R_{i+1} - R_i) = h\, [1 + \eta\, (R_{i+1} - R_i)]\,.$$

And now because R_{i+1}, $R_i \in (-0.5, 05)$ we can conclude that

$$(1 - \eta)\, h < x_{i+1} - x_i < (1 + \eta)\, h\,.$$

Comparing this equation with 2.3, we can conclude that

$$C_{min} = 1 - \eta = 0.8 \quad, C_{max} = 1 + \eta = 1.2\,.$$

2.1.2 Smooth and Non-Smooth Grids

For the practical application using a non-uniform grid, it is very useful and important to understand the difference between two essentially different types of non-uniform grids.

The first type of non-uniform grid is the *smooth grid*. To explain this notion assume that there is a smooth function $x = x(\xi)$ which transforms segment $[0, 1]$ into itself. This transformation must be a one-to-one transformation. For the 1-D case a sufficient condition for this is that function $x(\xi)$ must be a monotone increasing function. Let us assume that in segment $0 \leq \xi \leq 1$ there is a uniform grid with nodes $\xi_i = \frac{i-1}{M-1}$, $i = 1, \ldots, M$. In this assumption, grid $\{x_i; \ i = 1, \ldots, M\}$ on segment $0 \leq x \leq 1$ is called a *smooth grid* if the coordinates of this node are given as follows:

$$x_i = x(\xi_i) \tag{2.4}$$

and function $x(\xi)$ is a *smooth function.* (Note that the space of variable ξ is usually called *logical space,* and the space of variable x is called *physical space.*)

Theoretical properties of finite-difference schemes (for example, the order of truncation error) are usually investigated with the assumption that parameter $h \rightarrow 0$. And consequently we have not a single grid but a family of grids when $h \rightarrow 0$. Therefore in definition 2.4 it is very important that function $x(\xi)$ and its properties *do not depend on parameter h.* It is important to understand this fact because for each single grid (for any given h) it is very easy to construct a function $x(\xi)$ but when $h \rightarrow 0$, derivatives of this transformation may be unbounded and the transformation will not have smooth derivatives.

An example of a smooth grid given by the transformation $x(\xi) = \xi^2$ is given in Figure 2.3. A wider class of a non-uniform grid is the *non-smooth* grid. The *non-smooth grid* is a grid that cannot be constructed by using a smooth transformation. An example of such a grid is a grid in which the length between nodes alternates as h, $2h$, h, $\ldots$. This grid is presented in Figure 2.4

For the *non-smooth* grid it is usually assumed that there are some regularity conditions like 2.3 which provide us with some information about the distribution of nodes. For example, the grid that is represented in Figure 2.4, $C_{min} = 1$ and $C_{max} = 2$.

The difference between smooth and non-smooth grids will also be demonstrated in the section related to the investigation of truncation error and on numerical tests.

2.1.3   Different Types of Grid Functions

There are two main types of functions for discrete arguments in a 1-D case. In the first case, the values of the function correspond to the nodes. This is *nodal discretisation.* In this case grid function u^h is a set (vector) of M numbers:

$$u^h = \{u_i^h, \ i = 1, \ldots, M\}.$$

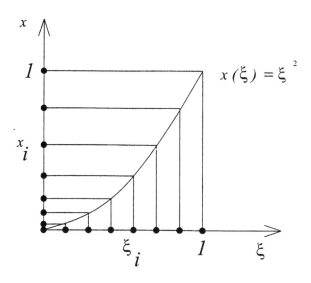

Figure 2.3: *Smooth non-uniform grid.*

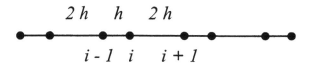

Figure 2.4: *Smooth non-uniform grid.*

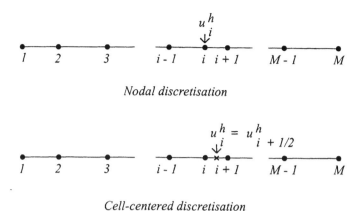

Figure 2.5: *Nodal and cell-centered discretisation.*

Using the interpretation of grid functions as a result of the measurements, we can say that in the case of nodal discretisation, the instruments measure values exactly in the nodes.

If it is possible to measure only the value averaged over cells (x_i, x_{i+1}) then the value of the grid functions corresponds to the cells. This is *cell-valued* discretisation. It is important to note that in this case, the value of a function does not correspond to a specific point in the cell but corresponds to the cell as a whole geometrical object. There are two ways to denote values for cell-valued grid functions. In the first case, half indices are used. This means that the notation $u^h_{i+1/2}$ is used for the value of the grid function in cell (x_i, x_{i+1}). In the second case, this value is denoted by using the index u^h_i of the left node of the cell.

As we mentioned above, one of the goals of this book is that the formulas in the text of this book be as close as possible to the text of the program in a programming language. It is known that it is not possible to use half indices in many programming languages. Therefore, in this book we will use integer indices. For cell-valued discretisation, index i varies in limits from 1 to $M-1$.

The difference in nodal and cell-valued discretisation is explained in Figure 2.5.

In cases where using integer indices for values of nodal and cell-valued functions can lead to a misunderstanding, we will write $u^h \in HN$ for nodal discretisation and $u^h \in HC$ for cell-valued discretisation. In general, if we do not want to concretely define the space of grid functions, we will write them as $u^h \in H_h$.

For comparison of grid functions as in a continuous case, the norm is introduced. The discrete analogs of norms 2.1, 2.2 for $u^h \in HC$ are

$$\| u^h \| = \max_{1 \leq i \leq M-1} | u_i^h | \qquad (2.5)$$

or

$$\| u^h \| = \left(\sum_{i=1}^{N-1} (u_i^h)^2 * (x_{i+1} - x_i) \right)^{1/2} . \qquad (2.6)$$

Let $u(x)$ be the function of a continuous argument. For example, $u(x)$ is the exact solution of the original partial differential equation. And let $u^h = \{u_i^h\}$ be a function that is the result of a finite-difference algorithm. The main interest for practice is how close u^h is to u. There are two ways to compare these functions.

1. Using the values of grid function u^h, we can determine the function of a continuous argument (for example, by using linear interpolation). As a result, we get the function $\tilde{u}(x, h)$. Then we can compare the difference $\tilde{u}(x, h) - u(x)$ in a continuous norm $\| \bullet \|_H$:

$$\| \tilde{u}(x, h) - u(x) \|_H .$$

2. Using a projection operator, space H is projected to space H_h. For each continuous function $u(x) \in H$ there is a corresponding grid function $\hat{u}^h$, that is $\hat{u}^h = P_h u \in H_h$, where P_h is a projection operator. For example, for the continuous function $u(x)$ and $H_h = HN$, we can define

$$\hat{u}_i^h = u(x_i).$$

Sometimes it is useful to consider P_h as the operator of the integral average. For example,

$$\hat{u}_i^h = \frac{1}{0.5 \, (x_{i+1} - x_{i-1})} \int_{(x_i+x_{i-1})/2}^{(x_i+x_{i+1})/2} u(x) dx .$$

Then if $u(x)$ is an exact solution of the original differential equation, $\hat{u}^h$ is its projection to the space of grid functions, and u^h is the approximate solution, we can define the grid function

$$z^h = u^h - \hat{u}^h = u^h - P_h u \in H_h .$$

Grid function z^h is called *error*. The accuracy of the finite-difference algorithm can be characterized by

$$\| z^h \|_{H_h} = \| u^h - \hat{u}^h \|_{H_h} ,$$

where $\| \bullet \|_{H_h}$ is some norm in H_h.

Everywhere in this book we will use a second possibility in which we will compare exact solutions of differential equations and approximate solutions obtained by some finite difference algorithm in *some norm in space of the grid functions*.

2.1.4 Finite-Difference Operators, Truncation Error

Now let's consider the second stage of constructing finite-difference schemes. That is, how to approximate differential operators by using difference schemes. Let $u(x)$ be the differentiable function on segment $[0, 1]$ and du/dx be the first derivative. In other words d/dx is the operator of the first derivative, which establishes the correspondence between the function and its derivative. Let's assume that the derivative is also a continuous function. Then operator d/dx acts from the space of continuous functions to itself.

The definition of du/dx is

$$\frac{du}{dx}\bigg|_x = \lim_{h \to 0} \frac{u(x+h) - u(x)}{h},$$

therefore it is natural to use this definition for construction of a finite-difference analog of the first derivative. Namely, for grid functions from HN (grid functions whose values are defined in *nodes*) the finite-difference analog of the first derivative can be defined as follows:

$$\left(D_x(u^h)\right)_i = \left(\frac{\delta u^h}{\delta x}\right)_i = \frac{u_{i+1}^h - u_i^h}{x_{i+1} - x_i}. \tag{2.7}$$

The *stencil* of this operator contains two nodes $i+1$ and i.

It is clear that operator $D_x(u^h)$ is a grid function. Because u_i^h is determined only for $i = 1, \ldots, M$ and the expression for $D_x(u^h)$ contains the term u_{i+1}, therefore $D_x(u^h)$ can be defined only for $i = 1, \ldots, M-1$. It means that the dimension of space grid functions, which results in action of operator D_x, is different from the dimension of space HN. It is natural to consider $D_x(u^h)$ as an element of space HC, then

$$D_x : HN \to HC.$$

Figure 2.6 helps in understanding how operator D_x acts.

Next, the important question is: *What is the relationship between differential operators and its difference analog?*

In the simplest case the statement of this question is: let grid function $u_i^h \in HN$ be the projection of the continuous function $u(x)$ and let, for simplicity, $u_i^h = u(x_i)$. Then, what we can say about *truncation error or error of approximation for a differential operator by a difference operator* is

$$\psi = \frac{du}{dx}\bigg|_{x^*} - \frac{u_{i+1}^h - u_i^h}{x_{i+1} - x_i} = \frac{du}{dx}\bigg|_{x^*} - \frac{u(x_{i+1}) - u(x_i)}{x_{i+1} - x_i}.$$

It is important to note that the value of ψ depends on the choice of point x^*, and in the general case, instead of $\frac{du}{dx}\big|_{x^*}$ we must consider projection of the continuous function du/dx to the space of grid functions HC, because $D_x(u^h)$ belongs to HC. To investigate the truncation error ψ in the simplest

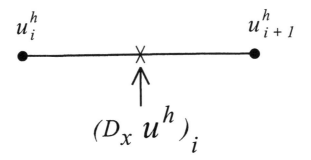

Figure 2.6: *Operator D_x.*

case, let's assume that $x_i \leq x^* \leq x_{i+1}$, and use the Taylor expansion of $u(x_{i+1})$ and $u(x_i)$ in point x^*. Using the expressions

$$
\begin{aligned}
u(x_{i+1}) &= u(x^*) + \frac{du}{dx}(x^*)(x_{i+1} - x^*) \\
&\quad + \frac{d^2u}{dx^2}(x^*)\frac{(x_{i+1} - x^*)^2}{2} \\
&\quad + O(x_{i+1} - x^*)^3,
\end{aligned}
$$

$$
\begin{aligned}
u(x_i) &= u(x^*) + \frac{du}{dx}(x^*)(x_i - x^*) \\
&\quad + \frac{d^2u}{dx^2}(x^*)\frac{(x_i - x^*)^2}{2} \\
&\quad + O(x_i - x^*)^3,
\end{aligned}
$$

we get

$$
\begin{aligned}
\psi &= \frac{du}{dx}\bigg|_{x^*} - \bigg[u(x^*) + \frac{du}{dx}(x^*)(x_{i+1} - x^*) + \\
&\quad + \frac{d^2u}{dx^2}(x^*)\frac{(x_{i+1} - x^*)^2}{2} + O(x_{i+1} - x^*)^3 - \\
&\quad - u(x^*) - \frac{du}{dx}(x^*)(x_i - x^*) \\
&\quad - \frac{d^2u}{dx^2}(x^*)\frac{(x_i - x^*)^2}{2} - O(x_i - x^*)^3 \bigg] / (x_{i+1} - x_i) \\
&= \frac{d^2u}{dx^2}\bigg|_{x^*} * \frac{(x_{i+1} - x^*)^2 - (x_i - x^*)^2}{2(x_{i+1} - x_i)} \\
&\quad + \frac{O(x_{i+1} - x^*)^3 - O(x_i - x^*)^3}{x_{i+1} - x_i}.
\end{aligned}
$$

The last term in the previous equation can be estimated by using assumption 2.3 as $O(h^2)$. And finally,

$$\psi = \left.\frac{d^2 u}{dx^2}\right|_{x^*} * \left[\frac{(x_{i+1} - x^*) + (x_i - x^*)}{2}\right] + O(h^2).$$

In the general case, the expression in square brackets is $O(h)$ and consequently $\psi = O(h)$. Therefore, for any x^*, the truncation error is the first order with respect to h. It is easy to see that the expression in square brackets is equal to zero only in the case when

$$x^* = \frac{x_{i+1} + x_i}{2}.$$

For this choice of x^*, the truncation error $\psi = O(h^2)$ for the general non-uniform grid. In other words, if we define the projection operator from the space of continuous functions to the space of grid functions HC as

$$(p_h u)_i = u\left(\frac{x_{i+1} + x_i}{2}\right)$$

we get a second-order truncation error.

In the *general case*, if Lu is the differential operator and $L_h u^h$ is the corresponding difference operator then the *truncation error* is the grid function which can be defined as follows:

$$\psi^h = \mathcal{P}_h(Lu) - L_h(p_h u)$$

where $\mathcal{P}_h$ and p_h are projection operators from the space of continuous functions H to the space of difference functions H_h. It is important that operators $\mathcal{P}_h$ and p_h can be different. Sometimes these operators must be different because u can be a scalar function and Lu can be a vector function.

We will say that L_h approximates L if

$$\| \psi^h \|_{H_h} \to 0 \quad when \quad h \to 0,$$

and that L_h approximates L with order m if

$$\| \psi^h \|_{H_h} \leq Rh^m$$

where R is a constant which is not dependent on h.

Let us demonstrate that the value of the truncation error strongly depends on the definition of operators $\mathcal{P}_h$ and p_h. For example, for the approximation of the first derivative we have a function which is determined in nodes and the approximation value of the derivative is determined somewhere in the cell. Therefore, for projection of a continuous function to a node, we can use the formula

$$(p_h u)_i = u(x_i)$$

but for projection into the cell we will not use a value in the middle of the cell, but an integral average value, which can be given by the operator $\mathcal{P}_h$:

$$(\mathcal{P}_h\, v)_i = \frac{1}{x_{i+1} - x_i} \int_{x_i}^{x_{i+1}} v\, dx\,.$$

Let us prove that in the case of these projection operators, operator D_x gives us the exact formula for the first derivative. In fact, we get the following expression for truncation error in a cell:

$$\psi_i^h = \left(\mathcal{P}_h\, \frac{du}{dx}\right)_i - \frac{(p_h\, u)_{i+1} - (p_h\, u)_i}{x_{i+1} - x_i}\,.$$

Using the definition of operator $\mathcal{P}_h$ we get

$$\left(\mathcal{P}_h\, \frac{du}{dx}\right)_i = \frac{1}{x_{i+1} - x_i} \int_{x_i}^{x_{i+1}} \frac{du}{dx}\, dx = \frac{u(x_{i+1}) - u(x_i)}{x_{i+1} - x_i}$$

and from the last two equations and the definition of p_h, we can conclude that

$$\psi_i^h \equiv 0\,.$$

Here we note that the goal of this book is not to formulate a precise and rigorous definition of all notions related to finite difference algorithms. The reader must be able to understand the basic notions and begin to solve the problems. In this context it is important to understand that the definition and value of truncation errors strongly depends on the meaning in which we intend to compare continuous and grid functions.

2.1.5   Two Approaches for the Investigation of Truncation Errors

The first approach is directly using a Taylor expansion as it is demonstrated in the previous section. This investigation can be done without any assumption as to how this grid is constructed. The second approach can be applied if we assume that the given grid is a smooth grid. Therefore we can assume that the function which gives us this grid from a grid in logical space has smooth derivatives.

Let us demonstrate the difference in the two approaches in the example of *a central difference approximation* of the first derivative on a non-uniform grid in a node. That is

$$(\hat{D}_x\, u^h)_i = \frac{u_{i+1}^h - u_{i-1}^h}{x_{i+1} - x_{i-1}}\,.$$

Using the Taylor expansion, we get

$$\psi_i^h = \left.\frac{du}{dx}\right|_{x_i} - (\hat{D}_x\, (p_h\, u))_i = 0.5 \left.\frac{d^2 u}{dx^2}\right|_{x_i} [(x_{i+1} - x_i) - (x_i - x_{i-1})] + O(h^2)\,.$$

This equation shows us that ψ^h is the first order in h. Therefore if we do not have any additional information about the grid we can only say that the central difference approximation is the first-order approximation for the first derivative. Let us now assume that there is some smooth transformation $x(\xi)$ which transforms logical space into physical space in our case segment $[0, 1]$ into itself. Then we can consider function u as a composite function of ξ, that is

$$u(\xi) = u(x(\xi)).$$

Then the first derivative is

$$\frac{du}{dx} = \frac{du}{d\xi}\frac{d\xi}{dx} = \frac{du}{d\xi}\left(\frac{dx}{d\xi}\right)^{-1}. \tag{2.8}$$

The central difference approximation can be written in a form that is similar to the previous equation for the first derivative, that is

$$(\hat{D}_x u^h)_i = \frac{u_{i+1}^h - u_{i-1}^h}{\xi_{i+1} - \xi_{i-1}}\left(\frac{x_{i+1} - x_{i-1}}{\xi_{i+1} - \xi_{i-1}}\right)^{-1}.$$

It is easy to show that by using the usual Taylor expansion in logical space, the central difference approximation on a uniform grid in logical space is the second order in $\xi_{i+1} - \xi_i = h$. Therefore we get

$$\frac{u_{i+1}^h - u_{i-1}^h}{\xi_{i+1} - \xi_{i-1}} = \left.\frac{du}{d\xi}\right|_{\xi_i} + O(h^2),$$

$$\frac{x_{i+1} - x_{i-1}}{\xi_{i+1} - \xi_{i-1}} = \left.\frac{dx}{d\xi}\right|_{\xi_i} + O(h^2),$$

therefore

$$(\hat{D}_x u^h)_i = \left(\left.\frac{du}{d\xi}\right|_{\xi_i} + O(h^2)\right)\left(\left.\frac{dx}{d\xi}\right|_{\xi_i} + O(h^2)\right)^{-1}.$$

From this relation and from formula

$$(a + O(h^2))^{-1} = \frac{1}{a + O(h^2)} = \frac{1}{a} + O(h^2)$$

we can conclude that

$$(\hat{D}_x u^h)_i = \left.\frac{du}{d\xi}\right|_{\xi_i}\left(\left.\frac{dx}{d\xi}\right|_{\xi_i}\right)^{-1} + O(h^2).$$

And finally from this equation and equation 2.8 we get

$$(\hat{D}_x u^h)_i = \left.\frac{du}{dx}\right|_{x_i} + O(h^2).$$

Therefore for a smooth grid we can conclude that the central difference approximation is a second order in h.

2.1.6   Approximation of Second-Order Derivatives

The approximation of finite-difference analogs of second-order derivatives is a more complicated task than approximation of first derivatives. The general principles for constructing finite-difference schemes will be described in chapter 2. In this section we consider only one expression for the approximation of second derivatives. By definition of $d^2 u/d x^2$ is

$$\frac{d^2 u}{dx^2}\bigg|_x = \lim_{h \to 0} \frac{\frac{du}{dx}(x+h) - \frac{du}{dx}(x)}{h}.$$

For our purposes it is more convenient to rewrite the previous definition in the following form:

$$\frac{d^2 u}{dx^2}\bigg|_x = \lim_{h \to 0} \frac{\frac{du}{dx}(x+\frac{h}{2}) - \frac{du}{dx}(x-\frac{h}{2})}{h}. \tag{2.9}$$

Then for grid function $u^h \in HN$, the finite-difference analog of formula 2.9 is

$$(D_{xx}u^h)_i = \frac{(D_x u^h)_i - (D_x u^h)_{i-1}}{0.5\,(x_{i+1}+x_i) - 0.5\,(x_i+x_{i-1})} = \frac{(D_x u^h)_i - (D_x u^h)_{i-1}}{0.5\,(x_{i+1}-x_{i-1})}, \tag{2.10}$$

where the finite-difference analog of the first derivative D_x is defined by formula 2.7. The formula 2.9 gives the finite-difference analog of the second derivative in node i. To understand this formula one can compare it with definition 2.9 and take into account that $(D_x u^h)_i$ approximates the first derivative du/dx at point $(x_{i+1}+x_i)/2$ with second order.

The explicit form of operator D_{xx} on a non-uniform grid is

$$(D_{xx}u^h)_i = \frac{\frac{u^h_{i+1} - u^h_i}{x_{i+1} - x_i} - \frac{u^h_i - u^h_{i-1}}{x_i - x_{i-1}}}{0.5\,(x_{i+1}-x_{i-1})}. \tag{2.11}$$

For a uniform grid we get

$$(D_{xx}u^h)_i = \frac{u^h_{i+1} - 2u^h_i - u^h_{i-1}}{h^2}.$$

The *stencil* of these operators contains the three points $i-1$, i, and $i+1$. Figure 2.7 helps in understanding how operator D_{xx} works.

Using the Taylor expansion, it can be proved that for a non-uniform grid *truncation error*

$$\psi^h = \frac{d^2 u}{dx^2}(x_i) - (D_{xx}u^h)_i \tag{2.12}$$

is

$$\psi^h = O(h), \quad \text{non-uniform grid,}$$
$$\psi^h = O(h^2), \quad \text{uniform grid}.$$

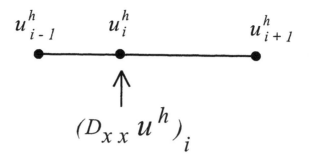

Figure 2.7: *Operator D_{xx}.*

2.1.7   Finite-Difference Schemes for the 1-D Poison Equation

To explain some ideas of finite-difference algorithms let's consider the 1-D Poisson equation

$$Lu = \frac{d^2 u}{dx^2} = -f(x), \quad 0 < x < 1 \tag{2.13}$$

with Dirichlet boundary condition at $x = 0$,

$$lu = u = \mu_1 \tag{2.14}$$

and Robin boundary condition at $x = 1$,

$$\frac{du}{dx} + \mu_2 (u - \mu_3) = 0, \tag{2.15}$$

or

$$lu = \frac{du}{dx} + \mu_2 u = \mu_4 \tag{2.16}$$

where $\mu_4 = \mu_2 \mu_3$.

To construct a finite-difference scheme on a general non-uniform grid in this section we will use nodal discretisation, that is, $u^h \in HN$. Therefore we must approximate the differential equation 2.13 and boundary conditions 2.14 and 2.15. To do this we will use operators D_{xx} and D_x from the previous section.

Differential equation 2.13 corresponds to difference equations in internal nodes $i = 2, \cdots, M - 1$:

$$\left(L_h u^h\right)_i = (D_{xx} u^h)_i = \frac{\dfrac{u^h_{i+1} - u^h_i}{x_{i+1} - x_i} - \dfrac{u^h_i - u^h_{i-1}}{x_i - x_{i-1}}}{0.5\,(x_{i+1} - x_{i-1})} - \varphi^h_i, \tag{2.17}$$

where φ_i^h is a difference function which approximates function $f(x)$ in node x_i.

The approximation of Dirichlet boundary condition 2.14 at $x = 0$ is

$$(l_h u^h)_1 = u_1^h = \chi_1 \qquad (2.18)$$

where χ_1 is an approximation for μ_1.

One possible approximation of Robin boundary condition 2.15 at $x = 1$ is

$$(l_h u^h)_M = (D_x u^h)_{M-1} + \chi_2 u_M^h = \frac{u_M^h - u_{M-1}^h}{x_M - x_{M-1}} + \chi_2 u_M^h = \chi_4, \qquad (2.19)$$

where χ_2 and χ_4 approximate μ_2 and μ_4.

The systems of difference equations 2.17, 2.18, and 2.19 give us the finite-difference scheme for differential problems 2.13, 2.14, and 2.15.

The difference equations 2.17, 2.18, 2.19 are the system of linear algebraic equations and the first question is: *Does this system have a solution and, if so, is this solution unique?*

To answer this question, we must investigate the properties of a related matrix problem. This investigation has to be done for each finite-difference scheme.

Assume that the system of difference equations has a unique solution. Then the question is what can we say about *error;* that is, the difference between the solution of difference equations and the solution of the original boundary value problem

$$z^h = u^h - p_h u.$$

We will say that the solution for FDS converges with the solution of BVP *(FDS is convergent)* if

$$\| z^h \|_{1_h} \to 0, \quad when \quad h \to 0.$$

FDS is the *n-th order accurate* if

$$\| z^h \|_{1_h} \leq R h^n$$

where R does not depend on h.

If we express u^h from this equation and substitute it for a difference equation we get the following equation for error in internal points:

$$L_h z^h = \Psi^h$$

where

$$\Psi^h = \varphi^h - L_h(p_h u).$$

Difference function Ψ^h is called *residual or error of approximation of differential equations by difference equations in the solution of the differential boundary value problem.*

For boundary conditions we get

$$(l_h z^h)_1 = \nu_1^h$$

and

$$(l_h z^h)_M = (D_x z^h)_{M-1} + \chi_2 z_M^h = \nu_2^h .$$

The ν_1^h and ν_2^h are called *residual or error of approximation of differential boundary conditions by the difference boundary conditions in the solution of the differential boundary value problem.* The expressions for ν_1^h and ν_2^h are

$$\nu_1^h = \chi_1 - (l_h(p_h u))_1 = \chi_1 - (p_h u)_1 ,$$

$$\nu_2^h = (D_x z^h)_{M-1} + \chi_2 z_M^h = \chi_4 - (l_h(p_h u))_M .$$

Therefore for error we have the similar problem as for u^h but with other right-hand sides and we can estimate the value of z^h by the value of residuals.

We define some norm for residuals Ψ^h, ν_1^h, and ν_2^h, that is $\| \Psi^h \|_{1_h}$ and $\| \nu \|_{2_h}$. The FDS is *consistent* with BVP if

$$\| \Psi^h \|_{1_h} \to 0 \quad when \quad h \to 0$$

and

$$\| \nu \|_{2_h} \to 0 \quad when \quad h \to 0 .$$

Therefore *consistency* means that residuals are going to zero when h is going to zero.

We will say that the FDS is the $n-th$ *order of approximation* if

$$\| \Psi^h \|_{1_h} = O(h^n)$$

and

$$\| \nu \|_{2_h} = O(h^n) .$$

It is important to note that residuals can be divided into two parts, one related to approximation of differential operators by difference one and another part related to approximation of the right-hand side. Because $Lu = -f$ we get $p_h(Lu + f) = p_h(Lu) + p_h f = 0$ and Ψ^h can be written as follows:

$$
\begin{aligned}
\Psi^h &= \varphi^h - L_h(p_h u) = \varphi^h - L_h(p_h u) + p_h(Lu) + p_h f \\
&= (-L_h(p_h u) + p_h(Lu)) + (\varphi^h - p_h f) \\
&= \psi_1^h + \psi_2^h,
\end{aligned}
$$

where

$$\psi_1^h = (-L_h(p_h u) + p_h(Lu))$$

is the *truncation error or error of approximation of the differential operator by the difference operator* and

$$\psi_2^h = \varphi^h - p_h f$$

is the *error of approximation of the right-hand side.*

As Ψ^h is the error of approximation in the solution of differential equations, then $\Psi^h = O(h^n)$ can be valid when each ψ_1^h and ψ_2^h are not $O(h^n)$. This effect is called *improving the order of approximation in the solution of differential equations.* In this context it means that when we construct the FDS the approximation of the differential operator and the right-hand side must be consistent with each other.

In our concrete case of FDS 2.17, 2.18, and 2.19 from 2.12 it follows that $\psi_1^h = \dot{O}(h)$. If we define the right-hand side as $\varphi^h = p_h f$ then $\psi_2^h = 0$ and finally $\Psi^h = O(h)$. In a similar way it can be proved that the error of approximation of boundary conditions is $O(h)$.

To find the relationship between the value of error and values of residuals one must find a relationship between the solution of FDS and input data of FDS or right-hand sides φ^h and χ.

The important notion which helps to do this is *stability.* This notion means continuous dependence of the solution of finite-difference schemes on input data. The stability is the internal property of finite-difference schemes. If we consider problems 2.17, 2.18, and 2.19 and denote their solution as u^h then the finite-difference scheme is called *stable* if

$$
\begin{aligned}
\| u^h \|_{H_h^1} \quad &\leq \quad M_1 \| \varphi^h \|_{H_h^1} \\
&+ \quad M_2 |\mu_1| \\
&+ \quad M_3 |\mu_2| \\
&+ \quad M_4 |\mu_3| .
\end{aligned}
$$

This relation means that a small change of data gives us a small change in the solution.

In this book we do not have enough room to consider questions related to investigating stability properties of finite-difference schemes so each single case will refer the reader to books or articles where he can find more information. We will also present some conditions that are sufficient for stability without any proof.

It is important to note for the elliptic equation method of support operators (which we use for construction of finite-difference schemes) that they automatically give us stable finite-difference schemes.

Assuming that our FDS is stable, we get

$$\| z^h \|_{1_h} \leq R(\| \Psi^h \|_{2_h} + \| \nu^h \|_{3_h}).$$

From this equation we can conclude that if the FDS is *stable* and *consistent* then it is *convergent.* It is important to note that error and residuals can be estimated in different norms.

2.1.8 Finite-Difference Operators and the Related Matrix

Let's consider the finite-difference operator D_x which is defined by equation 2.7. As we know, this operator is acting from the space of nodal grid functions HN to the space of cell-centered grid functions HC,

$$D_x : HN \rightarrow HC$$

and

$$\left(D_x(u^h)\right)_i = \frac{u^h_{i+1} - u^h_i}{x_{i+1} - x_i}.$$

Let us consider the grid function $u^h \in HN$ as a vector-column

$$u^h = \begin{pmatrix} u^h_1 \\ u^h_2 \\ \vdots \\ u^h_M \end{pmatrix}.$$

The result of the action of operator D_x on grid function $u^h \in HN$ is the grid function from HC, that is, $D_x u^h \in HC$. We can also consider this function as vector-column

$$D_x u^h = \begin{pmatrix} (D_x u^h)_1 \\ (D_x u^h)_2 \\ \vdots \\ (D_x u^h)_{M-1} \end{pmatrix}.$$

The relationship between vectors u^h and $D_x u^h$ can be written in the matrix form

$$D_x u^h = A u^h$$

where

$$A = \begin{pmatrix} \frac{-1}{x_2-x_1} & \frac{1}{x_2-x_1} & 0 & 0 & \cdots & 0 \\ 0 & \frac{-1}{x_3-x_2} & \frac{1}{x_3-x_2} & 0 & \cdots & 0 \\ \cdots & \cdots & \cdots & \cdots & \cdots & \cdots \\ 0 & \cdots & \frac{-1}{x_{i+1}-x_i} & \frac{1}{x_{i+1}-x_i} & \cdots & 0 \\ \cdots & \cdots & \cdots & \cdots & \cdots & \cdots \\ 0 & 0 & \cdots & 0 & \frac{-1}{x_M-x_{M-1}} & \frac{1}{x_M-x_{M-1}} \end{pmatrix}.$$

2.1.9   Finite-Difference Schemes and Related Matrix Problems

Now let's consider the matrix problem that corresponds to FDS 2.17, 2.18, and 2.19. Because u_1^h is given by equation 2.18, the vector of unknowns is

$$U = \begin{pmatrix} u_2^h \\ u_3^h \\ \vdots \\ u_M^h \end{pmatrix} .$$

At first we can multiply all equations in interior points $i = 2, \ldots, M-1$ by $(x_{i+1} - x_{i-1})/2$. The equation in point $i = 2$ contains given quantity $u_1^h = \chi_1$ and to formulate the matrix problem, we must transfer all given values to the right-hand side. As a result in point $i = 2$ we get

$$\left(\frac{1}{x_3 - x_2} + \frac{1}{x_2 - x_1} \right) u_2^h - \left(\frac{1}{x_3 - x_2} \right) u_3^h = \tag{2.20}$$
$$\left(\frac{1}{x_2 - x_1} \right) \chi_1 + \varphi_2^h \left(\frac{x_3 - x_1}{2} \right) .$$

For nodes $i = 3, \ldots, M-1$ we get

$$- \left(\frac{1}{x_i - x_{i-1}} \right) u_{i-1}^h + \left(\frac{1}{x_{i+1} - x_i} + \frac{1}{x_i - x_{i-1}} \right) u_i^h - \tag{2.21}$$
$$\left(\frac{1}{x_{i+1} - x_i} \right) u_{i+1}^h = \varphi_i^h \left(\frac{x_{i+1} - x_{i-1}}{2} \right) . \tag{2.22}$$

And finally, boundary condition 2.19 can be rewritten as follows

$$- \left(\frac{1}{x_M - x_{M-1}} \right) u_{M-1}^h + \left(\frac{1}{x_M - x_{M-1}} + \chi_2 \right) u_M^h = \chi_4 \tag{2.23}$$

The matrix form of these equations is

$$A U = \mathcal{F} \tag{2.24}$$

where matrix A is presented in Table 2.1 and

$$\mathcal{F} = \begin{pmatrix} \frac{\chi_1}{x_2 - x_1} + \varphi_2^h \frac{x_3 - x_1}{2} \\ \varphi_3^h \frac{x_4 - x_2}{2} \\ \vdots \\ \varphi_i^h \frac{x_{i+1} - x_{i-1}}{2} \\ \vdots \\ \varphi_{M-1}^h \frac{x_M - x_{M-2}}{2} \\ \chi_4 \end{pmatrix} . \tag{2.25}$$

It is easy to see that matrix A is symmetric and it can be proved that it is positive (see the next chapter related to elliptic equations). This means that the matrix problem and consequently FDS has a unique solution.

$$A = \begin{pmatrix}
\dfrac{1}{x_3-x_2}+\dfrac{1}{x_2-x_1} & -\dfrac{1}{x_3-x_2} & 0 & \cdots & 0 & 0 \\[2ex]
-\dfrac{1}{x_3-x_2} & \dfrac{1}{x_4-x_3}+\dfrac{1}{x_3-x_2} & -\dfrac{1}{x_4-x_3} & \cdots & 0 & 0 \\[2ex]
0 & -\dfrac{1}{x_i-x_{i-1}} & \dfrac{1}{x_i-x_{i-1}}+\dfrac{1}{x_{i+1}-x_i} & \cdots & \cdots & \cdots \\[2ex]
\vdots & \vdots & \vdots & & \vdots & \vdots \\[2ex]
0 & 0 & \dfrac{-1}{x_M-x_{M-1}} & \cdots & -\dfrac{1}{x_M-x_{M-1}}+\dfrac{1}{x_{M+1}-x_M} & -\dfrac{1}{x_M-x_{M-1}} \\[2ex]
0 & 0 & \cdots & \cdots & -\dfrac{1}{x_M-x_{M-1}} & \dfrac{1}{x_M-x_{M-1}}+\chi_2
\end{pmatrix}$$

Table 2.1: *Matrix A*.

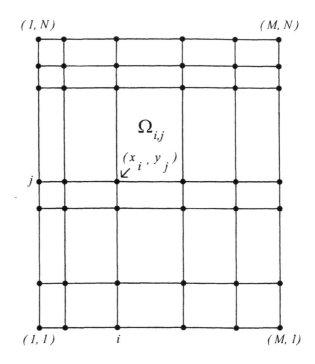

Figure 2.8: *Tensor product grid.*

2.2   2-D case

2.2.1   Grid in 2-D

The simplest example of a grid in 2-D is called the *tensor product* grid in a
rectangle. Let the domain of the variables x, y be the rectangle

$$R = \{0 \leq x \leq a\} \times \{0 \leq y \leq b\}.$$

On intervals $\{0 \leq x \leq a\}$ and $\{0 \leq y \leq b\}$ we construct the 1-D grids like
in the previous section:

$$0 \; = \; x_1 \leq x_2 < \cdots < x_i < x_{i+1} < \cdots < x_{M-1} \leq x_M = a,$$
$$0 \; = \; y_1 \leq y_2 < \cdots < y_j < x_{j+1} < \cdots < y_{N-1} \leq y_N = b.$$

The set of nodes with coordinates (x_i, y_j) on the plane is called the tensor
product grid or non-uniform rectangular grid. Clearly, the grid consists of
points intersecting the lines $x = x_i$ and $y = y_j$ (see Figure 2.8). In the case
of the tensor-product grids, density can be characterized by parameter

$$h = \max(h_x, h_y) = \max(\frac{a}{M - 1}, \frac{b}{N - 1}),$$

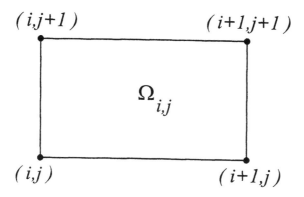

Figure 2.9: *Typical mesh of tensor product grid.*

which tends to go to zero when the number of nodes increase.

The grid is called *uniform* in some direction (x, y or both) if the 1-D grid in this direction is uniform. If the grid is uniform in both directions and steps h_x and h_y are equal then the grid is called a *square grid*.

For identification of nodes one can use two indices i, j. The rectangle with vertices (i, j), $(i+1, j)$, $(i+1, j+1)$, $(i, j+1)$ (called *cells*) (see Figure 2.8) is used for identification of this cell. We usually use the index of its lower left vertex - (i, j) and denote this cell by $\Omega_{i,j}$ (see Figure 2.9). A more complicated example of a grid in 2-D is called a *logically rectangular* grid. This grid has the same structure as the tensor product grid that makes it possible to use two indices for identification of the nodes. An example of this type of grid is given in Figure 2.10.

Logically rectangular grids play a central role in this book. The methods of construction or generation of logically rectangular grids are the subject of many books and articles. See for example, [70], [13], [135].

In this book we give just some examples of a real 2-D grid. For example, real tensor product grids can be obtained as follows. Suppose we have a square grid in a unit square $(0 < \xi < 1) \times (0 < \eta < 1)$ and

$$\xi_{i,j} = (i-1) h,$$
$$\eta_{i,j} = (j-1) h$$

are the coordinates of the nodes in this square grid. Then the coordinates of the nodes in the unit square $(0 < x < 1) \times (0 < y < 1)$ can be obtained as a result of the transformation

$$x_{i,j} = X(\xi_{i,j}, \eta_{i,j})$$
$$y_{i,j} = Y(\xi_{i,j}, \eta_{i,j}),$$

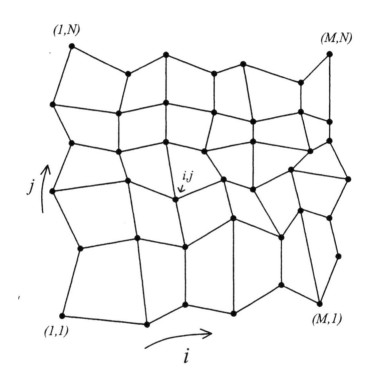

Figure 2.10: *Logically rectangular grid.*

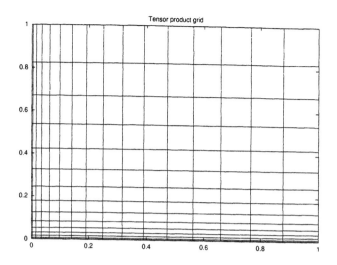

Figure 2.11: *Tensor product grid.*

where functions $X(\xi, \eta)$, $Y(\xi, \eta)$ determine the transformation. If

$$X(\xi, \eta) = \xi^2$$
$$Y(\xi, \eta) = \eta^3,$$

then we obtain a tensor product grid that is presented in Figure 2.11. Another example of a smooth curvilinear grid in a unit square can be obtained by the following transformation:

$$X(\xi, \eta) = \xi + \alpha \xi (1 - \xi)(0.5 - \xi) \eta (1 - \eta)$$
$$Y(\xi, \eta) = \eta + \alpha \eta (1 - \eta)(0.5 - \eta) \xi (1 - \xi).$$

The grid related to these functions is presented in Figure 2.12.

Both of these grids are examples of using mapping of a uniform grid to obtain a grid in a given domain. To do this we must know the analytical function which determines the transformation.

Another example of a grid can be presented using the *algebraic grid generation* technique (see [70] for details). Let us demonstrate this technique in a simple example of grid generation for domain with one curvilinear boundary. The domain in which we will generate a grid is presented in Figure 2.13, where the top boundary is the following curve:

$$Y(x) = 1 + 0.5 \sin(2 \pi x).$$

We define coordinates $x_{i,j}$ as

$$x_{i,j} = x_i = \frac{i - 1}{M - 1} ;$$

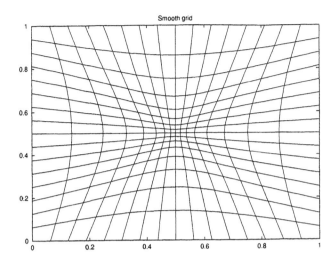

Figure 2.12: *Smooth grid.*

that is, it does not depend on index j, and line $i = const$ is a straight line that is parallel to the y axis. For each i we divide the segment of the straight line between the top and bottom boundaries on $M - 1$ that equal intervals with size

$$hy_i = \frac{Y(x_i)}{M - 1}$$

and then

$$y_{i,j} = (j - 1)\, hy_i.$$

The resulting grid is presented in Figure 2.13.

In practice we have situations where the grid in a computational domain is given for us, and is the result of some computation. For example, a grid can present fluid particles that are moved with fluid. An example of such a type of grid can be done using random perturbation of a uniform grid. Suppose we have a square grid in a unit square

$$\xi_i = (i - 1)\, h, \quad i = 1, \ldots, M$$
$$\eta_j = (j - 1)\, h, \quad i = 1, \ldots, M$$
$$h = \frac{1}{M - 1}.$$

We can then determine the coordinates of the nodes of the new grid as follows:

$$x_{i,j} = \xi_i - 0.25\, h + 0.5\, h\, R_x$$
$$y_{i,j} = \eta_j - 0.25\, h + 0.5\, h\, R_y,$$

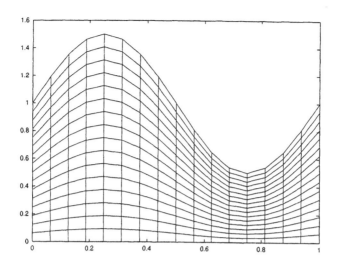

Figure 2.13: *Grid in domain with curvilinear boundaries.*

where R_x, R_y are random numbers from interval $(0,1)$. The resulting grid is presented in Figure 2.14 In the chapter relating to construction of finite-difference schemes for elliptic problems, we will demonstrate the accuracy of finite-difference schemes for different types of grids.

Now let's introduce some notation related to 2-D logically rectangular grids. As we already noted Ω_{ij} denotes a quadrangle with vertices (i,j), $(i+1,j)$, $(i+1,j+1)$, $(i,j+1)$ and are called *cells* (see Figure 2.15). The side of the quadrangle that connects vertex (i,j) and $(i+1,j)$ is denoted as $l\xi_{ij}$, and the side of the quadrangle that connects vertex (i,j) and $(i,j+1)$ is denoted as $l\eta_{ij}$. To denote the angles of Ω_{ij} we will use upper and lower indices. For example, $\varphi_{i+1,j}^{i,j}$ denotes the angle with vertex in node $i+1,j$ and is the angle of quadrangle Ω_{ij}; that is, the upper index is the index of the node and the lower index is the index of the cell (see Figure 2.15). To simplify notations we will use the same symbols for quadrangles and their area, for sides of quadrangles and their lengths.

For non-uniform grids it is usually assumed that there exists constants C_{max}^1 and C_{min}^1 that are not dependent on h, where

$$C_{min}^1 \, h^2 \le \Omega_{ij} \le C_{max}^1 \, h^2, \qquad (2.26)$$

and that there exists constants C_{max}^2 and C_{min}^2 that are not dependent on h, where

$$C_{min}^2 \, h \le l\xi_{ij}, l\eta_{ij} \le C_{max}^2 \, h. \qquad (2.27)$$

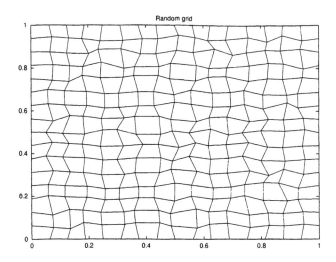

Figure 2.14: *Random grid.*

It is also usually assumed that

$$\sin \varphi_{ij}^{kl} \geq \delta > 0 \,,$$

where δ is a constant that does not depend on h. Let us note that this type of condition for *grid regularity* is also usual for finite-element methods [131].

2.2.2   Grid Functions and Difference Operators in 2-D

Scalar Functions

In this book we will consider two main types of scalar functions of a discrete argument. In the first case, the values of the function correspond to the nodes. This is *nodal discretisation*. In this case, grid function u^h is a set (vector) of $M \times N$ numbers :

$$u^h = \{u_{ij}^h \,, i = 1, \ldots, M \,; j = 1, \ldots, N\} \,.$$

The second possibility is *cell-valued discretisation*. To denote the values of a cell-valued function, we will use the same procedure as for the mesh of a grid; that is, u_{ij}^h as it is related to cell Ω_{ij}. For cell-centered discretisation, index i varies in limits from 1 to $M-1$ and $j = 1, \ldots, N-1$. The difference in nodal and cell-valued discretisation is explained in Figure 2.16.

When using integer indices for values of nodal and cell-centered functions, they may look the same. In cases where it can lead to a misunderstanding, we will write $u^h \in HN$ for nodal discretisation and $u^h \in HC$

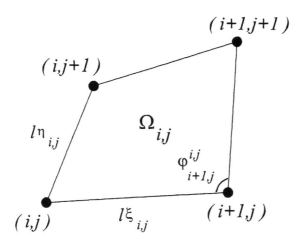

Figure 2.15: *Typical mesh of a logically rectangular grid.*

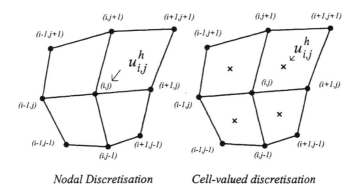

Nodal Discretisation   Cell-valued discretisation

Figure 2.16: *Nodal and cell-valued Discretisation*

for cell-centered discretisation. In general, if we do not want to concretely define the space of grid functions, we will write them as $u^h \in H_h$.

For comparison of grid functions as in a continuous case, the norm is introduced. The discrete analogs of norms in C and L_2 for $u^h \in HC$ are

$$\| u^h \| = \max_{\substack{1 \le i \le M-1 \\ 1 \le j \le N-1}} | u_{ij}^h | \qquad (2.28)$$

or

$$\| u^h \| = \left(\sum_{i,j=1}^{M-1,N-1} (u_{ij}^h)^2 * \Omega_{ij} \right)^{1/2} . \qquad (2.29)$$

Difference Operators

Let's consider an example of difference operators that act on scalar functions from HN and approximate the differential operator of the first partial derivative of u with respect to x - $\partial u / \partial x$.

To construct the difference operator we will use the Green formula,

$$\frac{\partial u}{\partial x} = \lim_{S \to 0} \frac{\oint_l u \, dy}{S} , \qquad (2.30)$$

where S is the area bounded by contour l.

In a discrete case, the role of S is played by the grid cell Ω_{ij}. Therefore l is the union of sides $l\xi_{ij}$, $l\eta_{i+1,j}$, $l\xi_{i,j+1}$, $l\eta_{ij}$. For approximation of the contour integral in the right-hand side of 2.30, we divide the contour integral into four integrals each over the corresponding side of quadrangle Ω_{ij} and for the approximate evaluation of each integral, we use the trapezium rule. As a result, we obtain the following expression for the difference analog of derivative $\partial u / \partial x$:

$$(D_x u^h)_{ij} = \frac{1}{\Omega_{ij}} \cdot$$

$$\left(\frac{u_{i+1,j} + u_{ij}}{2}(y_{i+1,j} - y_{ij}) + \frac{u_{i+1,j+1} + u_{i+1,j}}{2}(y_{i+1,j+1} - y_{i,j+1}) \right.$$

$$\left. \frac{u_{i,j+1} + u_{i+1,j+1}}{2}(y_{i,j+1} - y_{i+1,j+1}) + \frac{u_{ij} + u_{i,j+1}}{2}(y_{ij} - y_{i,j+1}) \right) .$$

Both here and below we will use the same notation for the quadrangle and its area, and for the sides and its length.

From this definition, it is clear that the natural domain of the values of operator D_x is the space HC. That is, D_x acts on discrete functions from HN and gives the discrete functions from HC:

$$D_x : HN \to HC.$$

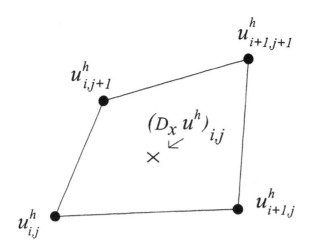

Figure 2.17: *Stencil for operator D_x in 2-D.*

In the example of D_x we can demonstrate the notion *stencil of difference operators* in 2-D. The stencil is a set of nodes that participate in the formula for discrete operators. Therefore, the stencil of operator D_x in cell (i, j) contains nodes (i, j), $(i + 1, j)$, $(i + 1, j + 1)$ $(i, j + 1)$ (see Figure 2.17).

Matrix Form of Difference Operators

The form of the matrix that corresponds to the difference operator depends on the way we number the component u_{ij}. Let us represent the discrete function u_{ij}^h as a vector U

$$U = \begin{pmatrix} u_{11} \\ u_{12} \\ \vdots \\ u_{1j} \\ \vdots \\ u_{1N} \\ u_{21} \\ u_{22} \\ \vdots \\ u_{2j} \\ \vdots \\ u_{2N} \\ \vdots \end{pmatrix}.$$

Then operator D_x corresponding to matrix A has the following structure:

$$A = \begin{pmatrix} \times & \times & 0 & 0 & \cdots & 0 & \times & \times & 0 & 0 & \cdots & 0 \\ 0 & \times & \times & 0 & \cdots & 0 & 0 & \times & \times & 0 & \cdots & 0 \\ 0 & 0 & \times & \times & \cdots & 0 & 0 & 0 & \times & \times & \cdots & 0 \\ \cdots & \cdots & \cdots & \cdots & \cdots & \cdots & \cdots & \cdots & \cdots & \cdots & \cdots & \cdots \end{pmatrix},$$

where $\times$ denotes elements that are not a zero.

This type of matrix is called the *sparse banded matrix* and matrices of this structure are typical for finite-difference algorithms. Here it is important to note that the structure of the matrix depends on the approach of numbering u_{ij}.

The examples for the approximation of second derivatives will be covered in the next chapter.

Vector Functions

In this book we will use Cartesian components of vector functions $\vec{A} = (AX, AY)$. As for scalar functions, we will use two examples. The first will be to measure these components in nodes. The notations are similar to the case of scalar functions, namely where AX_{ij}, AY_{ij} are the values for corresponding components in node (i, j). Space for this type of discrete vector functions is denoted as $\mathcal{HN}$ and the second possibility will be to measure these components in cells. Space for this type of discrete vector functions is denoted as $\mathcal{HC}$.

Because the vector function from $\mathcal{HN}$ is a combination of two scalar functions, each from HN, we can write $\mathcal{HN} = HN \oplus HN$ and similarly $\mathcal{HC} = HC \oplus HC$.

Block Difference Operators

To work with vector difference functions we must introduce block difference operators. This notion can be explained by an example of the finite-difference analog of operator divergence. In a continuous case, operator div acts on the vector function and as a result, gives the scalar function

$$\text{div}\,\vec{A} = \frac{\partial AX}{\partial x} + \frac{\partial AY}{\partial y}.$$

Let D_x and D_y be the analogs of derivatives $\partial/\partial x$ and $\partial/\partial y$ that act from HN to HC. Then we can construct the block difference operator $\text{DIV} : \mathcal{HN} \to HC$ as a matrix (2×1)

$$\text{DIV} = (D_x, D_y).$$

This operator acts as follows:

$$\text{DIV}\,\vec{A} = (D_x, D_y)\begin{pmatrix} AX \\ AY \end{pmatrix} = D_x(AX) + D_y(AY).$$

That is, we can work with block operators as we do with matrices. In a similar way we can construct the finite-difference analog of operator grad. In a differential case grad acts on the scalar function and gives us the vector function. In a discrete case we can define operator $GRAD : HN \rightarrow \mathcal{H}C$ as follows:

$$GRAD = \begin{pmatrix} D_x \\ D_y \end{pmatrix}$$

and

$$GRAD\, u^h = \begin{pmatrix} D_x \\ D_y \end{pmatrix} u^h = \begin{pmatrix} D_x(u^h) \\ D_y(u^h) \end{pmatrix}.$$

2.3 Methods for the Solution of the System of Linear Algebraic Equations

An important step towards implementation of finite-difference algorithms is the solution of the system of linear algebraic equations. Let us demonstrate this stage on the matrix problem 2.24 from the previous section. There are many effective methods for solving matrix problems with symmetric and positive matrix A. For examples, see [111].

There are two main classes of methods: direct methods and iteration methods. We will demonstrate the direct method in the example of the *Gauss elimination* method and the iteration method in the example of the *Gauss-Seidel* method.

The general system of linear algebraic equations can be written as follows:

$$
\begin{aligned}
a_{11}x_1 + a_{12}x_2 + \cdots + a_{1m}x_m &= f_1, \\
a_{21}x_1 + a_{22}x_2 + \cdots + a_{2m}x_m &= f_2, \\
\cdots \quad \cdots \quad \cdots \quad \cdots \quad \cdots & \qquad \cdots \\
\cdots \quad \cdots \quad \cdots \quad \cdots \quad \cdots & \qquad \cdots \\
a_{m1}x_1 + a_{m2}x_2 + \cdots + a_{mm}x_m &= f_m,
\end{aligned}
\tag{2.31}
$$

where $X = (x_1, x_2, \ldots x_m)^T$ is the vector of unknowns, a_{ij} are the elements of the matrix, and $F = (f_1, f_2, \ldots f_m)^T$ is the vector of the right-hand side.

The method of the *Gaussian elimination* consists in successive elimination of unknowns $x_1, x_2, \ldots, x_m$ from the system of equations. Let's assume that a_{11} is not equal to zero. Then we can divide the first equation by a_{11}. As a result we get

$$x_1 + c_{12}x_2 + \cdots\cdots + c_{1m}x_m = y_1, \tag{2.32}$$

where

$$c_{1j} = \frac{a_{1j}}{a_{11}}, \quad j = 2, \ldots, m, \quad y_1 = \frac{f_1}{a_{11}}.$$

Let us consider the remaining equations

$$a_{i1}x_1 + a_{i2}x_2 + \cdots + a_{im}x_m = f_i, \quad i = 2, 3, \ldots, m. \tag{2.33}$$

Multiply the 2.32 equation by a_{i1} and subtract the resulting equation from the i-th equation of 2.33, $i = 2, 3, \ldots, m$. As a result we get

$$
\begin{aligned}
x_1 + c_{12}x_2 + \cdots + c_{1j} + \cdots + c_{1m}x_m &= y_1, \\
a_{22}^{(1)}x_2 + \cdots + a_{mj}^{(1)} + \cdots + a_{2m}^{(1)}x_m &= f_2^{(1)}, \\
\cdots \quad \cdots \quad \cdots \quad \cdots \quad \cdots \quad &\quad \cdots \\
\cdots \quad \cdots \quad \cdots \quad \cdots \quad \cdots \quad &\quad \cdots \\
a_{m2}^{(1)}x_2 + \cdots + a_{mj}^{(1)} + \cdots + a_{mm}^{(1)}x_m &= f_m^{(1)},
\end{aligned}
\tag{2.34}
$$

where

$$a_{ij}^{(1)} = a_{ij} - c_{1j}, f_i^{(1)} = f_i - y_1 a_{i1}, \quad i, j = 2, 3, \ldots, m.$$

In system 2.34 only the first equation contains the unknown x_1; therefore, we can consider only the reduced system which contains unknowns $x_2, \ldots, x_m$:

$$
\begin{aligned}
a_{22}^{(1)}x_2 + \cdots + a_{mj}^{(1)} + \cdots + a_{2m}^{(1)}x_m &= f_2^{(1)}, \\
\cdots \quad \cdots \quad \cdots \quad \cdots \quad &\quad \cdots \\
a_{m2}^{(1)}x_2 + \cdots + a_{mj}^{(1)} + \cdots + a_{mm}^{(1)}x_m &= f_m^{(1)}.
\end{aligned}
\tag{2.35}
$$

If we found unknowns $x_2, \ldots, x_m$ from this system, we can compute the value of x_1 directly from the first equation of 2.35. With reduced system 2.35 we can do the same procedure as we did with the original system and as a result we get the system that contains only $x_3, \ldots, x_m$. Finally, we get the system

$$
\begin{aligned}
x_1 + c_{12}x_2 + \cdots\cdots\cdots + c_{1m}x_m &= y_1, \\
x_2 + \cdots\cdots + c_{1m}x_m &= y_2, \\
\cdots \quad \cdots \quad \cdots \quad \cdots \quad \cdots \quad &\cdots \\
x_{m-1} + c_{m-1,m}x_m &= y_{m-1}, \\
x_m &= y_m.
\end{aligned}
\tag{2.36}
$$

The process of obtaining system 2.36 is called an *elimination* in the Gauss method and *back substitution* is to find unknowns from 2.36. Because the matrix of this system is a triangular matrix, it is possible to successively find values of all unknowns beginning with x_m. Namely, $x_m = y_m$, $x_{m-1} = y_{m-1} - c_{m-1}x_m$, and so on. The general formula for back substitution is

$$x_m = y_m, \quad x_i = y_i - \sum_{j=i+1}^{m} c_{ij}x_j, \quad i = m - 1, \ldots, 1.$$

The main assumption which we made was the assumption that all elements $a_{kk}^{(k-1)}$ are not equal to zero. It is important to note that it is possible to use the Gauss elimination method as it described here for the solution of difference equations 2.17, 2.18, and 2.19 because diagonal elements of the related matrix do not equal zero. And in general if the matrix is symmetric and positive definite, that is

$$a_{ij} = a_{ji}$$

and

$$\sum_{i,j=1}^{m} a_{ij} \xi_i \, \xi_j > 0 \text{ , for } \vec{\xi} = (\xi_1, \xi_2, \ldots, \xi_m) \neq 0,$$

it is possible to prove that all diagonal elements are positive.

In a general situation we have to use the Gauss elimination method with *pivoting* (see, for example [51]). In this variant of the Gauss elimination method, on the k-th step of elimination we exclude not x_k, but x_j, $j \neq k$, that $|a_{kj}^{(k-1)}| > |a_{ki}^{(k-1)}|$, $i \neq j$.

The simplest example of iterative methods is the *Gauss-Seidel* method. The main idea of *iterative methods* can be explained as follows. Assume we want to solve system 2.31. The exact solution of this system is vector x and assume that we know some vector $y_1^{(0)}, y_2^{(0)}, \ldots, y_m^{(0)}$ which is some approximation to the exact solution. Next, we try to compute the new vector $y_1^{(1)}, y_2^{(1)}, \ldots, y_m(1)$ which is closer to the exact solution than the previous approximation. There are many different iterative methods (see [111]). The simplest example of the iterative methods is the *Gauss-Seidel* method. Assume that we know the $k-$th iterative approximation. Then the definition of $(k + 1)-$ st iteration begins by computation of the first component

$$a_{11} y_1^{(k+1)} = -\sum_{j=2}^{m} a_{1j} y_j^{(k)} + f_1.$$

Assuming that $a_{11} \neq 0$, we can then use this equation to find $y_1^{(k+1)}$. For $i = 2$ we obtain

$$a_{22} y_2^{(k+1)} = -a_{21} y_1^{(k+1)} - \sum_{j=3}^{m} a_{1j} y_j^{(k)} + f_2.$$

Therefore, in obtaining the second component of the new approximation we can use the second equation of the original system and the already computed first component of the new approximation.

Suppose that $y_1^{(k+1)}, y_2^{(k+1)}, \ldots, y_{i-1}^{(k+1)}$ have been found. Then y_i^{k+1} is found from the equation

$$a_{ii} y_i^{(k+1)} = -\sum_{j=1}^{i-1} a_{ij} y_j^{(k+1)} - \sum_{j=i+1}^{m} a_{ij} y_j^{(k)} + f_i.$$

From this formula it is clear that the algorithm for the Gauss-Seidel method is extremely simple. The value of $y_i^{(k+1)}$ found using this equation is over-written on $y_i^{(k)}$. It can be proved (see for example [111]) that if the matrix of the system of linear equations is symmetric and positive-definite then the Gauss-Seidel method is convergent.

Practical criterion for termination of the iteration is

$$\max_{1 \leq i \leq m} |y_i^{(k+1)} - y_i^{(k)}| < \varepsilon \tag{2.37}$$

where ε is some given tolerance. It is important to note here that, in general, when we do not have information about the spectrum of the matrix of a system of linear equations, there is no formula which gives us the estimation of accuracy of the Gauss-Seidel method in terms of ε. It is only clear that if $\varepsilon \to 0$ then we increase accuracy.

2.3.1 Application of the Gauss-Seidel Method for a Model Problem

In this section we describe how to use the Gauss-Seidel method for solving linear algebraic equations related to FDS for 1-D Poisson equations.

$$\frac{d^2u}{dx^2} = -f(x), \quad 0 < x < 1$$

with Dirichlet boundary condition at $x = 0$ and Robin boundary condition at $x = 1$:

$$u|_0 = \mu_1$$
$$\frac{du}{dx} + \mu_2\, u = \mu_4.$$

The finite-difference scheme for this problem is given by equations 2.20, 2.21, and 2.23. Let us be reminded that the vector of unknowns for this problem is

$$\begin{pmatrix} u_2^h \\ u_3^h \\ \vdots \\ u_M^h \end{pmatrix},$$

value u_1^h is known from boundary condition at $x = 0$ and equal to μ_1. To simplify the notations, we will drop index h in the notation of values for difference functions.

The formulas for the Gauss-Seidel method are as follows. For $i = 2$ we get

$$u_2^{(k+1)} = \frac{\dfrac{u_3^{(k)}}{x_3 - x_2} + \dfrac{\chi_1}{x_2 - x_1} + \varphi_2^h \dfrac{x_3 - x_1}{2}}{\dfrac{1}{x_3 - x_2} + \dfrac{1}{x_2 - x_1}}. \tag{2.38}$$

For $i = 3, \ldots, M - 1$ we get

$$u_i^{(k+1)} = \frac{\dfrac{u_{i-1}^{(k+1)}}{x_i - x_{i-1}} + \dfrac{u_{i+1}^{(k)}}{x_{i+1} - x_i} + \varphi_i^h \dfrac{x_{i+1} - x_{i-1}}{2}}{\dfrac{1}{x_{i+1} - x_i} + \dfrac{1}{x_i - x_{i-1}}}. \tag{2.39}$$

And finally, for node $i = M$ we get

$$u_M^{(k+1)} = \frac{\chi_4 + \dfrac{u_{M-1}^{(k+1)}}{x_M - x_{M-1}}}{\dfrac{1}{x_M - x_{M-1}} + \chi_2}. \tag{2.40}$$

These formulas give us the algorithm for computing all values $u_i^{(k+1)}$. The computational procedure is as follows. At first we must choose some initial value for vector u, that is, vector $u^{(0)}$. The Gauss-Seidel method will give the solution for difference equations independently of the choice of the initial value, but certainly the number of iterations needed to achieve the same accuracy will be different for different initial values and will be smaller if the initial value is closer to the exact solution of the system for difference equations. Suppose we know all values $u_i^{(k)}$ on some iteration k. Then the value $u_2^{(k+1)}$ can be determined from equation 2.38. The equation for determination of $u_3^{(k+1)}$ contains the value $u_2^{(k+1)}$ in the right-hand side which we already computed and value $u_4^{(k)}$ from the previous iteration. Therefore in computing any $u_i^{(k+1)}$ we use the already-computed value $u_{i-1}^{(k+1)}$ in the new iteration and value $u_{i+1}^{(k)}$ from the previous iteration. To terminate the iteration process we can use criteria 2.37.

2.4 Methods for the Solution of the System of Non-Linear Equations

In this book we will also consider finite-difference schemes for the gas dynamic equation in Lagrangian form. This is the system of non-linear equations. The resulting finite-difference scheme will also give us the system of non-linear equations and we must solve this system. In this section we briefly describe the Newton and Newton-Kantorovich methods for solving systems of non-linear equations. Readers who want to know more about methods for the solution of non-linear equations can find information in [94]. To demonstrate the main ideas we begin with the problem of solving one non-linear equation.

The Newton Method for One Non-Linear Equation

Let us consider the function $f(x)$ of one variable x. The problem is to find x^* which satisfies the equation

$$f(x^*) = 0.$$

The example of function $f(x)$ is presented in figure 2.18. To find x^* is the same as finding the point of intersection of curve $f(x)$ with the x axis of the coordinate system. Let us suppose that the initial approximation for x^* is given and is equal to $x^{(0)}$. Then if we approximately replace the original function

$$y = f(x)$$

with linear function

$$y = f'(x^{(0)})(x - x^{(0)}) + f(x^{(0)}),$$

which is the equation for the tangent line to curve $f(x)$ in point $x^{(0)}$, find the point of intersection of this tangent line with the x axis and it will give us the next approximation $x^{(1)}$ for x^*. The formula for $x^{(1)}$ is

$$x^{(1)} = x^{(0)} - \frac{f(x^{(0)})}{f'(x^{(0)})}.$$

And in general,

$$x^{(k+1)} = x_k - \frac{f(x^{(k)})}{f'(x^{(k)})}. \tag{2.41}$$

Let us note that replacing the function $f(x)$ with its tangent line is the same as replacing the function $f(x)$ by the first two terms of its Taylor series

$$f(x) \approx f(x^{(0)}) + (x - x^{(0)}) f'(x^{(0)}).$$

This interpretation will be useful for the system of non-linear equations.

Graphically the Newton iteration process is presented in Figure 2.18. We do not discuss here what conditions in function $f(x)$ and choice of initial value $x^{(0)}$ must be satisfied for convergence of the process or what the convergence rate is. Readers who are interested in these subjects can find more information in [94]. Let us note that problems which we will solve by the Newton method are convergent because we have the initial values that are close enough to the exact solution. From equation 2.41 we can see that for every iteration we must recompute not only the function in the new point $x^{(k)}$ but also the derivative $f'(x^{(k)})$ and the computation of this derivative can sometimes be a very time consuming procedure. To avoid computation of the derivative for each iteration, we can use a modification of the Newton method. One possible modification is described below.

The Newton-Kantorovich Method for One Non-Linear Equation

The formula for a modification of the Newton method which is called the *Newton-Kantorovich method* is

$$x^{(k+1)} = x^{(k)} - g \, f'(x^{(k)}), \tag{2.42}$$

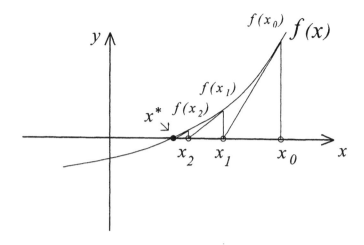

Figure 2.18: *Newton method.*

where g is a constant. For example, if

$$g = \frac{1}{f'(x^{(0)})},$$

then this method is called the modified Newton method. Generally, the choice of g depends on the choice of the initial value and in some sense g must be close to $1/f'$. In general, modifications of the Newton method have a smaller rate of convergence, but the total amount of computational time can be less than the cost of time saved for computing derivatives on each iteration. Graphically the Newton-Kantorovich method is illustrated in Figure 2.19.

The Newton and Newton-Kantorovich Methods for the System of Non-Linear Equations

The system of non-linear equations can be written as follows:

$$
\begin{aligned}
f_1(x_1, x_2, \ldots, x_m) &= 0, \\
f_2(x_1, x_2, \ldots, x_m) &= 0, \\
&\cdots\cdots\cdots\cdots\cdots \\
&\cdots\cdots\cdots\cdots\cdots \\
f_m(x_1, x_2, \ldots, x_m) &= 0,
\end{aligned}
\tag{2.43}
$$

where f_i, $i = 1, 2, \ldots, m$ is a function of variables $x_1, x_2, \ldots, x_m$. Let us describe the Newton method for these systems of equations. Let us suppose that we know values $x_i^{(k)}$ on iteration k. Then we can write the Taylor series

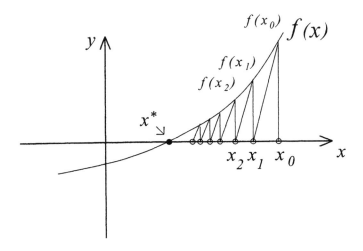

Figure 2.19: *Newton-Kantorovich method.*

for functions $f_i(x_1, x_2, \ldots, x_m)$ in point $x^{(k)} = (x_1^{(k)}, x_2^{(k)}, \ldots, x_m^{(k)})^T$:

$$f_i(x_1, x_2, \ldots, x_m) = f_i(x_1^{(k)}, x_2^{(k)}, \ldots, x_m^{(k)}) + (x_1 - x_1^{(k)}) \frac{\partial f_i(x^{(k)})}{\partial x_1} +$$

$$(x_2 - x_2^{(k)}) \frac{\partial f_i(x^{(k)})}{\partial x_2} + \ldots + (x_m - x_m^{(k)}) \frac{\partial f_i(x^{(k)})}{\partial x_m} + O(|x - x^{(k)}|^2).$$

If we now omit the terms of the second-order infinitesimal, then the original system can be replaced by the system of equations

$$\sum_{j=1}^{m}(x_j - x_j^{(k)}) \frac{\partial f_i(x^{(k)})}{\partial x_j} + f_i(x^{(k)}) = 0, \quad i = 1, 2, \ldots, m \qquad (2.44)$$

which is linear with respect to increments $x_j - x_j^{(k)}$. Then we take the solution of system 2.44 as the next approximation for the exact solution and denote it as

$$x^{(k+1)} = (x_1^{(k+1)}, x_2^{(k+1)}, \ldots, x_m^{(k+1)})^T.$$

Therefore, Newton's method for the system of non-linear equations can be written as follows

$$\sum_{j=1}^{m}(x_j^{(k+1)} - x_j^{(k)}) \frac{\partial f_i(x^{(k)})}{\partial x_j} + f_i(x^k) = 0, \quad i = 1, 2, \ldots, m. \qquad (2.45)$$

If we use notations

$$F(x) = (f_1(x),\ f_2(x),\ \ldots,\ f_m(x))^T$$

and

$$F'(x) = \begin{bmatrix} \frac{\partial f_1(x)}{\partial x_1} & \frac{\partial f_1(x)}{\partial x_2} & \cdots & \frac{\partial f_1(x)}{\partial x_m} \\ \frac{\partial f_2(x)}{\partial x_1} & \frac{\partial f_2(x)}{\partial x_2} & \cdots & \frac{\partial f_2(x)}{\partial x_m} \\ \cdots & \cdots & \cdots & \cdots \\ \frac{\partial f_m(x)}{\partial x_1} & \frac{\partial f_m(x)}{\partial x_2} & \cdots & \frac{\partial f_m(x)}{\partial x_m} \end{bmatrix}$$

then the matrix form of 2.45 looks as follows:

$$F'(x^{(k)})(x^{(k+1)} - x^{(k)}) + F(x^{(k)}) = 0. \tag{2.46}$$

To find $x^{(k+1)}$ we must solve the system of linear equations or, in other words, we must invert matrix $F'(x^{(k)})$ on each iteration. Therefore, at least the inverse matrix must exist. Each particular problem will need special investigation. Also, the existence of effective methods for the solution of linear systems depends on the properties of this matrix.

Similar to the case of one non-linear equation there is a modification of the Newton method called the Newton-Kantorovich method for the system of non-linear equations. In this method, instead of formula 2.46, values of $x_i^{(k+1)}$ on a new iteration satisfy the equation

$$G(x^{(k+1)} - x^{(k)}) + F(x^{(k)}) = 0, \tag{2.47}$$

where matrix G is close to matrix F' in some sense. Let us note that the Newton-Kantorovich method can have some advantages in comparison to the original Newton methods because the problem of solving the system of linear equations with a non-standard matrix $F'(x^{(k)})$ can be very difficult. We will demonstrate the use of the Newton-Kantorovich method in an example of non-linear finite-difference equations that correspond to gas dynamic equations in the Lagrangian form.

Chapter 2

Method of Support-Operators

There are two basic types of restrictions imposed on FDS, the first being approximation and stability restrictions stipulating the convergence of the approximate solution when the mesh size is small enough. The second type of requirement refers to the retention by the FDS of the important properties of the original differential equations. When computing the numerical solution to partial differential equations, difference operators that mimic the crucial properties of the differential operators are usually more accurate than those that do not. Properties such as symmetry, conservation, and duality relationships and identities between gradient, curl, and divergence are all important.

The author's opinion is that the most important of these properties is conservation of FDS. The conservation of FDS is especially important when computing the numerical solution using actual meshes.

There are several general constructive approaches such as the finite-volume or integral-interpolation method [137], the mapping method [70], and the variational method [138], [139] which enable us to find the FDS having the foregoing properties.

Independently of the approach used, the determination of FDS can be considered to be the construction of finite-difference analogs of basic invariant differential operators of the first order like grad, div, and curl of mathematical physics.

The advantages of the operational representation of FDS become clearer in the case of general logically rectangular grids and unstructured grids, in which the properties of the difference operators ensuring conservation of corresponding FDS become important.

In this book we demonstrate the possibility of constructing FDS by using a new method called the *support-operators method*. This method has been developed by A. Favorskii, A. Samarskii, M. Shashkov, and V. Tishkin [112], [113] over a number of years. Here a brief description of the method is given. A detailed description of the application of this method to construct finite-difference schemes for equations described in the introduction is given in the following special sections.

1 Main Stages

There are five main steps in the support-operators method of constructing conservative difference schemes for equations of mathematical physics.

1. First, write the original equations in terms of the invariant first-order differential operators div, grad, and curl.

2. It is important to understand what properties of the invariant differential operators imply the conservation laws and other main features of the original system of differential equations. In this connection, two different classes of operator properties are distinguished. The first class contains the properties of a single operator. The second class contains the properties that are connected with two or more operators. The second class of properties plays a primary role in this development. The main relations of the second type that are used for construction of finite-difference schemes are:

$$\int_V \varphi \operatorname{div} \vec{A} \, dV + \int_V (\vec{A}, \operatorname{grad} \varphi) \, dV = \oint_S \varphi (\vec{A}, \vec{n}) \, dS \,,$$

$$\int_V (\vec{A}, \operatorname{curl} \vec{B}) \, dV + \int_V (\vec{B}, \operatorname{curl} \vec{A}) \, dV = \oint_S (\vec{n}, \vec{A} \times \vec{B}) \, dS \,,$$

$$\int_V \varphi (\vec{C}, \operatorname{curl} \vec{A}) \, dV - \int_V (\vec{C}, \vec{A} \times \operatorname{grad} \varphi) \, dV =$$

$$\oint_S \varphi (\vec{C}, \vec{n} \times \vec{A}) \, dS \,,$$

$$\int_V \varphi \psi \operatorname{div} \vec{A} \, dV + \int_V \varphi (\vec{A}, \operatorname{grad} \psi) \, dV + \qquad (1.1)$$

$$\int_V \psi (\vec{A}, \operatorname{grad} \varphi) \, dV = \oint_S \varphi \psi (\vec{A}, \vec{n}) \, dS \,,$$

$$\int_V \nabla \vec{A} \cdot \cdot \psi \hat{\sigma}^* \, dV + \int_V (\vec{A}, \psi \operatorname{div} \hat{\sigma}) \, dV +$$

$$\int_V (\vec{A}, (\operatorname{grad} \psi \cdot \hat{\sigma})) \, dV = \oint_S (\psi \vec{n}, (\hat{\sigma} \cdot \vec{A})) \, dS$$

where V is the arbitrary volume with surface S and the outer unit normal $\vec{n}$; φ and ψ are arbitrary scalar functions; $\vec{A}$ and $\vec{B}$ are arbitrary vector functions; $\hat{\sigma}$ is the arbitrary tensor function; $\nabla \vec{A}$ is the gradient of vector function; $\cdot\cdot$ denotes double convolution of two tensors; $\cdot$ denotes convolution of tensor and vector; and $\vec{C}$ is a vector such that curl $\vec{C} = 0$.

In this book the first identity in (1.1) is used to derive consistent difference operators approximations for div and grad that are used in the construction of the difference schemes for all types of equations described in the introduction.

3. Choose where the scalar, vector, and tensor functions are to be located in the grid. In this book we consider two possibilities. The first possibility is to use nodal discretisation for the scalar function and a cell-centered discretisation of the Cartesian component vector function. A second possibility is to use nodal discretisation for the Cartesian component of the vector function and cell-centered discretisation for the scalar functions.

4. The choice of the *prime* operator must be made. This operator must be one of a first-order differential operator that is used in the formulation of original equations. A difference analog for this operator is given by definition. Then all other difference operators are derived from the definition of the discretisation of the prime operator and difference analogs of integral identities. In the case of nodal discretisation of the vector function, it is natural to use the divergence as the prime operator. In the case of nodal discretisation of the scalar function, it is natural to use the operator grad as the prime.

5. The formulas for other difference operators, called *derived* operators, are derived using the expression for the prime operator and the form of difference analogs of the integral identities.

6. Finally, if we have discrete analogs of differential operators which form original differential equations, then we can derive the FDS by substituting the discrete operators for differential operators.

2 The Elliptic Equations

Here we show how to use the support-operators method for the elliptic problem

$$-\operatorname{div} K \operatorname{grad} u = f \, .$$

This equation can be interpreted as a stationary heat equation. Then there is only one conservation law. That is the law of heat conservation

$$\frac{dQ}{dt} = -\oint_S (\vec{w}, \vec{n}) \, dS$$

where

$$Q = \int_V u \, dV, \quad \vec{w} = -K \operatorname{grad} u,$$

are the total amount of heat and heat flux.

From a formal viewpoint, this relationship can be obtained as follows:

$$\frac{dQ}{dt} = \frac{d}{dt} \int_V u \, dV = -\int_V \operatorname{div} \vec{w} \, dV = -\oint_S (\vec{w}, \vec{n}) \, dS.$$

Therefore to obtain conservative finite-difference scheme, the property

$$\int_V \operatorname{div} \vec{w} \, dV = \oint_S (\vec{w}, \vec{n}) \, dS \tag{2.1}$$

must be preserved in the discrete case. We will call property 2.1, *the divergence property* of operator divergence.

The differential operator $-\operatorname{div} K \operatorname{grad}$ is self-adjoint and positive. That is, if we introduce inner product

$$(u, v)_H = \int_V u \, v \, dV$$

and consider functions which are equal to zero on the boundary, then

$$(-\operatorname{div} K \operatorname{grad} u, v)_H = (u, -\operatorname{div} K \operatorname{grad} v)_H,$$

and

$$(-\operatorname{div} K \operatorname{grad} u, u)_H \geq 0.$$

From a formal viewpoint, this property follows from the first integral identity in (1.1),

$$\int_V \varphi \operatorname{div} \vec{A} \, dV + \int_V (\vec{A}, \operatorname{grad} \varphi) \, dV = \oint_S \varphi (\vec{A}, \vec{n}) \, dS, \tag{2.2}$$

which connects the operators div and grad, and properties of matrix K. In fact if for simplicity we consider the Dirichlet boundary condition, when $\varphi = 0$ on the boundary, we can write

$$\int_V (-\operatorname{div} K \operatorname{grad} u) v \, dV = \int_V (K \operatorname{grad} u, \operatorname{grad} v) \, dV \tag{2.3}$$

and because K is a symmetric positive definite matrix we can conclude that $-\operatorname{div} K \operatorname{grad}$ is self-adjoint and positive.

Therefore when deriving a finite-difference scheme, it is important to preserve the difference analog of the integral identity in (2.2). Namely, this integral identity is the basis for deriving the self-consistent system of difference operators DIV and GRAD, which are analogs of the differential operators divergence and gradient.

The choice of a prime operator depends on the type of discretisation and can be either DIV or GRAD. For example, for nodal discretisation of the vector function, operator DIV is used as prime because it is natural for the discretisation. The invariant definition of the prime operator div is used to derive its approximation DIV:

$$\operatorname{div} \vec{w} = \lim_{V \to 0} \frac{1}{V} \oint_{\partial V} (\vec{w}, \vec{n}) \, dV \,,$$

which gives a difference operator DIV preserving the divergence property of the differential operator div.

The derived operator GRAD is constructed using the difference analog of the integral identity in (2.2). Then the finite-difference scheme will preserve the symmetry and positive properties of the differential operator div K grad.

The finite-difference scheme can be written in a form that is similar to the differential case. That is

$$\operatorname{DIV} K \operatorname{GRAD} U = F \,,$$

where K is some discretisation for matrix K, F is the approximation for f, and U is the approximate solution.

3 Gas Dynamic Equations

One possible form of a Gas Dynamic Equation is the *Lagrangian form*

$$\frac{d\rho}{dt} = -\rho \operatorname{div} \vec{W} \,,$$

$$\rho \frac{d\vec{W}}{dt} = -\operatorname{grad} p \,,$$

$$\rho \frac{d\varepsilon}{dt} = -p \operatorname{div} \vec{W} \,,$$

where ρ is density, $\vec{W}$ is the velocity vector, p is pressure, and ε is specific internal energy.

For this equation, we have three main laws of conservation: conservation of mass, conservation of momentum, and conservation of total energy.

If we write an equation of continuity in the form

$$\frac{\partial \rho}{\partial t} + \operatorname{div} \rho \vec{W} = 0$$

then it is easy to see that conservation of mass will take place in the discrete case, if the difference analog of operator div, operator DIV will be conservative.

Conservation of momentum can be derived from the momentum equation by multiplying it by the element of volume and integrating it over the computational domain

$$\int_V \rho \frac{d\vec{W}}{dt}\, dV = \int_V \operatorname{grad} p\, dV.$$

Taking into account that for the Lagrange approach $d(\rho\, dV)/dt = 0$ we can rewrite the previous equation as follows:

$$\frac{d}{dt} \int_V \rho \vec{W}\, dV = \int_V \operatorname{grad} p\, dV.$$

The left-hand side is changing in total momentum and it must depend only on the work of surface forces. This means that the integral in the right-hand side must be transformed to the surface integral. This is another property that we want to preserve in the discrete case. That is, we want to construct the difference operator GRAD that will be divergent.

Conservation of total energy can be derived from the momentum equation and the equation for specific internal energy. Namely, if we form the scalar product of the momentum equation with vector $\vec{W}$, then add the resulting equation with the equation for specific internal energy and integrate the resulting equation over V we obtain

$$\frac{d}{dt}\left[\int_V \left(\frac{|\vec{W}|^2}{2} + \varepsilon\right) \rho\, dV\right] = \tag{3.1}$$

$$-\left[\int_V (\operatorname{grad} p, \vec{W})\, dV + \int_V p \operatorname{div} \vec{W}\, dV\right] = -\oint_{\partial V} p\,(\vec{W}, \vec{n})\, dS\,.$$

The last equality is the consequence of the integral identity relating to the operators div and grad. Equation 3.1 implies that when there is no exterior influence, the change in the total energy is equal to the work done by surface forces. This is a very important conservation law and we want to preserve it in the discrete case. It is clear that if difference analogs of operators div and grad are constructed independently, the analog of the integral identity in general fails to hold, and correspondence FDS will not be conservative. Therefore, if we want to construct conservative FDS, finite-difference analogs of div and grad must satisfy the difference analog of the integral identity.

Therefore for gas dynamic equations we have to construct operators DIV and GRAD with the following properties: DIV has divergent form, GRAD has divergent form, DIV and GRAD satisfy the difference form of the integral identity in 2.2.

Here we need to say that this terminology: "divergent form", "operator has divergent from", "operator is divergent" will be used throughout the book. For discrete operators it means that the discrete analog of integral over volume of these operators can be reduced to some sum over the boundary.

4 System of Consistent Difference Operators in 1-D

Let us consider how to construct a system of difference analogs of operators div and grad in a one-dimensional case that will satisfy conditions formulated at the end of the last section.

We will assume that $V = [0, 1]$ and $\vec{A} = \{AX, 0, 0\}$. That is, $\vec{A}$ has only one component. All scalar functions and components of vectors depend only on the variable x.

In the case where grad and div are both equal to du/dx, we must construct two different difference analogs for the first derivative.

In 1-D the first identity in 1.1 looks as follows:

$$\int_0^1 \varphi \frac{dAX}{dx}\,dx + \int_0^1 AX \frac{d\varphi}{dx}\,dx = \varphi(1)\,AX(1) - \varphi(0)\,AX(0).$$

For simplicity, assume that function φ is equal to zero at $x = 0$ and $x = 1$. Then the right-hand side in the previous equation is equal to zero and this equation can be rewritten as follows:

$$\int_0^1 \varphi \frac{dAX}{dx}\,dx + \int_0^1 AX \frac{d\varphi}{dx}\,dx = 0. \tag{4.1}$$

Therefore in the 1-D case, difference analogs of the first derivative must mimic some difference analog of this identity.

At first we consider the case where the difference scalar function belongs to HN, $u^h \in HN$ and $u_1^h = u_M^h = 0$. It means that in this case we will consider discrete functions $u^h = (u_2^h, u_3^h, \ldots, u_{M-1}^h)$. It is natural in this case to choose D_x as the prime operator (which is the analog of the first derivative) and $D_x : HN \to HC$. It's value in cell i is (see Introduction)

$$\left(D_x u^h\right)_i = \frac{u_{i+1}^h - u_i^h}{x_{i+1} - x_i}.$$

Because function u^h is equal to zero on the boundary we have to modify the definition of operator D_x in the cells $i = 1$ and $i = M - 1$:

$$\left(D_x u^h\right)_1 = \frac{u_2^h}{x_2 - x_1},$$

$$\left(D_x u^h\right)_{M-1} = \frac{-u_{M-1}^h}{x_M - x_{M-1}}.$$

Therefore, to have the ability to approximate the second derivative, we must construct another difference analog for the first derivative $\mathcal{D}$ where $\mathcal{D}_x : HC \to HN$.

To construct the difference operator $\mathcal{D}_x$, we will use the difference analog of the identity 4.1.

For difference functions $u^h \in HN$ and $v^h \in HC$ it looks as follows:

$$\sum_{i=2}^{M-1} u_i^h \left(\mathcal{D}_x v^h\right)_i VN_i + \sum_{i=1}^{M-1} \left(\mathcal{D}_x u^h\right) v_i^h VC_i = 0, \qquad (4.2)$$

where VC_i is the length (volume in 3-D) of cell i, VN_i is a length that corresponds to the node (near node volume in 3-D). There is a natural restriction for VC and VN, namely that it must be greater than zero and its sum must approximate the total volume

$$VC_i, VN_i > 0, \qquad \sum_{i=1}^{M-1} VC_i \approx 1; \sum_{i=2}^{M-1} VN_i \approx 1.$$

In this section we will use the following expressions for VC:

$$VC_i = x_{i+1} - x_i, \quad i = 1, \ldots, M - 1,$$

and

$$VN_i = \begin{cases} \frac{x_2 - x_1}{2} & \text{if } i = 1 \\ \frac{x_{i+1} - x_{i-1}}{2}, & i=2, \ldots, \text{M-1} \\ \frac{x_M - x_{M-1}}{2} & \text{if } i = M \end{cases} .$$

Let us transform the second sum in the difference identity to select a coefficient near u_i^h:

$$\sum_{i=1}^{M-1} \left(\mathcal{D}_x u^h\right) v_i^h VC_i =$$

$$\frac{u_2^h}{x_2 - x_1} v_1^h (x_2 - x_1) +$$

$$\sum_{i=2}^{M-2} \frac{u_{i+1}^h - u_i^h}{x_{i+1} - x_i} v_i^h (x_{i+1} - x_i) +$$

$$\frac{-u_{M-1}^h}{x_M - x_{M-1}} v_{M-1}^h (x_M - x_{M-1}) =$$

$$\sum_{i=1}^{M-2} u_{i+1}^h v_i^h - \sum_{i=2}^{M-1} u_i^h v_i^h.$$

To transform the first sum in the right-hand side we introduce a new summation index $i' = i + 1$, then we get

$$\sum_{i'=2}^{M-1} u_{i'}^h v_{i'-1}^h.$$

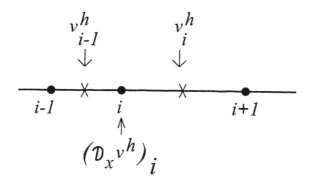

Figure 4.1: *Stencil for operator $\mathcal{D}_x$.*

Because i' is the index of the summation, we can denote it as i again, then from the previous consideration we get

$$\sum_{i=1}^{M-1} \left(D_x u^h \right)_i v_i^h \, V C_{ij} = \sum_{i=2}^{M-1} u_i^h \left(v_{i-1}^h - v_i^h \right).$$

Finally from this equation and the difference analog of the integral identity 4.2, we can conclude that

$$\left(\mathcal{D}_x v^h \right)_i = \frac{v_i^h - v_{i-1}^h}{V N_i}$$

and using the formula for $V N_i$,

$$\left(\mathcal{D}_x v^h \right)_i = \frac{v_i^h - v_{i-1}^h}{0.5 \left(x_{i+1} - x_{i-1} \right)} \quad i = 2, \ldots, M - 1.$$

The stencil for operator $\mathcal{D}_x$ contains two cells: i and $i - 1$ (see Figure 4.1).

Thus we can construct two difference operators D_x and $\mathcal{D}_x$ which correspond to derivative d/dx to satisfy the difference identity 4.2.

Operators D_x and $\mathcal{D}_x$ act as follows:

$$\begin{aligned} D_x &: & HN &\to HC \\ \mathcal{D}_x &: & HC &\to HN. \end{aligned}$$

We have to recall here that we actually consider the subspace of HN when discrete functions are equal to zero on the boundary.

Because the domain of one operator is the same as the range of the values of another operator the difference analogs of the second-order differential operation can be constructed. In particular, two different approximations of the second derivative can be constructed:

$$D_x(\mathcal{D}_x) \quad : \quad HC \to HC$$
$$\mathcal{D}_x(D_x) \quad : \quad HN \to HN.$$

Details about these operators will be discussed in the chapter relating to construction of finite-difference schemes for elliptic equations.

4.1   Inner Product in Spaces of Discrete Functions and Properties of Difference Operators

If we introduce *inner products* in spaces of difference functions then we can give a more elegant interpretation of the results from the previous section.

In the continuous case for two functions u, v on segment $[0, 1]$ the *inner product* can be determined as follows:

$$(u, v)_H = \int_0^1 u\,v\,dx. \tag{4.3}$$

Then 4.1 can be rewritten as

$$\left(\varphi, \frac{dAX}{dx} \right)_H + \left(AX, \frac{d\varphi}{dx} \right)_H = 0$$

or

$$\left(\varphi, \frac{dAX}{dx} \right)_H = \left(AX, -\frac{d\varphi}{dx} \right)_H. \tag{4.4}$$

Using the notion of the inner product, we can introduce the notion of the *adjoint* operator. For a given operator $A : H_1 \to H_2$, operator B is called *adjoint* to operator A if

$$(A\,u, v)_{H_2} = (u,\, B\,v)_{H_1}, \quad \text{for all } u \in H_1 \text{ and } v \in H_2.$$

The operator that is adjoint to A is denoted by A^*.

Then identity 4.4 means that in the continuous case we get

$$\left(\frac{d}{dx} \right)^* = -\frac{d}{dx}. \tag{4.5}$$

In the discrete case for space HC, the inner product can be determined as follows:

$$\left(u^h, v^h \right)_{HC} = \sum_{i=1}^{M-1} u_i^h\, v_i^h\, VC_i,$$

and in a similar way for HN:

$$\left(u^h, v^h\right)_{HN} = \sum_{i=1}^{M} u_i^h v_i^h VN_i.$$

Using the notation for inner products, the difference analog of integral identity 4.2 can be rewritten as follows:

$$\left(D_x u^h,\, v^h\right)_{HC} + \left(u^h,\, \mathcal{D}_x v^h\right)_{HN} = 0$$

or

$$\left(D_x u^h,\, v^h\right)_{HC} = \left(u^h,\, -\mathcal{D}_x v^h\right)_{HN}.$$

According to the definition of an adjoint operator it means that

$$\mathcal{D}_x = -\left(D_x\right)^*.$$

Therefore in the difference case, our difference operators mimic a very important property of the derivative in 4.5. This property plays a very important role and we will return to its consideration in the next chapter.

5   System of Consistent Discrete Operators in 2-D

Let us consider how to construct a system of difference analogs of operators div and grad in the two-dimensional case for logically rectangular grids. In this book, we will mainly use projections of vectors to Cartesian axes for the description of vector fields. Then

$$\operatorname{grad} u = \begin{pmatrix} \frac{\partial u}{\partial x} \\[6pt] \frac{\partial u}{\partial y} \end{pmatrix}$$

and

$$\operatorname{div} \vec{A} = \frac{\partial AX}{\partial x} + \frac{\partial AY}{\partial y}.$$

Thus grad and div contain only the first derivative with respect to x and y.

Therefore to construct the finite-difference analogs of operators grad and div, we must construct the finite-difference analogs of derivatives $\partial/\partial x$ and $\partial/\partial y$.

For this type of description of vector fields, the first identity 1.1 can be rewritten in terms of the first derivatives. If the scalar function is equal to zero on the boundary, then

$$\int_V \varphi \left(\frac{\partial AX}{\partial x} + \frac{\partial AY}{\partial y}\right) dx\, dy + \int_V \left(AX \frac{\partial \varphi}{\partial x} + AY \frac{\partial \varphi}{\partial y}\right) dx\, dy = 0. \quad (5.1)$$

As for the 1-D case, we can give the interpretation of this identity in terms of inner products. Let us introduce inner products in spaces of scalar and vector functions as follows:

$$(u, v)_H = \int_V u\, v\, dx\, dy,$$

$$(\vec{A}, \vec{B})_{\mathcal{H}} = \int_V (\vec{A}, \vec{B})\, dx\, dy,$$

where

$$(\vec{A}, \vec{B}) = AX\, BX + AY\, BY$$

is the scalar product of two vectors. Then identity 5.1 can be rewritten as follows:

$$(\varphi, \operatorname{div}\vec{A})_H + (\operatorname{grad}\varphi, \vec{A})_{\mathcal{H}} = 0$$

or

$$(\varphi, \operatorname{div}\vec{A})_H = (-\operatorname{grad}\varphi, \vec{A})_{\mathcal{H}}.$$

This means that

$$(\operatorname{div})^* = -\operatorname{grad}.$$

Because identity 5.1 must hold for any vector $\vec{A} = (AX, AY)$, we can consider two cases $\vec{A} = (AX, 0)$ and $\vec{A} = (0, AY)$. For these cases, identity 5.1 looks as follows:

$$\int_V \varphi\left(\frac{\partial AX}{\partial x}\right) dx\, dy + \int_V \left(AX\, \frac{\partial \varphi}{\partial x}\right) dx\, dy = 0 \qquad (5.2)$$

$$\int_V \varphi\left(\frac{\partial AY}{\partial y}\right) dx\, dy + \int_V \left(AY\, \frac{\partial \varphi}{\partial y}\right) dx\, dy = 0. \qquad (5.3)$$

It is evident that identity 5.1 follows from equations 5.2 and 5.3.

Thus if we use projections to Cartesian axes for the description of a vector field, we can use identities 5.2 and 5.3 instead of 5.1.

Identities 5.2 and 5.3 can be rewritten in terms of inner products

$$\left(\varphi, \frac{\partial AX}{\partial x}\right)_H = \left(AX, -\frac{\partial \varphi}{\partial x}\right)_H,$$

$$\left(\varphi, \frac{\partial AY}{\partial y}\right)_H = \left(AY, -\frac{\partial \varphi}{\partial y}\right)_H.$$

Let us go now to the discrete case. From the previous consideration, we can conclude that for the construction of a consistent system of difference analogs of operators div and grad, we must preserve in difference form some analog of the identity 5.1. That is, we must construct finite-difference analogs of the first derivatives which will satisfy the difference analogs of 5.2 and 5.3.

Nodal Discretisation of Scalar Functions and Cell-Centered Discretisation of Vector Functions

In this section we consider the case where the scalar functions belong to space HN, $u^h \in HN$ and vector functions (described by their projections to Cartesian axes) belong to HC, $\vec{A} = (AX, AY)$; $AX, AY \in HC$, that is, $\vec{A} \in \mathcal{HC}$.

For this type of discretisation, we must construct finite-difference operators

$$D_x, D_y : HN \to HC$$

which will be the analogs of the derivatives $\partial/\partial x$ and $\partial/\partial y$. Using these operators we can form the *prime* operator GRAD : $HN \to \mathcal{HC}$:

$$\text{GRAD}\, u^h = \begin{pmatrix} D_x(u^h) \\ D_y(u^h) \end{pmatrix}. \tag{5.4}$$

To construct operators D_x and D_y we will use the *Green* formula (see, for example, [103]):

$$\frac{\partial u}{\partial x} = \lim_{S \to 0} \frac{\oint_{\partial S} u\, dy}{S}, \tag{5.5}$$

$$\frac{\partial u}{\partial y} = -\lim_{S \to 0} \frac{\oint_{\partial S} u\, dx}{S}, \tag{5.6}$$

where S is some area and ∂S its boundary. We already used formula 5.5 in the previous chapter. Here we briefly repeat the consideration from this chapter. In the discrete case, the role of S is shown in grid cell Ω_{ij} (see Figure 5.2). Therefore ∂S is a union of sides $l\xi_{ij}$, $l\eta_{i+1,j}$, $l\xi_{i,j+1}$, $l\eta_{ij}$. For an approximation of the contour integral in the right-hand side of 5.5 and 5.6, we divide the contour integral into four integrals each over the corresponding side of quadrangle Ω_{ij}. Because each side of the quadrangle is a segment of the line, the integral over each side is a one-dimensional integral and to approximate it, we use the trapezium rule. As a result, we obtain the following expression for the difference analog of derivative $\partial u/\partial x$:

$$(D_x u^h)_{ij} = \left(\frac{u_{i+1,j} + u_{ij}}{2}(y_{i+1,j} - y_{ij}) + \right.$$

$$\frac{u_{i+1,j+1} + u_{i+1,j}}{2}(y_{i+1,j+1} - y_{i+1,j}) +$$

$$\frac{u_{i,j+1} + u_{i+1,j+1}}{2}(y_{i,j+1} - y_{i+1,j+1}) +$$

$$\left. \frac{u_{ij} + u_{i,j+1}}{2}(y_{ij} - y_{i,j+1}) \right) / \Omega_{ij}. \tag{5.7}$$

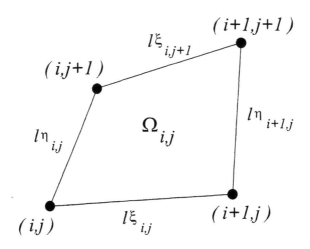

Figure 5.2: *Stencil for operators D_x and D_y.*

The domain of difference operator D_x is HN and the range of values is HC:

$$D_x : HN \rightarrow HC.$$

Collecting the coefficients, we can transform the formula for D_x to a more simple form:

$$(D_x u^h)_{ij} = \qquad (5.8)$$
$$\frac{(u_{i+1,j+1} - u_{ij})(y_{i,j+1} - y_{i+1,j}) - (u_{i,j+1} - u_{i+1,j})(y_{i+1,j+1} - y_{i,j})}{2\,\Omega_{ij}}.$$

In a similar way using 5.6 we derive the formula for operator D_y:

$$(D_y u^h)_{ij} = - \qquad (5.9)$$
$$\frac{(u_{i+1,j+1} - u_{ij})(x_{i,j+1} - x_{i+1,j}) - (u_{i,j+1} - u_{i+1,j})(x_{i+1,j+1} - x_{i,j})}{2\,\Omega_{ij}}.$$

The stencils of operators D_x and D_y in cell (i, j) are the same and contain nodes (i, j), $(i+1, j)$, $(i+1, j+1)$ $(i, j+1)$ (see Figure 5.2). Using operators D_x and D_y we can determine the block operator GRAD : $HN \rightarrow \mathcal{HC}$:

$$\text{GRAD} = \left(\begin{array}{c} D_x \\ D_y \end{array} \right).$$

Now we must construct another finite-difference analog for the first derivative $\partial/\partial x$ that goes from HC to HN. We denote this operator as $\mathcal{D}_x$. To do this we will use the difference analog of identity 5.2:

$$\sum_{i,j=1}^{M,N} \varphi_{ij} \left(\mathcal{D}_x AX \right)_{ij} VN_{ij} + \sum_{i,j=1}^{M-1,N-1} \left(\mathcal{D}_x \varphi \right)_{ij} AX_{ij} VC_{ij} = 0, \qquad (5.10)$$

where VC_{ij} is the area (volume in 3-D) relating to the cell Ω_{ij} and VN_{ij} is the area (volume in 3-D) relating to the node (i,j). The approximation properties of operator $\mathcal{D}_x$ are strongly dependent on the choice of VC_{ij} and VN_{ij}. Now we can assume that $VC_{ij} = \Omega_{ij}$ and we will not use the concrete expressions for VN_{ij}. Details relating to the choice of VN_{ij} will be considered in the next chapter.

To obtain the formula for operator $\mathcal{D}_x$ we must transform the second sum in 5.10 to find a coefficient near φ_{ij}. To do this we will use a technique that is similar to the 1-D case. That is, changing the indices of the summation. Substituting the explicit form of operator $\mathcal{D}_x$ to the second sum in 5.10 we get

$$\sum_{i,j=1}^{M-1,N-1} \left(\mathcal{D}_x \varphi \right)_{ij} AX_{ij} VC_{ij} =$$

$$\sum_{i,j=1}^{M-1,N-1} \varphi_{i+1,j+1} \frac{y_{i,j+1} - y_{i+1,j}}{2} AX_{ij} -$$

$$\sum_{i=1;j=1}^{M-1;N-1} \varphi_{ij} \frac{y_{i,j+1} - y_{i+1,j}}{2} AX_{ij} +$$

$$\hspace{9cm} (5.11)$$

$$\sum_{i,j=1}^{M-1,N-1} \varphi_{i,j+1} \frac{y_{i+1,j+1} - y_{i,j}}{2} AX_{ij} -$$

$$\sum_{i,j=1}^{M-1,N-1} \varphi_{i+1,j} \frac{y_{i+1,j+1} - y_{i,j}}{2} AX_{ij}.$$

Let us, for example, transform the first sum

$$\sum_{i,j=1}^{M-1,N-1} \varphi_{i+1,j+1} \frac{y_{i,j+1} - y_{i+1,j}}{2} AX_{ij}.$$

By changing the indices of the summation by introducing new indices $i' = i+1$, $j' = j+1$, we get

$$\sum_{i',j'=2}^{M,N} \varphi_{i',j'} \frac{y_{i'-1,j'} - y_{i',j'-1}}{2} AX_{i'-1,j'-1}.$$

We then denote indices of the summation as i, j and obtain

$$\sum_{i,j=2}^{M,N} \varphi_{i,j} \frac{y_{i-1,j} - y_{i,j-1}}{2} AX_{i-1,j-1}.$$

If we transform all four terms in a similar way and take into account that $\varphi = 0$ on the boundary $(i = 1, j = 1, i = M, j = N)$, we get

$$\sum_{i,j=2}^{M-1,N-1} (D_x \varphi)_{ij} AX_{ij} VC_{ij} =$$

$$\sum_{i,j=2}^{M-1,N-1} \varphi_{i,j} \left(\frac{y_{i-1,j} - y_{i,j-1}}{2} AX_{i-1,j-1} - \frac{y_{i,j+1} - y_{i+1,j}}{2} AX_{ij} + \right.$$

$$\left. \frac{y_{i+1,j} - y_{i,j-1}}{2} AX_{i,j-1} - \frac{y_{i,j+1} - y_{i-1,j}}{2} AX_{i-1,j} \right).$$

(5.12)

By comparison of 5.10 and the last equation, we can conclude that

$$(\mathcal{D}_x AX)_{ij} =$$

$$- \left(\frac{y_{i-1,j} - y_{i,j-1}}{2} AX_{i-1,j-1} - \frac{y_{i,j+1} - y_{i+1,j}}{2} AX_{ij} + \right.$$

(5.13)

$$\left. \frac{y_{i+1,j} - y_{i,j-1}}{2} AX_{i,j-1} - \frac{y_{i,j+1} - y_{i-1,j}}{2} AX_{i-1,j} \right) / V N_{ij}.$$

Similarly for $\mathcal{D}_y$ we get

$$(\mathcal{D}_y AY)_{ij} =$$

$$\left(\frac{x_{i-1,j} - x_{i,j-1}}{2} AY_{i-1,j-1} - \frac{x_{i,j+1} - x_{i+1,j}}{2} AY_{ij} + \right.$$

(5.14)

$$\left. \frac{x_{i+1,j} - x_{i,j-1}}{2} AY_{i,j-1} - \frac{x_{i,j+1} - x_{i-1,j}}{2} AY_{i-1,j} \right) / V N_{ij}.$$

The stencils for operators $\mathcal{D}_x$ and $\mathcal{D}_y$ are the same and are shown in Figure 5.3

If we introduce inner products in spaces HC and HN as

$$(u^h, v^h)_{HC} = \sum_{i,j=1}^{M-1,N-1} u_{ij}^h v_{ij}^h VC_{ij},$$

$$(u^h, v^h)_{HN} = \sum_{i,j=1}^{M,N} u_{ij}^h v_{ij}^h V N_{ij},$$

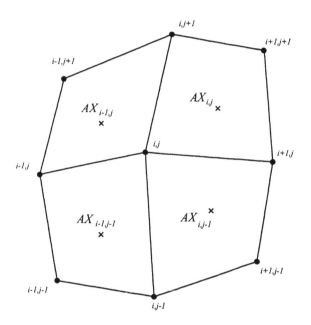

Figure 5.3: *Stencil for operators $\mathcal{D}_x$ and $\mathcal{D}_y$.*

then the difference identity 5.10 can be rewritten as

$$(\varphi,\, \mathcal{D}_x AX)_{HN} + (D_x \varphi,\, AX)_{HC} = 0.$$

This means that similar to the continuous case we get

$$\mathcal{D}_x = -\left(D_x\right)^*.$$

And in a similar way

$$\mathcal{D}_y = -\left(D_y\right)^*.$$

Now using operators $\mathcal{D}_x$ and $\mathcal{D}_y$ we can make up operator $\mathcal{DIV}$:

$$\mathcal{DIV} = \left(\mathcal{D}_x \mid \mathcal{D}_y\right).$$

Using properties of operators D_x, $\mathcal{D}_x$, D_y, $\mathcal{D}_y$ and the definition of the adjoint operator, we can conclude that

$$\mathcal{DIV} = -\mathrm{GRAD}^*.$$

Nodal Discretisation of Vector Functions and Cell-Centered Discretisation of Scalar Functions

Now let's consider a second case when vector functions belong to $\mathcal{HN}$, $\vec{A} \in \mathcal{HN}$. That is, $AX \in HN$ and $AY \in HN$ and scalar functions belonging to HC, $\varphi \in HC$.

To construct the system of consistent finite-difference analogs of operators grad and div we can use already constructed operators D_x, $\mathcal{D}_x$, D_y, $\mathcal{D}_y$. Namely, we will use as the *prime* operator

$$\mathrm{DIV} = (\mathrm{D_x} \mid \mathrm{D_y}) : \mathcal{HN} \to \mathrm{HC}.$$

Then the *derived* operator $\mathcal{GRAD}$ is adjoint to DIV:

$$\mathcal{GRAD} = -\mathrm{DIV}^* = - \begin{pmatrix} \mathcal{D}_x \\ \\ \mathcal{D}_y \end{pmatrix}.$$

Thus for both types of discretisation we have constructed consistent systems of finite-difference analogs for operators grad and div. All details relating to these operators will be considered in the next chapter.

Chapter 3

The Elliptic Equations

1   Introduction

This chapter considers the problem of accurately solving boundary value problems that consist of a general elliptic partial-differential equation

$$- \operatorname{div} K \operatorname{grad} u = f, \quad (x, y) \in V \qquad (1.1)$$

given in some region V, and Dirichlet

$$u(x, y) = \varphi, \quad (x, y) \in \partial V \qquad (1.2)$$

or a general mixed *Robin* boundary condition

$$(\vec{n}, K \operatorname{grad} u) + \alpha u = \gamma \quad (x, y) \in \partial V \qquad (1.3)$$

given on the boundary ∂V of V. The problem is solved for the function u when K is a given matrix that is symmetric and positive definite, f is a source term, $\vec{n}$ is a unit outward normal to ∂V, and φ, α, and γ are given functions given on ∂V.

In the next two sections the properties of the continuum boundary value problem for Dirichlet and Robin boundary conditions are explicitly derived in preparation for the discretisation process. It is assumed that all functions are smooth enough that all definitions make sense. Consequently, precise definitions of function spaces are not given.

1.1   Continuum Elliptic Problems. Dirichlet Boundary Conditions

Theoretical investigations of the Dirichlet problem usually begin with the homogeneous boundary condition case

$$\begin{aligned} -\operatorname{div} K \operatorname{grad} u = f, \quad & (x, y) \in V, \\ u = 0, \quad & (x, y) \in \partial V. \end{aligned} \qquad (1.4)$$

In other words, the problem is to find a solution for the equation

$$-\operatorname{div} K \operatorname{grad} u = f, \quad (x, y) \in V,$$

which belongs to space $\overset{0}{H}$ of functions defined on V and are the functions that are equal to zero on the boundary ∂V. The inner product in $\overset{0}{H}$ is

$$(u, v)_{\overset{}{H}} = \int_V u\, v\, dV. \tag{1.5}$$

The original equation can be rewritten in the operator form

$$\overset{0}{\mathcal{A}}\, u = f,$$

where the operator $\overset{0}{\mathcal{A}}$ is given by

$$\overset{0}{\mathcal{A}} : \overset{0}{H} \to \overset{0}{H}, \quad \overset{0}{\mathcal{A}}\, u = -\operatorname{div} K \operatorname{grad} u.$$

1.1.1 The Properties of the Differential Operator

Let us prove that the operator $\overset{0}{\mathcal{A}}$ has the following properties:

$$(\overset{0}{\mathcal{A}}\, u, v)_{\overset{}{H}} = (u, \overset{0}{\mathcal{A}}\, v)_{\overset{}{H}}, \quad (\overset{0}{\mathcal{A}}\, u, u)_{\overset{}{H}} > 0, \quad \text{for } u \neq 0. \tag{1.6}$$

The proof of these properties relies on the definition of the operator, the inner product, and the boundary conditions:

$$
\begin{aligned}
(\overset{0}{\mathcal{A}}\, u, v)_{\overset{}{H}} &= -\int_V (\operatorname{div} K \operatorname{grad} u)\, v\, dV \\
&= \int_V (K \operatorname{grad} u, \operatorname{grad} v)\, dV - \oint_{\partial V} (K \operatorname{grad} u, \vec{n})\, v\, dS \\
&= \int_V (K \operatorname{grad} u, \operatorname{grad} v)\, dV.
\end{aligned}
\tag{1.7}
$$

This relationship implies that $\overset{0}{\mathcal{A}}$ is a symmetric operator.

It is important to note that in the second step of this transformation, we first used the integral identity from 1.1 which connects the operators div and grad. That is,

$$\int_V u \operatorname{div} \vec{w}\, dV + \int_V \vec{w} \operatorname{grad} u\, dV = \oint_{\partial V} u\, (\vec{w}, \vec{n})\, dS. \tag{1.8}$$

If v is taken to be equal to u in the previous equation, then

$$(\overset{0}{\mathcal{A}}\, u, u)_{\overset{}{H}} = \int_V (K \operatorname{grad} u, \operatorname{grad} u)\, dV$$

and from the properties of the matrix K, we can conclude that $\overset{0}{\mathcal{A}}$ is positive.

1.1.2   The Properties of First-Order Operators and the Operator of the BVP

To investigate the properties of first-order operators div and grad, the space of vector functions must be defined. The notation $\mathcal{H}$ is used for this space and for the two vector functions $\vec{A}, \vec{B} \in \mathcal{H}$, the inner product is defined as:

$$(\vec{A}, \vec{B})_{\mathcal{H}} = \int_V \vec{A} \cdot \vec{B}\, dV .$$

For the next consideration, it is useful to rewrite the original equations in flux form:

$$\operatorname{div} \vec{w} = f, \quad (x, y) \in V,$$
$$\vec{w} = -K \operatorname{grad} u, \quad (x, y) \in V,$$
$$u = 0, \quad (x, y) \in \partial V.$$

Therefore we can consider operators div and grad as

$$\operatorname{div} : \mathcal{H} \to \overset{0}{H}$$
$$\operatorname{grad} : \overset{0}{H} \to \mathcal{H}.$$

The integral identity 1.8 for functions from $\overset{0}{H}$ is

$$\int_V u \operatorname{div} \vec{w}\, dV + \int_V \vec{w} \operatorname{grad} u\, dV = 0.$$

Now using the definitions of the inner product in spaces $\overset{0}{H}$ and $\mathcal{H}$, we can rewrite this identity as follows:

$$(u, \operatorname{div} \vec{w})_{\overset{0}{H}} + (\vec{w}, \operatorname{grad} u)_{\mathcal{H}} = 0$$

or

$$(u, \operatorname{div} \vec{w})_{\overset{0}{H}} = -(\vec{w}, \operatorname{grad} u)_{\mathcal{H}}.$$

This last equation means that

$$\operatorname{div} = -\operatorname{grad}^* .$$

Let us recall that to show symmetry and positivity properties of the operator div K grad, we used only the integral identity, which means that div $= -\operatorname{grad}^*$. Therefore, this property of the operators div and grad is responsible for symmetry and positivity properties of the operator from the original boundary value problem.

These considerations, in the continuous case, illuminate the properties of the first-order operators that must be preserved when the finite-difference schemes are constructed.

1.2   Continuum Elliptic Problems.   Robin Boundary Conditions

To better understand the properties of differential operators in the general case when the solution is not equal to zero on the boundary, let us consider the case of Robin boundary conditions.

The problem to be solved is

$$-\text{div } K \text{ grad } u = f, \qquad (x,y) \in V,$$
$$(K\text{grad } u, \vec{n}) + \alpha\, u = \psi, \qquad (x,y) \in \partial V, \quad \alpha \geq 0, \qquad (1.9)$$

where the function u belongs to space H with the following inner product:

$$(u,\, v)_H = \int_V u\, v\, dV + \oint_{\partial V} u\, v\, dS, \qquad u, v \in H. \qquad (1.10)$$

Let us note that there are no restrictions on the boundary for function u, but the definition of the inner product includes a boundary integral.

Equation (1.9) can be written in the operator form

$$\mathcal{A}u = \mathcal{F},$$

where the operator $\mathcal{A}$ is given by

$$\mathcal{A}: H \to H, \quad \mathcal{A}u = \begin{cases} -\text{div } K \text{ grad } u, & (x,y) \in V \\ (K\text{grad } u, \vec{n}) + \alpha u, & (x,y) \in \partial V \end{cases},$$

and the right-hand side has the form

$$\mathcal{F} = \begin{cases} f, & (x,y) \in V \\ \psi, & (x,y) \in \partial V \end{cases}.$$

1.2.1   The Properties of the Differential Operator

Now, let us prove that the operator $\mathcal{A}$ has the following properties:

$$(\mathcal{A}u,\, v)_H = (u,\, \mathcal{A}v)_H, \qquad (\mathcal{A}u,\, u)_H \geq 0,$$
$$(1.11)$$
$$(\mathcal{A}u,\, u)_H > 0, \quad \text{if } \alpha > 0.$$

The proof of these properties relies on the definition of the operator and the inner product:

$$(\mathcal{A}u,\, v)_H = -\int_V (\text{div } K \text{ grad } u)\, v\, dV$$
$$+ \oint_{\partial V} (K \text{ grad } u, \vec{n})\, v\, dS + \oint_{\partial V} \alpha\, u\, v\, dS$$

$$= \int_V (K \operatorname{grad} u, \operatorname{grad} v) \, dV - \oint_{\partial V} (K \operatorname{grad} u, \vec{n}) \, v \, dS$$

$$+ \oint_{\partial V} (K \operatorname{grad} u, \vec{n}) \, v \, dS + \oint_{\partial V} \alpha \, u \, v \, dS \qquad (1.12)$$

$$= \int_V (K \operatorname{grad} u, \operatorname{grad} v) \, dV + \oint_{\partial V} \alpha \, u \, v \, dS .$$

This relationship implies that $\mathcal{A}$ is a symmetric operator.

If v is taken to be equal to u in the previous equation (1.12), then

$$(\mathcal{A}u, u)_H = \int_V (K \operatorname{grad} u, \operatorname{grad} u) \, dV + \oint_{\partial V} \alpha \, u^2 \, dS \geq 0 .$$

Now because matrix K is symmetric and the positive definite first term in the right-hand side of this equation is greater than zero, the second term in the right-hand side also greater than zero if $\alpha > 0$, and therefore

$$(\mathcal{A}u, u)_H \geq 0 , \quad (\mathcal{A}u, u)_H > 0 , \text{if } \alpha > 0 .$$

That is, $\mathcal{A}$ is non-negative, and if $\alpha > 0$ then $\mathcal{A}$ is positive.

1.2.2 The Properties of First-Order Operators and the Operator of the BVP

Here we will use the same space of vector functions as for the case of Dirichlet boundary conditions. For the next consideration it is useful, as before, to rewrite the original equations in flux form:

$$\operatorname{div} \vec{w} = f , \qquad \vec{w} = -K \operatorname{grad} u ,$$
$$-(\vec{w}, \vec{n}) + \alpha \, u = \psi , \qquad (x, y) \in \partial V . \qquad (1.13)$$

The previous relation (1.13) shows that operator $\mathcal{A}$ can be represented in the form

$$\mathcal{A} = \mathcal{B} \mathcal{K} \mathcal{C} + \mathcal{D} ,$$

where the operators $\mathcal{B}$, $\mathcal{K}$, $\mathcal{C}$, and $\mathcal{D}$ have the following definitions:

$$\begin{aligned}
\mathcal{C}u &= -\operatorname{grad} u , \quad \text{for all } (x, y), \\
\mathcal{K}\vec{w} &= K \, \vec{w} , \quad \text{for all } (x, y), \\
\mathcal{B}\vec{w} &= \begin{cases} +\operatorname{div} \vec{w} , & (x, y) \in V \\ -(\vec{w}, \vec{n}) , & (x, y) \in \partial V , \end{cases} \\
\mathcal{D}u &= \begin{cases} 0 , & (x, y) \in V \\ \alpha \, u , & (x, y) \in \partial V , \end{cases}
\end{aligned} \qquad (1.14)$$

and

$$\mathcal{C} : H \to \mathcal{H}, \; \mathcal{K} : \mathcal{H} \to \mathcal{H}, \; \mathcal{B} : \mathcal{H} \to H, \; \mathcal{D} : H \to H . \qquad (1.15)$$

Let us prove that
$$\mathcal{B} = \mathcal{C}^* \, .$$

The definition of operator $\mathcal{B}$, the formula (1.10) for the inner product in space H, and the formula from (1.1) which connects div and grad, give the following result:

$$
\begin{aligned}
(\mathcal{B}\,\vec{w}, u)_H &= \int_V u \operatorname{div} \vec{w}\, dV - \oint_{\partial V} u\,(\vec{w}, \vec{n})\, dS \\
&= -\int_V (\vec{w}, \operatorname{grad} u)\, dV = (\vec{w}, \mathcal{C}\, u)_{\mathcal{H}} \, .
\end{aligned}
$$

The statement of the original problem gives
$$\mathcal{K} = \mathcal{K}^* > 0 \, .$$

It is also easy to derive the required relations for the operator $\mathcal{D}$:
$$(\mathcal{D}u, v)_H = \oint_{\partial V} \alpha\, u\, v\, dS = (u, \mathcal{D}v)_H \, ,$$

and
$$(\mathcal{D}u, u)_H = \oint_{\partial V} \alpha\, u^2\, dS \geq 0 \, ,$$

because $\alpha \geq 0$. This means that
$$\mathcal{A} = \mathcal{C}^*\, \mathcal{K}\, \mathcal{C} + \mathcal{D} \, .$$

Then the properties (1.11) of operator $\mathcal{A}$ follow from the properties of operators $\mathcal{B}, \mathcal{K}, \mathcal{C},$ and $\mathcal{D}$. The required properties are
$$\mathcal{B} = \mathcal{C}^*, \quad \mathcal{D} = \mathcal{D}^* \geq 0, \quad \mathcal{K} = \mathcal{K}^* > 0 \, .$$

These considerations, in the continuous case, illuminate the properties of the first-order operators that must be preserved when the finite-difference schemes are constructed.

2   One-Dimensional Support-Operators Algorithms

In this section, the support-operators method is used to construct a finite-difference scheme for elliptic equations with Dirichlet and general Robin boundary conditions in the 1-D case.

The Poisson equation in 1-D can be written as follows:

$$- \operatorname{div} K \operatorname{grad} u = -\frac{d}{dx}\left(K \frac{du}{dx}\right) = f, \quad a < x < b. \tag{2.1}$$

The *Dirichlet* problem is to solve equation 2.1 on the interval $[a, b]$ with the following boundary conditions:

$$u(a) = \psi_a, \quad u(b) = \psi_b. \tag{2.2}$$

The problem is to solve for the function u when K is a given positive function of coordinates and f is a source term, where ψ_a and ψ_b are given numbers.

The *Robin* problem is to solve equation 2.1 with the following boundary conditions:

$$-(K \frac{du}{dx}) + \alpha_a u = \psi_a, \quad x = a,$$

$$\tag{2.3}$$

$$(K \frac{du}{dx}) + \alpha_b u = \psi_b, \quad x = b,$$

where $\alpha_a > 0$, $\alpha_b > 0$, ψ_a, ψ_b are given numbers.

Again, the basic equation can be written as the first-order system

$$\text{div}\, \vec{w} = f, \quad \vec{w} = -K \,\text{grad}\, u,$$

or if $\vec{w} = (WX, 0, 0)$, then as

$$\frac{dWX}{dx} = f, \quad WX = -K \frac{du}{dx}.$$

Then the Robin boundary conditions can be rewritten in terms of $\vec{w}$ and u as follows:

$$WX + \alpha_a u = \psi_a, \quad x = a,$$

$$\tag{2.4}$$

$$-WX + \alpha_b u = \psi_b, \quad x = b.$$

Finite-difference schemes are constructed for two cases. The first case uses a *nodal* discretisation of scalar functions and a *cell-centered* discretisation of vector functions, while the second case uses a *cell-centered* discretisation for scalar functions and a *nodal* discretisation for vector functions.

2.1 Nodal Discretisation of Scalar Functions and Cell-Centered Discretisation of Vector Functions

Consider the segment $[a, b]$ and introduce the grid with nodes

$$\{x_i, i = 1, \cdots, M\},$$

$$a = x_1 < \cdots < x_i < x_{i+1} < \cdots < x_M = b$$

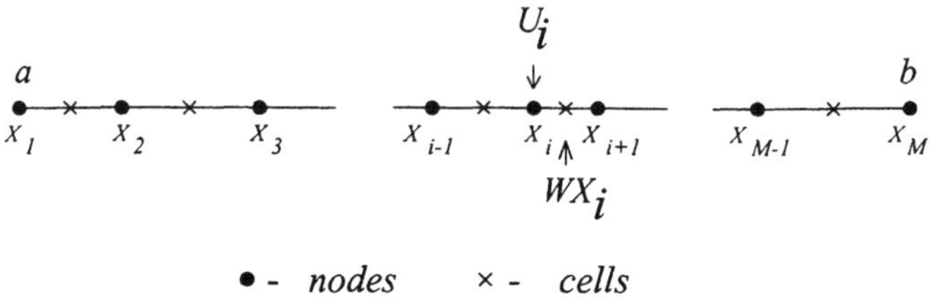

Figure 2.1: *N-C Discretisation.*

(see Figure 2.1). It is important that these nodes can be irregularly distributed.

Scalar functions are discretised using values at the nodes (see Figure 2.1). This is called *node* or *nodal* discretisation. We denote the discrete scalar function u^h, that is

$$u^h = (u_1^h, u_2^h, \cdots, u_M^h).$$

If it will not lead to misunderstanding we will drop index h.

A vector function $\vec{W} = (WX, 0, 0)$ is discretised using its values in the center of cells (see Figure 2.1) which is called *cell-centered* discretisation:

$$WX = (WX_1, WX_2, \cdots, WX_{M-1}).$$

2.1.1 Spaces of Discrete Functions

In accordance with the theory of support operators, spaces of discrete functions that are analogous of spaces H and $\mathcal{H}$ must be defined. The space HN of discrete scalar functions (see previous subsection) is given the inner product

$$(u, v)_{HN} = \sum_{i=2}^{M-1} u_i v_i VN_i + u_1 v_1 + u_M v_M,$$

where VN_i is the volume of the i-th node:

$$VN_i = 0.5 (x_{i+1} - x_{i-1}).$$

This inner product converges as mesh size goes to zero to the expression (1.10) for the inner product in the continuous case. The space of the vector function is denoted $\mathcal{H}C$ and has the inner product

$$(\vec{A}, \vec{B})_{\mathcal{H}C} = \sum_{i=1}^{M-1} (\vec{A}, \vec{B})_i VC_i,$$

where $(\vec{A}, \vec{B})_i$ is the scalar product of the two vectors in cell i. In 1-D,

$$(\vec{A}, \vec{B})_i = AX_i\, BX_i\,,$$

and VC_i is the volume of a cell:

$$VC_i = x_{i+1} - x_i, \quad i = 1, \cdots, M - 1\,. \tag{2.5}$$

2.1.2 The Prime Operator

For nodal discretisation of the scalar functions and cell-centered discretisation of the vector functions, it is natural to use the finite-difference analog of operator $\mathcal{C} = -\text{grad}$ as the prime operator. For the difference analog of operator $\mathcal{C}$, we will use the notation $\mathcal{C}^h$ and similar notations for other difference analogs of our differential operators. The considerations of the first chapter give us the following formulas for $\mathcal{C}^h = -\text{GRAD} : HN \to \mathcal{HC}$:

$$(\mathcal{C}^h u)_i = -(\text{GRAD}\, u)_i = -\frac{u_{i+1} - u_i}{VC_i}, \quad i = 1, \cdots, M - 1. \tag{2.6}$$

2.1.3 The Derived Operator

In accordance with the theory of support operators, see the considerations in the previous section, the derived operator is the finite-difference analog of the operator

$$\mathcal{B}\vec{w} = \begin{cases} \text{div}\,\vec{w}\,, & (x, y) \in V \\ -(\vec{w}, \vec{n})\,, & (x, y) \in \partial V \end{cases}, \tag{2.7}$$

which can be derived using the difference analog of the property $\mathcal{B} = \mathcal{C}^*$. For a given discretisation it means that

$$\sum_{i=2}^{M-1} (\mathcal{B}^h\, \vec{W})_i\, u_i\, VN_i + (\mathcal{B}^h\, WX)_M\, u_M + (\mathcal{B}^h\, WX)_1\, u_1 =$$

$$\sum_{i=1}^{M-1} (\vec{W}, \mathcal{C}^h\, u)_i\, VC_i\,.$$

The following chain of formulas gives the expression for operator $\mathcal{B}^h$:

$$(\text{GRAD}\, u, W)_{\mathcal{HC}} = \tag{2.8}$$

$$\sum_{i=1}^{M-1} (\text{GRAD}\, u)_i\, WX_i\, VC_i =$$

$$\sum_{i=1}^{M-1} (u_{i+1} - u_i)\, WX_i =$$

$$\sum_{i=1}^{M-1} u_{i+1}\, W X_i - \sum_{i=1}^{M-1} u_i\, W X_i =$$

$$\sum_{i=2}^{M} u_i\, W X_{i-1} - \sum_{i=1}^{M-1} u_i\, W X_i =$$

$$\sum_{i=2}^{M-1} u_i\, (W X_{i-1} - W X_i) + u_M\, (W X_{M-1}) + u_1\, (-W X_1).$$

The final expression for operator $\mathcal{B}^h$ is

$$(\mathcal{B}^h\, \vec{W})_i = \begin{cases} (\mathrm{DIV}\ \vec{W})_i = \frac{W X_i - W X_{i-1}}{V N_i} & i = 2, \ldots M - 1 \\ (\vec{W},\, \vec{n})_1 = -W X_1 & i = 1 \\ (\vec{W},\, \vec{n})_M = W X_{M-1} & i = M. \end{cases} \tag{2.9}$$

Thus we have an approximation for the operator div at the internal nodes and an approximation of the normal component for flux on the boundary. The approximation of the flux on the boundary will be used for constructing an FDS for the Poisson equation with Robin boundary conditions.

It is easy to check that, for homogeneous boundary conditions and by using the space of scalar functions with the inner product

$$(u,\, v)_{\underset{H}{0}} = \sum_{i=2}^{M-1} u_i\, v_i\, V N_i,$$

we get

$$\mathrm{DIV} = -\mathrm{GRAD}^*.$$

Therefore the difference operators DIV and GRAD mimic the important properties of the differential operators. It is evident from the formula for DIV that it is a divergence operator. This property also follows from the fact that GRAD is equal to zero for the function which is equal to 1 at all nodes.

2.1.4 Multiplication by a Matrix K

The definition for the difference analog of the multiplication of a vector by a matrix depends on the discretisation of the matrix elements. In 1-D, a matrix has only one entry:

$$\mathcal{K} = (KXX).$$

To shorten the notation, set $KXX = K$. If a cell-centered discretisation of the matrix element is used, then the discrete matrix multiplication operator $\mathcal{K}$ is given by

$$(\mathcal{K}^h\, WX)_i = K_i\, WX_i,$$

and then

$$\mathcal{K}^h : \mathcal{HC} \rightarrow \mathcal{HC}.$$

For other types of approximations of the matrix elements, a more complicated formula will be needed. Also, when entries of the matrix are discontinuous, a special averaging procedure related to the position of discontinuity is needed (see [95]).

2.1.5 The Finite-Difference Scheme for the Dirichlet Boundary Value Problem

Using the above considerations, the difference scheme at the internal nodes can be written in the following form:

$$(-\text{DIV} \cdot K \cdot \text{GRAD}\, u)_i = \varphi_i, \quad i = 2, \ldots, M - 1 \qquad (2.10)$$

where φ is some approximation for the function f at the nodes x_i; for example, we can use $\varphi_i = f(x_i)$.

The approximation of the Dirichlet boundary conditions is

$$u_1 = \psi_a, \quad u_M = \psi_b. \qquad (2.11)$$

The explicit form of difference equations in the internal nodes is

$$-\frac{K_i \left(\frac{u_{i+1}-u_i}{x_{i+1}-x_i}\right) - K_{i-1} \left(\frac{u_i-u_{i-1}}{x_i-x_{i-1}}\right)}{(x_{i+1} - x_{i-1})} = f(x_i)/2. \qquad (2.12)$$

From the Dirichlet boundary conditions (2.11) we know the values of the difference function u in nodes $i = 1$ and $i = M$. Thus, the vector of unknowns is

$$u = \begin{pmatrix} u_2 \\ u_3 \\ \vdots \\ u_{M-1} \end{pmatrix}.$$

Because known values u_1 and u_M are contained only in the equations for nodes $i = 2$ and $i = M - 1$, we can transform these equations by moving known values to the right-hand side of the equation:

$$-\frac{K_2 \left(\frac{u_3-u_2}{x_3-x_2}\right) - K_1 \left(\frac{u_2}{x_2-x_1}\right)}{0.5\,(x_3 - x_1)} = \varphi_2 + \frac{K_1 \left(\frac{\psi_a}{x_2-x_1}\right)}{0.5\,(x_3 - x_1)},$$

and

$$\frac{K_{M-1} \left(\frac{-u_{M-1}}{x_M-x_{M-1}}\right) - K_{M-2} \left(\frac{u_{M-1}-u_{M-2}}{x_{M-1}-x_{M-2}}\right)}{0.5\,(x_M - x_{M-2})} =$$

$$\varphi_{M-1} + \frac{K_{M-1} \left(\frac{\psi_b}{x_M-x_{M-1}}\right)}{0.5\,(x_M - x_{M-2})}.$$

By moving the known values to the right-hand side of the finite-differen-ce equations, we change not only the right-hand side but also the left-hand side of the equation. Changing the left-hand side is equivalent to consider-ing the operator GRAD on functions which are equal to zero on the bound-ary. Thus, the new finite-difference equations are similar to those for the homogeneous Dirichlet boundary value problem for the Poisson equation.

For the class of difference functions which are equal to zero on the boundary, we have $DIV = -GRAD^*$. Then, the basic equation (2.10) can be written in the form

$$(GRAD^* \cdot K \cdot GRAD\, u)_i = \tilde{\varphi}_i, \qquad (2.13)$$

where the formula for GRAD is changed near the boundary by using ho-mogeneous boundary conditions and $\tilde{\varphi}_i$ is the new right-hand side. That is, for $i = 1$ and $i = M - 1$, operator GRAD is given by

$$(GRAD\, u)_1 = \frac{u_2}{x_2 - x_1},$$

$$(GRAD\, u)_M = \frac{-u_{M-1}}{x_M - x_{M-1}},$$

and

$$\tilde{\varphi}_2 = f(x_2) + \frac{K_1\left(\frac{\psi_a}{x_2 - x_1}\right)}{0.5\,(x_3 - x_1)},$$

$$\tilde{\varphi}_{M-1} = f(x_{M-1}) + \frac{K_{M-1}\left(\frac{\psi_b}{x_M - x_{M-1}}\right)}{0.5\,(x_M - x_{M-2})}.$$

From 2.13 we can see that the finite-difference scheme for the Dirichlet problem can be written in the operator form

$$\mathcal{A}_h\, u = \tilde{\varphi},$$

where $\mathcal{A}_h : \overset{\scriptscriptstyle 0}{H}N \to \overset{\scriptscriptstyle 0}{H}N$ and

$$\mathcal{A}_h = GRAD^*\, K\, GRAD.$$

Using this representation, it is easy to see that $\mathcal{A}_h$ is self-adjoint and non-negative in the sense of the inner product on $\overset{\scriptscriptstyle 0}{H}N$, that is,

$$(\mathcal{A}_h\, u,\, v)_{\overset{\scriptscriptstyle 0}{H}N} = (GRAD^*\, K\, GRAD\, u,\, v)_{\overset{\scriptscriptstyle 0}{H}N} = (K\, GRAD\, u,\, GRAD\, v)_{\mathcal{H}C},$$

and if $v = u$, then

$$(\mathcal{A}_h\, u,\, u)_{\overset{\scriptscriptstyle 0}{H}N} = (K\, GRAD\, u,\, GRAD\, u)_{\mathcal{H}C} > 0.$$

Here it is important to note that self-adjointness and the non-negative properties of operator $\mathcal{A}_h$ in the sense of the inner product on $\overset{0}{H}N$ means that in the matrix sense, to obtain a symmetric and non-negative matrix, we must multiply the matrix of operator $\mathcal{A}_h$ by the diagonal matrix N, such that

$$(N\,u)_i = V N_i\, u_i.$$

This fact follows from the formula for the inner product.

2.1.6 The Matrix Problem

In accordance with the considerations in the previous section, to obtain a matrix problem with a symmetric and non-negative matrix, we multiply both sides of the i-th difference equations by $V N_i = 0.5\,(x_{i+1} - x_{i-1})$. Then the matrix form of the difference equation has the form

$$M\,u = F,$$

where

$$
F =
\begin{pmatrix}
f(x_2)\cdot 0.5\cdot(x_3 - x_1) + K_1\,\psi_a/(x_2 - x_1) \\
f(x_3)\cdot 0.5\cdot(x_4 - x_2) \\
\vdots \\
f(x_i)\cdot 0.5\cdot(x_{i+1} - x_{i-1}) \\
\vdots \\
f(x_{M-1})\cdot 0.5\cdot(x_M - x_{M-2}) + K_{M-1}\,\psi_b/(x_M - x_{M-1}).
\end{pmatrix}
$$

and M is the matrix. If for convenience we use the notation $V C_i = x_{i+1} - x_i$, then matrix M is

$$
M =
\begin{pmatrix}
\frac{K_2}{VC_2} + \frac{K_1}{VC_1} & -\frac{K_2}{VC_2} & 0 & 0 & \cdot \\
-\frac{K_2}{VC_2} & \frac{K_2}{VC_2} + \frac{K_3}{VC_3} & -\frac{K_3}{VC_3} & 0 & \cdots \\
0 & -\frac{K_3}{VC_3} & \frac{K_3}{VC_3} + \frac{K_4}{VC_4} & -\frac{K_4}{VC_4} & \cdots \\
\vdots & \vdots & \vdots & \vdots & \vdots \\
0 & 0 & 0 & 0 & \cdot
\end{pmatrix}.
$$

It is evident that matrix M is symmetric and the construction of the scheme ensures that this matrix is non-negative.

2.1.7 The Finite-Difference Scheme for the Robin Boundary Problem

In accordance with the considerations for the Robin boundary value problem in the continuous case, we must construct finite-difference analogs for the operators $\mathcal{C}, \mathcal{B}, \mathcal{D}, \mathcal{K}$. We have already constructed finite-difference analogs for $\mathcal{C}, \mathcal{B}, \mathcal{K}$ during the construction of FDS for the Dirichlet problem. Operator $\mathcal{D}$, in the 1-D case, is given by the very simple formula

$$(\mathcal{D}^h u)_1 = \alpha_a u_1, \quad (\mathcal{D}^h u)_M = \alpha_b u_M$$

$$(\mathcal{D}^h u)_i = 0, i = 2, \ldots M - 1,$$

and thus
$$\mathcal{D}^h : HN \to HN.$$

Now the finite-difference approximation in internal nodes $i = 2, \ldots, M - 1$ is the same as the Dirichlet problem and is given by the formula (2.12).

In accordance with the support-operators method, to approximate the Robin boundary conditions (2.4), we use the derived operator $\mathcal{B}^h$ on the boundary (which gives an approximation for the normal component of flux) and the operator $\mathcal{D}^h$:

$$(\mathcal{B}^h WX)_1 + (\mathcal{D}^h u)_1 = \psi_a, \qquad (2.14)$$

$$(\mathcal{B}^h WX)_M + (\mathcal{D}^h u)_M = \psi_b. \qquad (2.15)$$

Using the expressions for the operator $\mathcal{B}^h$ on the boundary (see (2.9)) and the definition for operator $\mathcal{D}^h$, we can rewrite (2.14) as follows:

$$WX_1 + \alpha_a u_1 = \psi_a,$$

$$\qquad (2.16)$$

$$-WX_{M-1} + \alpha_b u_M = \psi_b.$$

Finally, because $WX = -\text{GRAD}\, u$ we get

$$-K_1 \frac{u_2 - u_1}{x_2 - x_1} + \alpha_a u_1 = \psi_a,$$

$$\qquad (2.17)$$

$$K_{M-1} \frac{u_M - u_{M-1}}{X_M - X_{M-1}} + \alpha_b u_M = \psi_b.$$

2.1.8   The Matrix Problem for Robin Boundary Conditions

Unlike the Dirichlet boundary conditions for the case of Robin boundary conditions, the vector of unknowns contains values at all nodes including the boundary,

$$u = \begin{pmatrix} u_1 \\ u_2 \\ \vdots \\ u_M \end{pmatrix}.$$

Note that for the Robin boundary value problems, the vector of unknowns contains values in all nodes and therefore it is not necessary to move any terms to the right-hand side.

Similar to the case for Dirichlet boundary conditions, we multiply the finite-difference equations at the internal nodes, by $V N_i$, and then we obtain the following equations at the internal nodes $i = 2, \ldots, M - 1$:

$$
-\left(K_i \left(\frac{u_{i+1} - u_i}{x_{i+1} - x_i} \right) - K_{i-1} \left(\frac{u_i - u_{i-1}}{x_i - x_{i-1}} \right) \right) = f(x_i) \frac{(x_{i+1} - x_{i-1})}{2}.
\tag{2.18}
$$

We do not need to change the difference equations at the boundary points because there is no weight near these values in the expression of the inner product. Equations (2.18) and (2.17) can be written in matrix form as follows (the first is the equation for the left boundary, next is the equations at the internal nodes, and the last equation is for the right boundary):

$$
M u = F.
$$

The matrix M is

$$
M = \begin{pmatrix}
\frac{K_1}{V C_1} + \alpha_a & -\frac{K_1}{V C_1} & 0 & 0 & \cdots \\
-\frac{K_1}{V C_1} & \frac{K_2}{V C_2} + \frac{K_1}{V C_1} & -\frac{K_2}{V C_2} & 0 & \cdots \\
0 & -\frac{K_2}{V C_2} & \frac{K_3}{V C_3} + \frac{K_2}{V C_2} & \frac{-K_3}{V C_3} & \cdots \\
\cdots & \cdots & \cdots & \cdots & \cdots \\
\cdots & \cdots & \cdots & \cdots & \cdots
\end{pmatrix}.
\tag{2.19}
$$

The right-hand side is

$$
F = \begin{pmatrix}
\psi_a \\
f(x_2) V N_2 \\
\vdots \\
f(x_i) V N_i \\
\vdots \\
f(x_{M-1}) V N_{M-1} \\
\psi_b.
\end{pmatrix}.
$$

It is easy to see from (2.19) that matrix M is symmetric and the support-operators method implies that matrix M is non-negative, that is, $M \geq 0$.

2.1.9 Approximation Properties

Let us consider the approximation properties of prime and derived difference operators and the superposition of these operators which approximate the differential operator $d(K \, du/dx)/dx$.

To investigate the approximation properties, we will use the following projection operators:

$$(p_h(u))_i = u(x_i)$$

for scalar functions, and

$$(\mathcal{P}_h \vec{W})_i = WX\left(\frac{x_{i+1} + x_i}{2}\right)$$

for vector functions.

Let us be reminded that we can consider two cases. One case is of the smooth grid, where the grid is the result of a smooth transformation of a uniform grid, and the second case is a regular non-uniform grid when we assume that

$$C_{min}\, h \leq VC_i \leq C_{max}\, h.$$

Then, using the Taylor expansion, we can prove that

$$(\mathcal{P}_h(\operatorname{grad} u))_i - (\operatorname{GRAD}(p_h u))_i =$$
$$\frac{du}{dx}\bigg|_{\left(\frac{x_{i+1}+x_i}{2}\right)} - \frac{u(x_{i+1}) - u(x_i)}{x_{i+1} - x_i} = O(h^2),$$

that is, the truncation error for operator grad is of the second-order with respect to h. This will be true for both smooth and regular grids.

For operator div we get

$$\left(p_h(\operatorname{div}\vec{W})\right)_i - \left(\operatorname{DIV}(\mathcal{P}_h\vec{W})\right)_i =$$
$$\frac{dWX}{dx}\bigg|_{x_i} - \frac{WX_i - WX_{i-1}}{0.5\,(x_{i+1} - x_{i-1})} = O(h).$$

That is, the truncation error for operator div is of the first order with respect to h for regular non-uniform grids. For smooth grids we can show that the truncation error is h^2.

And finally, for the second-order differential operator we get

$$(p_h(\operatorname{div} K \operatorname{grad} u))_i - (\operatorname{DIV} K \operatorname{GRAD}(p_h u))_i =$$
$$(\operatorname{div} K \operatorname{grad} u)\big|_{x_i} - \frac{K_i\left(\frac{u(x_{i+1})-u(x_i)}{x_{i+1}-x_i}\right) - K_{i-1}\left(\frac{u(x_i)-u(x_{i-1})}{x_i-x_{i-1}}\right)}{0.5\,(x_{i+1} - x_{i-1})} =$$
$$O(h).$$

That is, the truncation error is $O(h)$ for regular non-uniform grids, and similar to operator div, we can prove that for smooth grids, we get a second-order truncation error.

Now we can evaluate the *residual on the solution of differential boundary value problems* (see the first chapter) for Dirichlet boundary conditions.

From this chapter we know that the residual is the sum of the truncation error for the operator of the differential equation and the error of approximation of the right-hand side. For computing the right-hand side, we use the exact value of the function in the right-hand side of the differential equation in nodes. Then the error for the right-hand side is equal to zero and the residual is the same as the truncation error for the second-order operator, that is $O(h)$. Because we use the exact approximation for the Dirichlet boundary conditions, it does not affect the residual. Therefore for the Dirichlet problem, we get the first-order residual for regular non-uniform grids and the second-order residual for smooth grids.

For Robin boundary conditions we must do an additional investigation of the approximation on the boundary. Let us, for example, consider the left boundary. The differential boundary condition is

$$-K(a)\frac{du}{dx}(a) + \alpha_a\, u(a) = \psi_a$$

and the corresponding finite-difference approximation has the form

$$-K_1\frac{u_2^h - u_1^h}{x_2 - x_1} + \alpha_a\, u_1^h = \psi_a\,. \qquad (2.20)$$

By comparison of these two equations, it is easy to see that in the difference case, we have an exact approximation for the second term in the left-hand side and for the right-hand side. Therefore, we must investigate the truncation error for the first term.

Using the technique similar to the one in Chapter 1, we can prove that this truncation error is of the first order with respect to h. Thus, to investigate the truncation error, we must use the Taylor expansion for all terms in (2.20) for point a and compare the resulting expression with the differential boundary condition. Because

$$K_1 = k\left(\frac{x_1 + x_2}{2}\right),$$

we get

$$K_1 = k(x_1) + \frac{x_2 - x_1}{2}\frac{dk}{dx}(x_1) + O(h^2)\,.$$

For the difference analog of the first derivative we get

$$\frac{u(x_2) - u(x_1)}{x_2 - x_1} = \frac{du}{dx}(x_1) + \frac{x_2 - x_1}{2}\frac{d^2u}{dx^2}(x_1) + O(h^2)\,.$$

From the two previous equations we can conclude that

$$-k\left(\frac{x_1 + x_2}{2}\right)\frac{u(x_2) - u(x_1)}{x_2 - x_1} =$$

$$\left(-k\frac{du}{dx}\right) - \frac{x_2 - x_1}{2}\left(\frac{dk}{dx}\frac{du}{dx}\right) - \frac{x_2 - x_1}{2}\left(k\frac{d^2u}{dx^2}\right) + O(h^2)$$

where all functions in the right-hand side are computed in point $x_1 = a$.

Using common terms from the previous equation and taking into account that

$$\frac{d}{dx}\left(k\,\frac{du}{dx}\right) = \frac{dk}{dx}\frac{du}{dx} + k\,\frac{d^2u}{dx^2},$$

we get

$$-k\left(\frac{x_1 + x_2}{2}\right)\frac{u(x_2) - u(x_1)}{x_2 - x_1} = -k\,\frac{du}{dx} - \frac{x_2 - x_1}{2}\left[\frac{d}{dx}\,k\,\frac{du}{dx}\right] + O(h^2).$$
$$(2.21)$$

It is easy to see from this equation that the truncation error for the Robin boundary condition, in the general case, is of the first order.

We can *improve* the order of the truncation error if we use information about the solution. In particular, if we take into account that $u(x)$ is not an arbitrary function but is the solution for the differential equation (2.1), then we can conclude that the term in square brackets in the right-hand side of 2.21 is equal to $-f(a)$, and consequently

$$-k\left(\frac{x_1 + x_2}{2}\right)\frac{u(x_2) - u(x_1)}{x_2 - x_1} - \frac{x_2 - x_1}{2}\,f(a) = -k\,\frac{du}{dx} + O(h^2).$$

Thus from this equation and equation 2.20 we get the difference equation

$$-K_1\,\frac{u_2^h - u_1^h}{x_2 - x_1} + \alpha_a\,u_1^h = \psi_a + \frac{x_2 - x_1}{2}\,f(a),\qquad (2.22)$$

which gives us the second-order truncation error of Robin boundary conditions for general non-uniform grids.

This is the simplest example for *improving the order of approximation on the solution of a differential equation*. We will demonstrate how this improvement affects the accuracy in the section relating to numerical solutions of test problems.

Here we do not have enough space to prove the convergence of the constructed finite-difference scheme. Readers who are interested in the theory of the convergence of finite-difference schemes can find this information in the following books: [1], [12], [34], [61], [90], [100], [101], [107].

To demonstrate the convergence properties, we consider examples for the solution to some test problems. Also, the reader can solve similar problems by using the subroutines that are a part of this book.

Using the Gauss-Seidel Method for the Solution to the System of Linear Equations

For the reader's convenience, we will provide the explicit form of the formula for the Gauss-Seidel method for the Dirichlet and Robin boundary value problems. Details about the Gauss-Seidel method and other more effective iteration methods can be found in [50], [53], [54], [89], [111], [146], and [152].

Dirichlet boundary conditions: At first we must specify the initial guess for the vector of unknowns, that is, we specify values

$$u_2^{(0)}, u_3^{(0)}, \ldots, u_{M-1}^{(0)}.$$

Now suppose we know the values on iteration s; then the values on the next iteration, $s + 1$, is determined as follows: for $i = 2$,

$$u_2^{(s+1)} = \frac{\frac{K_2}{VC_2} u_3^{(s)} + f(x_2) \frac{x_3 - x_1}{2} + K_1 \frac{\psi_a}{x_2 - x_1}}{\frac{K_2}{VC_2} + \frac{K_1}{VC_1}};$$

for $i = 3, \ldots, M - 2$,

$$u_i^{(s+1)} = \frac{\frac{K_{i-1}}{VC_{i-1}} u_{i-1}^{(s+1)} + \frac{K_i}{VC_i} u_i^{(s)} + f(x_i) \frac{x_{i+1} - x_{i-1}}{2}}{\frac{K_i}{VC_i} + \frac{K_{i-1}}{VC_{i-1}}};$$

and for $i = M - 1$,

$$u_{M-1}^{(s+1)} = \frac{\frac{K_{M-2}}{VC_{M-2}} u_{M-2}^{(s+1)} + f(x_{M-1}) \frac{x_M - x_{M-2}}{2} + K_{M-1} \frac{\psi_b}{x_M - x_{M-1}}}{\frac{K_{M-2}}{VC_{M-2}} + \frac{K_{M-1}}{VC_{M-1}}}.$$

The criteria for stopping the iteration process is

$$\max_i |u_i^{(s+1)} - u_i^{(s)}| < \varepsilon,$$

where ε is some given small number.

For the Robin boundary conditions, the vector of unknowns is

$$u_1, u_2, u_3, \ldots, u_{M-1}, u_M$$

and the formulas are as follows: for $i = 1$,

$$u_1^{(s+1)} = \frac{\frac{K_1}{VC_1} u_2^{(s)} + \psi_a}{\frac{K_1}{VC_1} + \alpha_a};$$

for $i = 2, \ldots, M - 1$,

$$u_i^{(s+1)} = \frac{\frac{K_{i-1}}{VC_{i-1}} u_{i-1}^{(s+1)} + \frac{K_i}{VC_i} u_i^{(s)} + f(x_i) \frac{x_{i+1} - x_{i-1}}{2}}{\frac{K_i}{VC_i} + \frac{K_{i-1}}{VC_{i-1}}};$$

for $i = M$,

$$u_M^{(s+1)} = \frac{\frac{K_{M-1}}{VC_{M-1}} u_{M-1}^{(s+1)} + \psi_b}{\frac{K_{M-1}}{VC_{M-1}} + \alpha_b}.$$

The criteria for finishing the iteration process is the same as for the case of Dirichlet boundary conditions.

2.1.10   The Numerical Solution of Test Problems

First, we present the results for the solution of the Dirichlet boundary value problem. The statement of the problem is:

$$a = 0,\ b = 1,$$
$$\psi_a = 0,\ \psi_b = 0,$$
$$K(x) = x^4 + 1,$$

$$f(x) = -4\,x^3\,2\,\pi\,\cos(2\,\pi\,x) + (x^4 + 1)\,4\,\pi^2\sin(2\,\pi\,x)\,.$$

The exact solution for this problem is

$$u(x) = \sin(2\,\pi\,x)\,.$$

To check the convergence properties of the constructed finite-difference scheme, we use the non-uniform smooth grid with the following coordinates of nodes:

$$x_i = \left(a + \frac{b-a}{M-1}(i-1)\right)^2.$$

This is the simplest case of using the transformation of a uniform grid in constructing a non-uniform grid.
    That is,

$$x_i = f(\xi_i),$$

where

$$f(\xi) = \xi^2$$

and

$$\xi_i = a + \frac{b-a}{M-1}(i-1)\,.$$

Distribution of the grid nodes for $M = 17$ is shown in Figure 2.2 (solid circles). Approximate and exact solutions are presented in Figure 2.2. To investigate the convergence properties of the constructed finite-difference scheme, we calculate the two norms of difference between the exact and approximate solutions.
    The first norm is

$$\| u^h - u \|_{max} = \max_{1 < i < M} |u_i^h - u(x_i)|,\qquad (2.23)$$

where u^h is the approximate solution, and $u(x)$ is the exact solution for the differential equation.
    The second norm is the difference analog of the L_2 norm:

$$\| u^h - u \|_{L_2} = \left(\sum_{i=2}^{M-1}(u_i^h - u(x_i))^2\,V N_i\right)^{\frac{1}{2}}.\qquad (2.24)$$

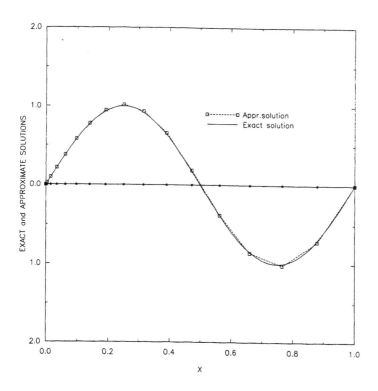

Figure 2.2: *Dirichlet boundary conditions, Smooth grid, $M = 17$.*

M	max norm	L_2 norm	Const max	Const L_2	R max	R L_2
17	0.257E-01	0.171E-01	6.585	4.386	-	-
33	0.696E-02	0.426E-02	7.131	4.370	3.694	4.370
65	0.174E-02	0.106E-02	7.148	4.365	3.990	4.004
129	0.436E-03	0.266E-03	7.156	4.364	3.995	4.000

Table 2.1: *Convergence analysis, Dirichlet boundary conditions, smooth-grid.*

By using the numerical experiment we can demonstrate that

$$\| u^h - u \|_{max} \leq C_1 h^2 \tag{2.25}$$

and

$$\| u^h - u \|_{L_2} \leq C_2 h^2 \tag{2.26}$$

where

$$h = \frac{1}{M-1},$$

and constants C_1, C_2 do not depend on h.

To show this, we compute the mentioned norms for $M = 2^r + 1$ and correspondingly, $h = \frac{1}{2^r}$. Therefore, if we have a second-order convergence, then the ratio between the norm for some r and the norm for $r+1$ must be

$$\frac{\left(\frac{1}{2^r}\right)^2}{\left(\frac{1}{2^{r+1}}\right)^2} = 4.$$

Table 2.1 demonstrates the second-order convergence of the finite-difference scheme in max and L_2 difference norms. In the first column we present the number of nodes, in the second and third columns we present the max and L_2 norms. Correspondingly, in the fourth and fifth columns, we present the numerically computed constants in inequalities 2.25 and 2.26. These constants are

$$\frac{\| u^h - u \|_{max}}{h^2}, \quad \frac{\| u^h - u \|_{L_2}}{h^2}.$$

In the last two columns we present a ratio between the error for the number of nodes given in this row and the error for the number of nodes given in the previous row. Then when $h \to 0$, this ratio must converge to 4 for the second-order method. The results presented in the table demonstrate the second-order convergence of the finite-difference scheme for Dirichlet boundary problems on a smooth grid.

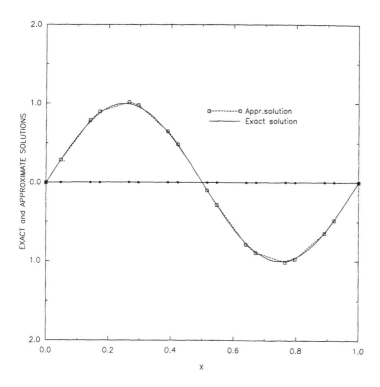

Figure 2.3: *Dirichlet boundary conditions, non-smooth grid, $M = 17$.*

To compare the quality of the solution on smooth and non-smooth grids, we present the results of the solution for the test problem on the following grid:

$$x_1 = 0; \; x_i = \xi_i + (-1)^{i-1} \frac{h}{4}, \; i = 2, \ldots, M - 1; \; x_M = 1.$$

This grid with the approximate and exact solutions are shown in Figure 2.3. The results of the convergence analysis are presented in Table 2.2. This table is arranged in the same way as the previous table for a smooth grid. Comparison of the results for smooth and non-smooth grids show us that accuracy in both cases are approximately the same, and for both cases, we have a second-order convergence. This means that the constructed finite-difference scheme can be used for a wide class of grids.

Therefore, we demonstrated numerically the very important property of conservative finite-difference schemes. In particular, conservative finite-

M	max norm	L_2 norm	Const max	Const L_2	R max	R L_2
17	0.238E-01	0.140E-01	6.092	3.584	-	-
33	0.608E-02	0.362E-02	6.225	3.706	3.914	3.867
65	0.153E-02	0.930E-03	6.266	3.809	3.973	3.892
129	0.388E-03	0.236E-03	6.356	3.886	3.943	3.940

Table 2.2: *Convergence analysis, Dirichlet boundary conditions, non-smooth grid.*

difference schemes for Dirichlet boundary conditions, which have a second-order truncation error on a smooth grid, will have a second-order accuracy on an arbitrary grid. More accurate formulation of this result can be found in [108] (see also [72]).

Now let us consider the results for solving the test problem for *Robin boundary* conditions 2.3. The statement of problem is:

$$a = 0, \ b = 1,$$
$$K(x) = x^4 + 1,$$
$$f(x) = -4\,x^3\,2\,\pi\,\cos(2\,\pi\,x) + (x^4 + 1)\,4\,\pi^2\,\sin(2\,\pi\,x),$$
$$\tag{2.27}$$

$$\alpha_a = 1, \ \alpha_b = 1,$$
$$\psi_a = -K(a)\,2\,\pi\,\cos(2\,\pi\,a) + \sin(2\,\pi\,a) = -2\,\pi,$$
$$\psi_b = K(b)\,2\,\pi\,\cos(2\,\pi\,b) + \sin(2\,\pi\,b) = 4\,\pi.$$

The exact solution of this problem is

$$u(x) = \sin(2\,\pi\,x).$$

First, we present the results for the first-order approximation of boundary conditions 2.17. In Figure 2.4, we present the approximate solution for $M = 17$ and $M = 33$ for a smooth grid with transformation $x = \xi^2$. The position of the nodes are shown only for $M = 17$.

The results of the convergence analysis are shown in Table 2.3. For the case of Robin boundary conditions, the definition of max-norm is the same as for the case of Dirichlet boundary conditions and is shown in formula 2.23. The difference analog of the L_2 norm includes boundary terms. That is,

$$\| u^h - u \|_{L_2} = \tag{2.28}$$

$$\left(\sum_{i=2}^{M-1} (u_i^h - u(x_i))^2 \, VN_i + (u_1^h - u(x_1))^2 + (u_M^h - u(x_M))^2 \right)^{\frac{1}{2}}.$$

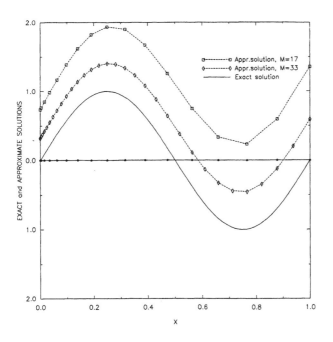

Figure 2.4: *Robin boundary conditions, first-order approximation, smooth grid.*

M	max norm	L_2 norm	Const max	Const L_2	R max	R L_2
17	0.136E+01	0.186E+01	21.76	29.76	-	-
33	0.596E+00	0.822E+00	19.07	26.30	2.281	2.262
65	0.276E+00	0.383E+00	17.66	24.51	2.159	2.146
129	0.133E+00	0.184E+00	17.02	23.55	2.075	2.081

Table 2.3: *Robin boundary conditions, first-order approximation, smooth grid.*

M	max norm	L_2 norm	Const max	Const L_2	R max	R L_2
17	0.369E+00	0.511E+00	94.46	130.8	-	-
33	0.927E-01	0.129E+00	94.92	132.0	3.980	3.961
65	0.232E-01	0.324E-01	95.02	132.7	3.998	3.981
129	0.580E-02	0.813E-02	95.02	133.2	4.000	3.985

Table 2.4: *Robin boundary conditions, second-order approximation, smooth grid.*

For the case of the first-order approximation of Robin boundary conditions, it is natural to expect the *first order of accuracy*. Then the constants in Table 2.3 are the ratio of the corresponding error and h, and for the first order convergence, these constants must converge to 2. The results from this table confirm the first-order convergence of the constructed finite-difference scheme with the first-order approximation of Robin boundary conditions.

Now let us see what will happen on a non-smooth grid. As an example, we will use the same non-smooth grid as we did for the Dirichlet boundary condition. The approximate and exact solutions are shown in Figure 2.5. The results of the convergence analysis confirm the first-order convergence for this case of grids also. The absolute values of error for non-smooth grids are even less than for smooth grids. This can be explained by the fact that near the right boundary, the mesh size for non-smooth grids is less than for smooth grids, and as we know, the error strongly depends on the truncation error on the boundary. Let us now consider formula 2.22 for the case of a second-order approximation of Robin boundary conditions. The exact and approximate solutions for a smooth grid are shown in Figure 2.6. For the second-order approximation of boundary conditions, it is natural to expect second-order convergence of the finite-difference scheme. The results of the convergence analysis, which confirm second-order convergence, are presented in Table 2.4. The results for computations on a non-smooth grid are presented in Figure 2.7 and in Table 2.5.

These results confirm second-order convergence of finite-difference schemes with second-order approximation of boundary conditions on a non-smooth grid.

2.2   Cell-Valued Discretisation of Scalar Functions and Nodal Discretisation of Vector Functions

As we have seen from the case of nodal discretisation of scalar functions and cell-valued discretisation of vector functions, to construct finite-difference schemes for Dirichlet and Robin boundary value problems, it is sufficient to

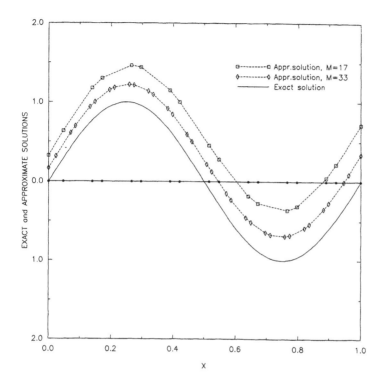

Figure 2.5: *Robin boundary conditions, first-order approximation, non-smooth grid.*

M	max norm	L_2 norm	Const max	Const L_2	R max	R L_2
17	0.782E-01	0.897E-01	20.01	22.96	-	-
33	0.226E-01	0.262E-01	23.14	26.82	3.460	3.423
65	0.607E-02	0.7061-02	24.86	28.91	3.723	3.711
129	0.157E-02	0.182E-02	25.72	29.81	3.866	3.879

Table 2.5: *Robin boundary conditions, second-order approximation, non-smooth grid.*

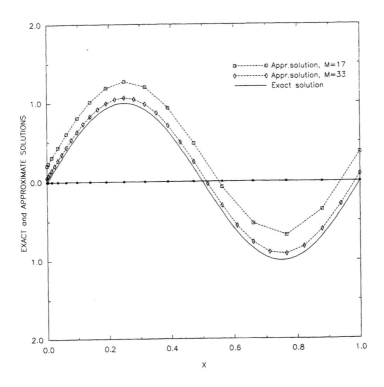

Figure 2.6: *Robin boundary conditions, second-order approximation, smooth grid.*

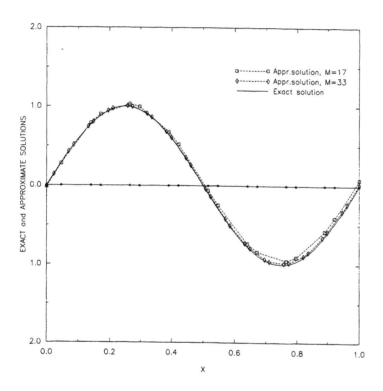

Figure 2.7: *Robin boundary conditions, second-order approximation, non-smooth grid.*

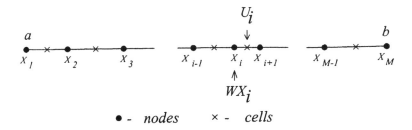

• - nodes   × - cells

Figure 2.8: *C-N discretisation.*

construct finite-difference analogs of operators $\mathcal{B}$, $\mathcal{C}$, $\mathcal{K}$, and $\mathcal{D}$. Therefore, in this section, we will construct these operators for the case of cell-valued discretisation of scalar functions and nodal discretisation of vector functions, then use them to construct the finite-difference schemes.

We will first consider the segment $[a, b]$ and the grid with nodes

$$\{x_i, \ i = 1, \cdots, M\}, \tag{2.29}$$

and vector function $\vec{W} = (WX, 0, 0)$ is discretised using its values at the nodes,

$$WX = \{WX_1, WX_2, \cdots, WX_M\}.$$

This is called *node* or *nodal* discretisation. Scalar functions are discretised using values at the cells $\{u_1, u_2, \cdots, u_{M-1}\}$ (see Figure 2.8).

For the Robin boundary condition, this discretisation must be expanded to include the values of function u_0 and u_M at the end points of the domain. Then the values of the scalar function are denoted by

$$u = (u_0, u_1, \cdots, u_{M-1}, u_M).$$

2.2.1   Spaces of Discrete Functions

In accordance with the theory of support operators, spaces of discrete functions must be defined. The space HC of discrete scalar functions (see previous subsection) is denoted by the inner product

$$(u, v)_{HC} = \sum_{i=1}^{M-1} u_i\, v_i\, VC_i + u_0\, v_0 + u_M\, v_M\,,$$

where VC_i is the volume of the i-th cell (see 2.5). This inner product converges to the expression (1.10) for the inner product in the continuous case. The space of the vector function is denoted by $\mathcal{HN}$ and has the inner product

$$(\vec{A}, \vec{B})_{\mathcal{HN}} = \sum_{i=1}^{M} (\vec{A}, \vec{B})_i \, VN_i \,,$$

where $(\vec{A}, \vec{B})_i$ is the scalar product of the two vectors at node i,

$$(\vec{A}, \vec{B})_i = AX_i \, BX_i \,,$$

and VN_i is the volume of the node:

$$\begin{cases} VN_1 = \frac{x_2 - x_1}{2} \\ VN_i = \frac{x_{i+1} - x_{i-1}}{2}, & i = 2, \cdots, M - 1 \,. \\ VN_M = \frac{x_M - x_{M-1}}{2} \end{cases} \qquad (2.30)$$

2.2.2 The Prime Operator

For cell-valued discretisation of scalar functions and nodal discretisation for vector functions, it is natural to use a finite-difference analog of operator $\mathcal{B}$ as the prime operator:

$$\mathcal{B}\vec{w} = \begin{cases} \operatorname{div} \vec{w}, & (x, y) \in V \\ -(\vec{w}, \vec{n}), & (x, y) \in \partial V \end{cases} \,.$$

The divergence operator is approximated using its invariant definition:

$$\operatorname{div} \vec{w}\big|_x = \lim_{V \to 0} \frac{\oint_{\partial V} (\vec{w}, \vec{n}) \, dS}{V} \,,$$

where the volume V, containing the point x; $\vec{n}$ is of the outer normal to the boundary ∂V. For the construction of the difference operator DIV, the segment $[X_i, X_{i+1}]$ is used as the volume and consequently,

$$(\operatorname{DIV} \vec{W})_i = \frac{WX_{i+1} - WX_i}{VC_i}, \qquad i = 1, \cdots, M - 1. \qquad (2.31)$$

The values WX_1 and WX_M are used to approximate the normal component of a vector $\vec{W}$ at the end points of the domain. This gives the following expression for the prime difference operator $\mathcal{B}^h$:

$$\mathcal{B}^h \vec{W} = \begin{cases} (\mathcal{B}^h \vec{W})_0 = -WX_1 \\ (\mathcal{B}^h \vec{W})_i = \frac{WX_{i+1} - WX_i}{VC_i}, & i = 1, \cdots, M - 1 \,, \\ (\mathcal{B}^h \vec{W})_M = WX_M \end{cases}$$

and

$$\mathcal{B}^h : \mathcal{HN} \to HC \,.$$

2.2.3   The Derived Operator

In accordance with the theory of support operators, the derived operator $C^h = -\text{GRAD}$ is derived using the difference analog of the integral identity (1.1):

$$\sum_{i=1}^{M-1} (\text{DIV } \vec{W})_i \, u_i \, VC_i + \sum_{i=1}^{N} (\vec{W}, \text{GRAD } u)_i \, VN_i = u_M \, WX_M - u_0 \, WX_1 \, .$$

Taking into account the definitions of the inner products, gives us

$$C^h = \left(B^h \right)^* .$$

The following chain of formulas gives the expression for operator C:

$$
\begin{aligned}
(B^h \, \vec{W}, u)_{HC} &= \sum_{i=1}^{M-1} u_i \, (B^h \, \vec{W})_i \, VC_i + u_0 \, (B^h \, \vec{W})_0 + u_N \, (B^h \, \vec{W})_N \\
&= \left[\sum_{i=1}^{M-1} u_i \, (WX_{i+1} - WX_i) \right] - u_0 \, WX_1 + u_M \, WX_M \\
&= \sum_{i=1}^{M} (C^h \, u, \vec{W})_i \, VN_i \quad\quad\quad (2.32) \\
&= \sum_{i=1}^{M} (C^h \, u)_i \, WX_i \, VN_i \\
&= (C^h u, \vec{W})_{\mathcal{HN}} \, .
\end{aligned}
$$

The final expression for operator C is

$$(C^h \, u)_i = -\frac{u_i - u_{i-1}}{VN_i} \, , \quad i = 1, M \, . \quad\quad\quad (2.33)$$

It is important to note that the expression for operator C is the same for all nodes, but the formula for the volume of the nodes is different in the internal and boundary nodes (see 2.30).

2.2.4   Multiplication by a Matrix and the Operator $\mathcal{D}$

The definition of a difference analog of the multiplication of a vector by a matrix depends on the discretisation of the matrix elements. In the 1-D case, a matrix has only one entry

$$\mathcal{K} = (KX X) \, .$$

To shorten the notation, set $KXX = K$. In this section a nodal discretisation of the matrix elements is used, so the discrete matrix multiplication operator $\mathcal{K}$ is denoted by

$$(\mathcal{K}^h \, WX)_i = K_i \, WX_i \,,$$

and then

$$\mathcal{K}^h : \mathcal{HN} \to \mathcal{HN} \,.$$

For other types of approximation of the matrix elements, a more complicated formula will be needed. Also, when the entries of the matrix are discontinuous, a special averaging procedure relevant to the position of discontinuity is needed.

In the 1-D case, operator $\mathcal{D}$ is given by a very simple formula:

$$(\mathcal{D}^h \, u)_0 = \alpha_a \, u_0 \,, \quad (\mathcal{D}^h \, u)_M = \alpha_b \, u_M \,,$$

$$(\mathcal{D}^h \, u)_i = 0 \,, i = 2, \ldots, M - 1 \,,$$

and thus,

$$\mathcal{D}^h : HC \to HC \,.$$

2.2.5 The Finite-Difference Scheme for Dirichlet Boundary Conditions

The finite-difference scheme for the case of Dirichlet boundary conditions is as follows:

$$(- \, \mathrm{DIV} \, K \, \mathrm{GRAD} \, u)_i = \varphi_i \,, \quad i = 1, \ldots, M - 1$$

$$\tag{2.34}$$

$$u_0 = \psi_a, \quad u_M = \psi_b,$$

where operators DIV and GRAD are shown in formulas 2.31 and 2.33. Keeping in mind that the equations in 2.34 are written for cells, the difference function U belongs to HC, and the right-hand side of the difference-equation is a projection of the function $f(x)$ to the cell, we will use following formula:

$$\varphi_i = f \left(\frac{x_i + x_{i+1}}{2} \right) \,.$$

The explicit form of 2.34 is

$$\frac{-K_{i+1} \frac{u_{i+1} - u_i}{VN_{i+1}} + K_i \frac{u_i - u_{i-1}}{VN_i}}{VC_i} = \varphi_i \,, \quad i = 1, \cdots, M - 1. \tag{2.35}$$

It is important to note the contrast to the case of nodal discretisation of the scalar function, whereas in this formula, $K_i = K(x_i)$.

2.2.6   The Matrix Problem

Because u_0 and u_M are known for Dirichlet boundary conditions, the vector of unknowns is

$$u = \begin{pmatrix} u_1 \\ u_2 \\ \vdots \\ u_{M-1} \end{pmatrix}.$$

Similar to the case of nodal discretisation of scalar functions, to obtain a symmetric and non-negative matrix, we must multiply equation 2.35 by VC_i and move the known values into the right-hand side. As a result, we obtain the following matrix problem:

$$M\,u = F,$$

where

$$M = \begin{pmatrix} \frac{K_2}{VN_2} + \frac{K_1}{VN_1} & -\frac{K_2}{VN_2} & 0 & 0 & \cdots \\ -\frac{K_2}{VN_2} & \frac{K_2}{VN_2} + \frac{K_3}{VN_3} & -\frac{K_3}{VN_3} & 0 & \cdots \\ 0 & -\frac{K_3}{VN_3} & \frac{K_3}{VN_3} + \frac{K_4}{VN_4} & -\frac{K_4}{VN_4} & \cdots \\ \vdots & \vdots & \vdots & \vdots & \cdots \\ \vdots & \vdots & \vdots & \vdots & \cdots \end{pmatrix}, \quad (2.36)$$

and for the right-hand side,

$$F = \begin{pmatrix} f(\frac{x_1+x_2}{2})\,VC_1 + K_1\,\psi_a/VN_1 \\ f(\frac{x_2+x_3}{2})\,VC_2 \\ \vdots \\ f(\frac{x_i+x_{i+1}}{2})\,VC_i \\ \vdots \\ f(\frac{x_{M-1}+x_M}{2})\,VC_{M-1} + K_{M-1}\,\psi_b/VN_M \end{pmatrix}.$$

Similar to the case of nodal discretisation, it is easy to see that M is symmetric and as it follows from the approach in the construction of the FDS, that matrix M is non-negative.

2.2.7   The Finite-Difference Scheme for Robin Boundary Conditions

Using the constructed operators, the difference scheme for Robin boundary conditions can be written in the following form:

$$\mathcal{A}^h\,u^h = (\mathcal{B}^h\,\mathcal{K}^h\,\mathcal{C}^h + \mathcal{D}^h)\,u^h = \mathcal{F}^h.$$

The explicit form for the difference equations is

$$-K_1 \frac{u_1^h - u_0^h}{VN_1} + \alpha_a u_0^h = \psi_a,$$

$$\frac{-K_{i+1} \frac{u_{i+1}^h - u_i^h}{VN_{i+1}} + K_i \frac{u_i^h - u_{i-1}^h}{VN_i}}{VC_i} = \varphi_i, \quad i = 1, \cdots, M-1,$$

$$+K_M \frac{u_M^h - u_{M-1}^h}{VN_M} + \alpha_b u_M^h = \psi_b.$$

2.2.8 The Matrix Problem

The difference equations can be written in the matrix form

$$M u = F,$$

where

$$u = (u_0, u_1, u_2, \cdots, u_M)^T,$$
$$F = (\psi_a, \phi_1 \cdot VC_1, \phi_2 \cdot VC_2, \cdots, \psi_b)^T,$$

and

$$
M = \begin{pmatrix}
\frac{K_1}{VN_1} + \alpha_a & -\frac{K_1}{VN_1} & 0 & 0 & \cdots \\
-\frac{K_1}{VN_1} & \frac{K_2}{VN_2} + \frac{K_1}{VN_1} & -\frac{K_2}{VN_2} & 0 & \cdots \\
0 & -\frac{K_2}{VN_2} & \frac{K_3}{VN_3} + \frac{K_2}{VN_2} & \frac{-K_3}{VN_3} & \cdots \\
\vdots & \vdots & \vdots & \vdots & \cdots
\end{pmatrix}.
$$

It is evident that matrix M is symmetric and the construction of the scheme ensures that this matrix is positive.

2.2.9 Approximation Properties

Let us consider the approximation properties of the prime and derived difference operators, and also the supposition of these operators which approximates the differential operator $d(K \, du/dx)/dx$.

To investigate the approximation properties, we will use the following projection operators:

$$(p_h(u))_i = u\left(\frac{x_i + x_{i+1}}{2}\right), \quad i = 1, \ldots, M-1,$$

$$(p_h(u))_0 = u(x_1), \ (p_h(u))_M = u(x_M)$$

for scalar functions, and

$$(\mathcal{P}_h \vec{W})_i = WX(x_i), \quad i = 1, \ldots, M$$

for vector functions.

Then using the Taylor expansion, we can prove that for $i = 2, \ldots, M-1$:

$$(\mathcal{P}_h(\mathrm{grad}\, u))_i - (\mathrm{GRAD}\,(p_h u))_i =$$

$$\left.\frac{du}{dx}\right|_{x_i} - \frac{u\left(\frac{x_{i+1}+x_i}{2}\right) - u\left(\frac{x_i+x_{i+1}}{2}\right)}{0.5\,(x_{i+1} - x_{i-1})}$$

$$= \frac{1}{4}\left.\frac{\partial^2 u}{\partial x^2}\right|_{x_i} [(x_{i+1} - x_i) - (x_i - x_{i-1})] + O(h^2).$$

That is, the truncation error for operator grad is generally only of the first order with respect to h, and it is of the second order only for a uniform grid when the coefficient near the second derivative is equal to zero. For a smooth grid, it is possible to prove that the coefficient near the second derivative is of the second order with respect to h and therefore, the truncation error is of the second-order.

For $i = 1$ we get

$$(\mathcal{P}_h(\mathrm{grad}\, u))_1 - (\mathrm{GRAD}\,(p_h u))_1 =$$

$$\left.\frac{du}{dx}\right|_{x_1} - \frac{u\left(\frac{x_2+x_1}{2}\right) - u(x_1)}{0.5\,(x_2 - x_1)} =$$

$$\frac{1}{4}\left.\frac{\partial^2 u}{\partial x^2}\right|_{x_1} (x_2 - x_1) + O(h^2).$$

From this formula we can conclude that the truncation error for operator *grad* is of the first-order with respect to h for *any* grid. For $i = M$ we get similar results.

For operator div we get

$$\left(p_h(\mathrm{div}\, \vec{W})\right)_i - \left(\mathrm{DIV}(\mathcal{P}_h \vec{W})\right)_i =$$

$$\left.\frac{dWX}{dx}\right|_{\left(\frac{x_i+x_{i+1}}{2}\right)} - \frac{WX_{i+1} - WX_i}{x_{i+1} - x_i} = O(h^2).$$

That is, the truncation error for operator div is of the second-order with respect to h for *any* non-uniform grid.

For the second-order differential operator and cells $i = 2, \ldots, M-2$ we get

$$(p_h(\mathrm{div}\, K\, \mathrm{grad}\, u))_i - (\mathrm{DIV}\, K\, \mathrm{GRAD}\,(p_h u))_i =$$

$$(\mathrm{div}\, K\, \mathrm{grad}\, u)|_{x_i} -$$

$$\left\{ K_{i+1} \frac{u\left(\frac{x_{i+2}+x_{i+1}}{2}\right) - u\left(\frac{x_{i+1}+x_i}{2}\right)}{0.5\,(x_{i+2}-x_i)} - \right.$$

$$\left. K_i \frac{u\left(\frac{x_{i+1}+x_i}{2}\right) - u\left(\frac{x_i+x_{i-1}}{2}\right)}{0.5\,(x_{i+1}-x_{i-1})} \right\} \Big/ (x_{i+1}-x_i) =$$

$$K\left(\frac{x_{i+1}+x_i}{2}\right) \frac{\partial^2 u}{\partial x^2}\left(\frac{x_{i+1}+x_i}{2}\right) \times$$

$$\left[1 - 0.25 \frac{(x_{i+2}-x_{i-1})+(x_{i+1}-x_i)}{x_{i+1}-x_i}\right] + O(h^2).$$

It is easy to see that the first term in the expression for the truncation error is $O(1)$ for a general non-uniform regular grid. Therefore, the truncation error is $O(1)$. Although for a smooth grid, we can prove that the truncation error is $O(h^2)$. This follows from the fact that

$$x_{i+2} - x_{i-1} = 3\,\Delta\xi\,\frac{\partial x}{\partial \xi}\left(\frac{\xi_{i+1}+\xi_i}{2}\right) + O\left((\Delta\xi)^2\right),$$

$$x_{i+1} - x_i = \Delta\xi\,\frac{\partial x}{\partial \xi}\left(\frac{\xi_{i+1}+\xi_i}{2}\right) + O\left((\Delta\xi)^2\right).$$

Thus, the expression in square brackets is $O(h^2)$ for a smooth grid.

For cells $i = 1$ and $i = M - 1$, the order of approximation is $O(1)$ for *any* grid. This fact is a consequence of the first-order approximation of the operator grad in nodes $i = 1$ and $i = M$. For example, the truncation error for $i = 1$ is

$$K\left(\frac{x_2+x_1}{2}\right) \frac{\partial^2 u}{\partial x^2}\left(\frac{x_2+x_1}{2}\right) \left[1 - 0.25 \frac{(x_3-x_1)+(x_2-x_1)}{0.5\,(x_2-x_1)}\right] +$$

$$O(h^2).$$

Now we can evaluate the *residual on the the solution of the differential boundary value problem* for Dirichlet boundary conditions. From this chapter we know that the residual is the sum of the truncation error for the operator of the differential equation and the error of approximation of the right-hand side. In computing the right-hand side, we use the exact value of the function in the right-hand side of the differential equation in nodes. Then the error for the right-hand side is equal to zero and the residual is the same as the truncation error for the second-order operator, namely $O(1)$, for a general regular non-uniform grid. Because we use the exact approximation of Dirichlet boundary conditions, it does not affect the residual.

For the Dirichlet problem, the residual for a regular non-uniform grid is $O(1)$, whereas for a smooth grid, it is $O(h^2)$.

For Robin boundary conditions, it is necessary to perform an additional investigation for an approximation on the boundary. Let us, for example, consider the left boundary. The differential boundary condition is

$$-K(a)\frac{du}{dx}(a) + \alpha_a\, u(a) = \psi_a.$$

Correspondingly, the finite-difference approximation is as follows:

$$- K_1\,\frac{u_1^h - u_0^h}{0.5\,(x_2 - x_1)} + \alpha_a\, u_0^h = \psi_a. \qquad (2.37)$$

From a comparison of these two equations, it is easy to see that in the difference case, we have an exact approximation for the second term in the left-hand side and for the right-hand side. The truncation error for the first term is $O(h)$, as we found in the investigation done in the beginning of this section. Hence, we have the first-order residual for Robin boundary conditions.

Now let us see what happens when we investigate the residual in the solution of the original equation.

If we integrate the original equation with the cell $[x_1, x_2]$, we get

$$-\left[\left(K\frac{du}{dx}\right)\bigg|_{x_2} - \left(K\frac{du}{dx}\right)\bigg|_{x_1}\right] = \int_{x_1}^{x_2} f\, dx\,.$$

Now we can divide both sides of this equality by $x_2 - x_1$:

$$-\left[\left(K\frac{du}{dx}\right)\bigg|_{x_2} - \left(K\frac{du}{dx}\right)\bigg|_{x_1}\right]\bigg/(x_2 - x_1) = \left(\int_{x_1}^{x_2} f\, dx\right)\bigg/(x_2 - x_1).$$

Because we know from the boundary condition at $x = a$ that

$$\left(K\frac{du}{dx}\right)\bigg|_{x_1} = \alpha_a\, u(a) - \psi_a$$

and φ_i approximates the integral average of the function f with the second order, we can write

$$-\left[\left(K\frac{du}{dx}\right)\bigg|_{x_2} - (\alpha_a\, u(a) - \psi_a)\right]\bigg/(x_2 - x_1) = \varphi_1 + 0(h^2). \qquad (2.38)$$

We can present the difference equation in cell $i = 1$ in a similar form if we express the term

$$K_1\,\frac{u_1^h - u_0^h}{0.5\,(x_2 - x_1)}$$

by using the difference equation for the boundary conditions

$$K_1 \frac{u_1^h - u_0^h}{0.5\,(x_2 - x_1)} = \alpha_a\, u_0^h - \psi_a.$$

Then the finite-difference equation in cell $i = 1$ will be denoted as:

$$-\left[K_2 \frac{u_2^h - u_1^h}{0.5\,(x_3 - x_1)} - (\alpha_a\, u(a) - \psi_a) \right] \Big/ (x_2 - x_1) = \varphi_1. \qquad (2.39)$$

From a comparison of equations 2.38 and 2.39, we can conclude that the order of the residual is equal to the order of the truncation error for the operator gradient in node $i = 2$. And as we know, it is the first order for a general regular non-uniform grid and the second order for a smooth grid.

Therefore for residuals in the solution for the original differential equation, we have the first-order approximation for general regular grids and the second order for smooth grids.

As we already noted for conservative finite-difference schemes, which has a second-order approximation on smooth grids, it is natural to expect second-order accuracy. The convergence of the second-order is demonstrated in the next section by numerical experiments both for Dirichlet and Robin boundary conditions.

2.3 Numerical Solution for Test Problems

In this section we present the results for the numerical solution to the same test problem as for the case of nodal discretisation of scalar functions and cell-valued discretisation of vector functions. That is, we solve problem 2.23 for the Dirichlet boundary conditions and problem 2.27 for the Robin boundary conditions. The numerical solution for Dirichlet boundary conditions is presented in Figure 2.9 (smooth grid) and in Figure 2.10 (non-smooth grid), where solid circles give us the distribution of nodes, and symbol "x" is the center of the cells.

Convergence analysis is presented in Tables 2.6 and 2.7. The results in these tables confirm second-order convergence for smooth and for non-smooth grids.

The results for Robin boundary conditions are presented for smooth grids in Figure 2.11 and Table 2.8, and for non-smooth grids in Figure 2.12 and Table 2.9. The presented results confirm the second-order convergence of finite-difference schemes for the case of Robin boundary conditions for smooth and non-smooth grids.

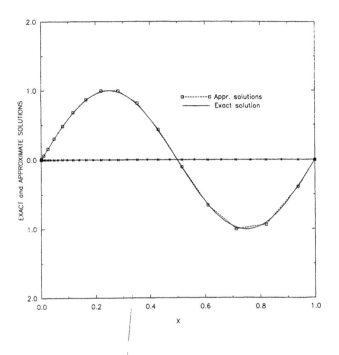

Figure 2.9: *Dirichlet boundary conditions, smooth grid, M=17.*

M	max norm	L_2 norm	Const max	Const L_2	R max	R L_2
17	0.322E-01	0.186E-01	8.243	4.761	-	-
33	0.778E-02	0.453E-02	7.966	4.638	4.138	4.105
65	0.194E-02	0.112E-02	7.946	4.587	4.010	4.044
129	0.485E-03	0.281E-03	7.946	4.603	4.000	3.985

Table 2.6: *Dirichlet boundary conditions, smooth grid.*

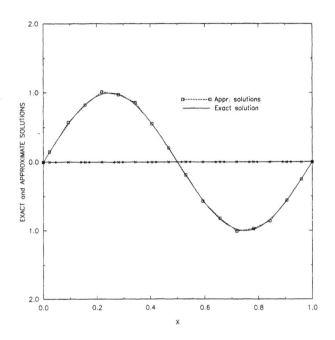

Figure 2.10: *Dirichlet boundary conditions, non-smooth grid, M=17.*

M	max norm	L_2 norm	Const max	Const L_2	R max	R L_2
17	0.359E-01	0.207E-01	9.190	5.299	-	-
33	0.876E-02	0.499E-02	8.970	5.109	4.098	4.148
65	0.217E-02	0.122E-02	8.888	4.997	4.036	4.090
129	0.543E-03	0.305E-03	8.749	4.997	3.996	4.000

Table 2.7: *Dirichlet boundary conditions, non-smooth grid.*

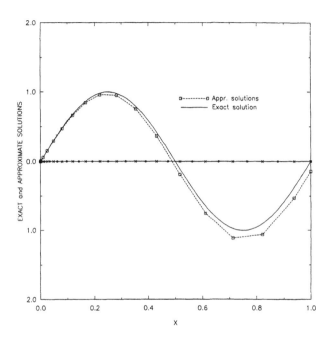

Figure 2.11: *Robin boundary conditions, smooth grid, M=17.*

M	max norm	L_2 norm	Const max	Const L_2	R max	R L_2
17	0.158E-00	0.174E-00	40.44	44.54	-	-
33	0.398E-01	0.435E-01	40.75	44.54	3.969	4.000
65	0.997E-02	0.108E-01	40.83	44.23	3.991	4.027
129	0.249E-02	0.272E-02	40.79	44.56	4.004	3.970

Table 2.8: *Robin boundary conditions, smooth grid.*

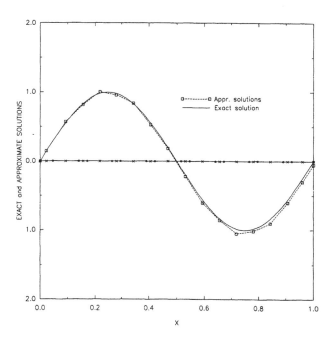

Figure 2.12: *Robin boundary conditions, non-smooth grid, M=17.*

M	max norm	L_2 norm	Const max	Const L_2	R max	R L_2
17	0.704E-01	0.631E-01	18.02	16.15	–	–
33	0.185E-01	0.172E-01	18.94	17.61	3.805	3.668
65	0.475E-02	0.448E-02	19.45	18.35	3.894	3.839
129	0.120E-02	0.114E-02	19.66	18.67	3.958	3.929

Table 2.9: *Robin boundary conditions, non-smooth grid.*

3 Two-Dimensional Support-Operators Algorithms

As it follows from the considerations for the continuous case in the beginning of this chapter and from the description of 1-D support-operators algorithms, to construct the finite-difference scheme for two-dimensional problems, we must construct the finite-difference analogs of operators $\mathcal{C}, \mathcal{K}, \mathcal{B}$, and $\mathcal{D}$, which are given by 1.14

$$
\begin{aligned}
\mathcal{C}u &= -\operatorname{grad} u, \quad \text{for all } (x,y) \\
\mathcal{K}\vec{w} &= K\,\vec{w}, \quad \text{for all } (x,y) \\
\mathcal{B}\vec{w} &= \begin{cases} +\operatorname{div}\vec{w}, & (x,y) \in V \\ -(\vec{w},\vec{n}), & (x,y) \in \partial V \end{cases} \\
\mathcal{D}u &= \begin{cases} 0, & (x,y) \in V \\ \alpha\,u, & (x,y) \in \partial V \end{cases},
\end{aligned}
$$

in two dimensions.

3.1 Nodal Discretisation of Scalar Functions and Cell-Valued Discretisation of Vector Functions

For this case of discretisation, similar to the 1-D case, it is natural to take the finite-difference analog of operator $\mathcal{C}$ as prime. That is, we must construct the finite-difference analog of operator grad with domain HN and range of values $\mathcal{HC}$:

$$\text{GRAD} : HN \to \mathcal{HC}.$$

Because we use the Cartesian components of vectors, operator grad is

$$
\operatorname{grad} u = \begin{pmatrix} \dfrac{\partial u}{\partial x} \\[2mm] \dfrac{\partial u}{\partial y} \end{pmatrix}.
$$

Therefore, we must construct finite-difference approximations for the first derivatives $\partial u/\partial x$, $\partial u/\partial y$. This problem was already solved in the introduction where we constructed a consistent system of finite-difference operators in 2-D. The formulas for these difference operators are given by formulas 5.8 and 5.9 in the introduction, and are repeated here for the convenience of the reader.

$$(D_x u)_{ij} = \tag{3.1}$$
$$\frac{(u_{i+1,j+1} - u_{ij})\,(y_{i,j+1} - y_{i+1,j}) - (u_{i,j+1} - u_{i+1,j})\,(y_{i+1,j+1} - y_{i,j})}{2\,\Omega_{ij}},$$

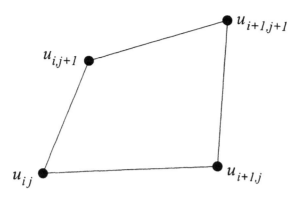

Figure 3.13: *Stencil for operator* GRAD.

$$(D_y u)_{ij} = -$$

$$\frac{(u_{i+1,j+1} - u_{ij})(x_{i,j+1} - x_{i+1,j}) - (u_{i,j+1} - u_{i+1,j})(x_{i+1,j+1} - x_{i,j})}{2\,\Omega_{ij}}.$$

$$(3.2)$$

Here and later where it will not lead to misunderstanding we will drop superscript h in the notation for discrete functions. The stencil for both of these operators is the same and is presented in Figure 3.13. Using operators D_x and D_y, we can form the *prime* operator GRAD : $HN \rightarrow \mathcal{H}C$ such that

$$\text{GRAD}\, u = \begin{pmatrix} D_x(u) \\ D_y(u) \end{pmatrix}.$$

Hence we get the approximation for the operator $C = -\text{grad}$. The next step is to construct the inner products in spaces of discrete functions.

For the space HN, the inner product must be approximate to the following inner product in a continuous space:

$$(u, v)_H = \int_V u\,v\,dV + \oint_{\partial V} u\,v\,dS.$$

We will define the inner product in discrete space HN as follows:

$$(u, v)_{HN} = \sum_{i,j=2}^{M-1,N-1} u_{ij}\, v_{ij}\, VN_{ij} +$$

$$\sum_{i=1}^{M-1} u_{i,1}\, v_{i,1}\, l_{i,1} + \sum_{j=1}^{N-1} u_{M,j}\, v_{M,j}\, l_{M,j} +$$

$$\sum_{i=1}^{M-1} u_{i,N}\, v_{i,N}\, l_{i,N} + \sum_{j=1}^{N-1} u_{1,j}\, v_{1,j}\, l_{1,j} +$$

$$u_{1,1}\, v_{1,1}\, l_{1,1} + u_{M,1}\, v_{M,1}\, l_{M-1,1} +$$

$$u_{M,N}\, v_{M,N}\, l_{M,N-1} + u_{1,N}\, v_{1,N}\, l_{1,M} \; ,$$

where VN_{ij} is some volume of the node ij and l_{ij} is the length which is associated with nodes on the boundary.

It is clear that the order of the approximation operator div in internal nodes will depend on the choice of volume $VN_{i,j}$ and the order of approximation of the normal component of the vectors $(\vec{W}, \vec{n})$ on the boundary will depend on the definition of the length $l_{i,j}$. Therefore, we will choose some concrete expression for these quantities after investigation of the approximation of the derived operators.

It is important to note that the discrete analog of property $B = C^*$ does not depend on the definition of $VN_{i,j}$ and $l_{i,j}$.

The inner product in discrete space $\mathcal{HC}$ can be determined as follows:

$$(\vec{A}, \vec{B})_{\mathcal{HC}} = \sum_{i,j=1}^{M-1,N-1} \left(AX_{ij}\, BX_{ij} + AY_{ij}\, BY_{ij} \right) \Omega_{ij} \, .$$

Let us be reminded that for space $\mathcal{HC}$, both Cartesian components of vector functions are given in cells.

Now let us construct the derived operator, which is the difference analog of operator B and will give us the approximation of the operator divergence in internal points and the approximation of the normal component of the vector on the boundary.

As in the 1-D case, we will regroup terms in the expression for

$$(C^h u, \vec{W})_{\mathcal{HC}} = (-\mathrm{GRAD}\, u, \vec{W})_{\mathcal{HC}}$$

to find coefficients near the components of vector $\vec{W}$. This will give us the formulas for the difference analog of operator B:

$$(-\mathrm{GRAD}\, u, \vec{W})_{\mathcal{HC}} =$$

$$- \sum_{i,j=1}^{M-1,N-1} \left((D_x(u))_{ij}\, WX_{ij} + (D_y(u))_{ij}\, WY_{ij} \right) =$$

$$- \sum_{i,j=1}^{M-1,N-1} \left[\left((u_{i+1,j+1} - u_{ij}) \frac{y_{i,j+1} - y_{i+1,j}}{2} - \right. \right.$$

$$\left. (u_{i,j+1} - u_{i+1,j}) \frac{y_{i+1,j+1} - y_{i,j}}{2} \right) WX_{ij} -$$

$$\left((u_{i+1,j+1} - u_{ij}) \frac{x_{i,j+1} - x_{i+1,j}}{2} - \right.$$

$$\left. (u_{i,j+1} - u_{i+1,j}) \frac{x_{i+1,j+1} - x_{i,j}}{2} \right) WY_{ij} \right] =$$

$$\sum_{i,j=2}^{M-1,N-1} \left[-\left(\frac{y_{i-1,j} - y_{i,j-1}}{2} WX_{i-1,j-1} - \right. \right.$$

$$\frac{y_{i,j+1} - y_{i+1,j}}{2} WX_{ij} -$$

$$\left. \frac{y_{i+1,j} - y_{i,j-1}}{2} WX_{i,j-1} + \frac{y_{i,j+1} - y_{i-1,j}}{2} WX_{i-1,j} \right) +$$

$$\left(\frac{x_{i-1,j} - x_{i,j-1}}{2} WY_{i-1,j-1} - \frac{x_{i,j+1} - x_{i+1,j}}{2} WY_{ij} - \right.$$

$$\left. \left. \frac{x_{i+1,j} - x_{i,j-1}}{2} WY_{i,j-1} + \frac{x_{i,j+1} - x_{i-1,j}}{2} WY_{i-1,j} \right) \right] u_{ij} +$$

$$\sum_{i=2}^{M-1} \left[\left(\frac{y_{i,2} - y_{i+1,1}}{2} WX_{i,1} - \frac{y_{i,2} - y_{i-1,1}}{2} WX_{i-1,1} \right) - \right.$$

$$\left. \left(\frac{x_{i,2} - x_{i+1,1}}{2} WY_{i,1} - \frac{x_{i,2} - x_{i-1,1}}{2} WY_{i-1,1} \right) \right] u_{i1} +$$

$$\sum_{j=2}^{N-1} \left[-\left(\frac{y_{M-1,j} - y_{M,j-1}}{2} WX_{M-1,j-1} + \right. \right.$$

$$\frac{y_{M,j+1} - y_{M-1,j}}{2} WX_{M-1,j} \right) +$$

$$\left(\frac{x_{M-1,j} - x_{M,j-1}}{2} WY_{M-1,j-1} + \right.$$

$$\left. \left. \frac{x_{M,j+1} - x_{M-1,j}}{2} WY_{M-1,j} \right) \right] u_{M,j} +$$

$$\sum_{i=2}^{M-1} \left[-\left(\frac{y_{i-1,N} - y_{i,N-1}}{2} WX_{i-1,N-1} - \right. \right.$$

$$\frac{y_{i+1,N} - y_{i,N-1}}{2} WX_{i,N-1} \right) +$$

$$\left(\frac{x_{i-1,N} - x_{i,N-1}}{2} WY_{i-1,N-1} - \right.$$

$$\left. \left. \frac{x_{i+1,N} - x_{i,N-1}}{2} WY_{i,N-1} \right) \right] u_{iN} +$$

$$\sum_{j=2}^{N-1} \left[\left(\frac{y_{1,j+1} - y_{2,j}}{2} WX_{1j} + \frac{y_{2,j} - y_{1,j-1}}{2} WX_{1,j-1} \right) - \right.$$

$$\left. \left(\frac{x_{1,j+1} - x_{2,j}}{2} WY_{1j} + \frac{x_{2,j} - x_{1,j-1}}{2} WY_{1,j-1} \right) \right] u_{1j} +$$

$$\left(\frac{y_{1,2} - y_{2,1}}{2} \, WX_{1,1} - \frac{x_{1,2} - x_{2,1}}{2} \, WY_{1,1}\right) u_{1,1} -$$

$$\left(\frac{y_{M,2} - y_{M-1,1}}{2} \, WX_{M-1,1} - \frac{x_{M,2} - x_{M-1,1}}{2} \, WY_{M-1,1}\right) u_{M,1} -$$

$$\left(\frac{y_{M-1,N} - y_{M,N-1}}{2} \, WX_{M-1,N-1} - \right.$$

$$\left. \frac{x_{M-1,N} - x_{M,N-1}}{2} \, WY_{M-1,N-1}\right) u_{M,N} +$$

$$\left(\frac{y_{2,N} - y_{1,N-1}}{2} \, WX_{1,N-1} - \frac{x_{2,N} - x_{1,N-1}}{2} \, WY_{1,N-1}\right) u_{1,N}$$

$$= \left(u, \left(C^h\right)^* \vec{W}\right)_{HN} = \left(u, \mathcal{B}^h \, \vec{W}\right)_{HN} .$$

Therefore, for the internal points $i = 2, \ldots, M - 1; j = 2, \ldots, N - 1$ operator $\mathcal{B}^h$, which is the same as operator DIV, is

$$(\text{DIV} \, \vec{W})_{ij} = \tag{3.3}$$

$$\left[-\left(\frac{y_{i-1,j} - y_{i,j-1}}{2} \, WX_{i-1,j-1} - \frac{y_{i,j+1} - y_{i+1,j}}{2} \, WX_{ij} + \right.\right.$$

$$\left. \frac{y_{i+1,j} - y_{i,j-1}}{2} \, WX_{i,j-1} - \frac{y_{i,j+1} - y_{i-1,j}}{2} \, WX_{i-1,j}\right) +$$

$$\left(\frac{x_{i-1,j} - x_{i,j-1}}{2} \, WY_{i-1,j-1} - \frac{x_{i,j+1} - x_{i+1,j}}{2} \, WY_{ij} + \right.$$

$$\left.\left. \frac{x_{i+1,j} - x_{i,j-1}}{2} \, WY_{i,j-1} - \frac{x_{i,j+1} - x_{i-1,j}}{2} \, WY_{i-1,j}\right)\right] / VN_{ij} .$$

The stencil for operator DIV is shown in Figure 3.14.

For $i = 2, \ldots, M - 1$ and $j = 1$, we get the following approximation for the normal component of vector $\vec{W}$:

$$(\mathcal{B}^h \, \vec{W})_{i,1} = \tag{3.4}$$

$$\left[\left(\frac{y_{i,2} - y_{i+1,1}}{2} \, WX_{i,1} - \frac{y_{i,2} - y_{i-1,1}}{2} \, WX_{i-1,1}\right) - \right.$$

$$\left.\left(\frac{x_{i,2} - x_{i+1,1}}{2} \, WY_{i,1} - \frac{x_{i,2} - x_{i-1,1}}{2} \, WY_{i-1,1}\right)\right] / l_{i,1} .$$

For $j = 2, \ldots, N - 1$ and $i = M$, we get

$$(\mathcal{B}^h \, \vec{W})_{M,j} =$$

$$\left[-\left(\frac{y_{M-1,j} - y_{M,j-1}}{2} \, WX_{M-1,j-1} + \right.\right.$$

$$\left. \frac{y_{M,j+1} - y_{M-1,j}}{2} \, WX_{M-1,j}\right) +$$

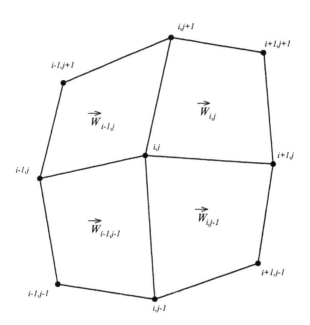

Figure 3.14: *Stencil for operator* DIV.

$$\left(\frac{x_{M-1,j} - x_{M,j-1}}{2} WY_{M-1,j-1} + \right.$$

$$\left. \frac{x_{M,j+1} - x_{M-1,j}}{2} WY_{M-1,j} \right) \Big] / l_{M,j} \ .$$

For $i = 2, \ldots, M - 1$ and $j = N$, we get

$$(\mathcal{B}^h \vec{W})_{i,N} =$$

$$\left[-\left(\frac{y_{i-1,N} - y_{i,N-1}}{2} WX_{i-1,N-1} - \frac{y_{i+1,N} - y_{i,N-1}}{2} WX_{i,N-1} \right) + \right.$$

$$\left. \left(\frac{x_{i-1,N} - x_{i,N-1}}{2} WY_{i-1,N-1} - \frac{x_{i+1,N} - x_{i,N-1}}{2} WY_{i,N-1} \right) \right] / l_{i,N} \ .$$

For $j = 2, \ldots, N - 1$ and $i = 1$, we get

$$(\mathcal{B}^h \vec{W})_{1,j} =$$

$$\left[\left(\frac{y_{1,j+1} - y_{2,j}}{2} WX_{1j} + \frac{y_{2,j} - y_{1,j-1}}{2} WX_{1,j-1} \right) - \right.$$

$$\left. \left(\frac{x_{1,j+1} - x_{2,j}}{2} WY_{1j} + \frac{x_{2,j} - x_{1,j-1}}{2} WY_{1,j-1} \right) \right] / l_{1,j} \ .$$

And finally, for corner nodes, we get

$$(\mathcal{B}^h \vec{W})_{1,1} = \tag{3.5}$$

$$\left(\frac{y_{1,2} - y_{2,1}}{2} WX_{1,1} - \frac{x_{1,2} - x_{2,1}}{2} WY_{1,1} \right) / l_{1,1} \ ,$$

$$(\mathcal{B}^h \vec{W})_{M,1} = \tag{3.6}$$

$$-\left(\frac{y_{M,2} - y_{M-1,1}}{2} WX_{M-1,1} - \frac{x_{M,2} - x_{M-1,1}}{2} WY_{M-1,1} \right) / l_{M,1} \ ,$$

$$(\mathcal{B}^h \vec{W})_{M,N} = \tag{3.7}$$

$$-\left(\frac{y_{M-1,N} - y_{M,N-1}}{2} WX_{M-1,N-1} - \right.$$

$$\left. \frac{x_{M-1,N} - x_{M,N-1}}{2} WY_{M-1,N-1} \right) \Big/ l_{M,N} \ ,$$

$$(\mathcal{B}^h \vec{W})_{1,N} = \tag{3.8}$$

$$\left(\frac{y_{2,N} - y_{1,N-1}}{2} WX_{1,N-1} - \frac{x_{2,N} - x_{1,N-1}}{2} WY_{1,N-1} \right) / l_{1,N} \ .$$

To complete our consideration, we must choose concrete formulas for the volume of the nodes $VN_{i,j}$ and the length associated with the boundary nodes $l_{i,j}$. To do this, we must use reasons related to the approximation properties of the constructed operator.

We can regroup terms in expression for DIV as follows:

$$(\text{DIV } \vec{W})_{ij} =$$
$$[((y_{i,j+1} - y_{i+1,j}) \, WX_{ij} - (x_{i,j+1} - x_{i+1,j}) \, WY_{ij}) +$$
$$((y_{i-1,j} - y_{i,j+1}) \, WX_{i-1,j} - (x_{i-1,j} - x_{i,j+1}) \, WY_{i-1,j}) +$$
$$((y_{i,j-1} - y_{i-1,j}) \, WX_{i-1,j-1} - (x_{i,j-1} - x_{i-1,j}) \, WY_{i-1,j-1}) +$$
$$((y_{i+1,j} - y_{i,j-1}) \, WX_{i,j-1} - (x_{i+1,j} - x_{i,j-1}) \, WY_{i,j-1})] / (2 \, VN_{ij}).$$

Now if we introduce the quadrangle $\diamondsuit_{i,j}$, which is presented in Figure 3.15 by dashed lines, and denote its sides as

$$l_{(i,j+1)(i+1,j)} \, , \, l_{(i-1,j)(i,j+1)} \, , \, l_{(i,j-1)(i-1,j)} \, , \, l_{(i+1,j)(i,j-1)} \, ,$$

and the outward normal to these sides as

$$\vec{n}_{(i,j+1)(i+1,j)} = \left(\frac{y_{i,j+1} - y_{i+1,j}}{l_{(i,j+1)(i+1,j)}} , \ -\frac{x_{i,j+1} - x_{i+1,j}}{l_{(i,j+1)(i+1,j)}} \right) ,$$

$$\vec{n}_{(i-1,j)(i,j+1)} = \left(\frac{y_{i-1,j} - y_{i,j+1}}{l_{(i-1,j)(i,j+1)}} , \ -\frac{x_{i-1,j} - x_{i,j+1}}{l_{(i-1,j)(i,j+1)}} \right) ,$$

$$\vec{n}_{(i,j-1)(i-1,j)} = \left(\frac{y_{i,j-1} - y_{i-1,j}}{l_{(i,j-1)(i-1,j)}} , \ \frac{x_{i,j-1} - x_{i-1,j}}{l_{(i,j-1)(i-1,j)}} \right) ,$$

$$\vec{n}_{(i+1,j)(i,j-1)} = \left(\frac{y_{i+1,j} - y_{i,j-1}}{l_{(i+1,j)(i,j-1)}} , \ -\frac{x_{i+1,j} - x_{i,j-1}}{l_{(i+1,j)(i,j-1)}} \right) ,$$

we can then rewrite the expression for DIV as follows:

$$(\text{DIV } \vec{W})_{ij} =$$
$$\left[\left(\vec{W}_{ij}, \vec{n}_{(i,j+1),(i+1,j)} \right) l_{(i,j+1)(i+1,j)} + \right.$$
$$\left(\vec{W}_{i-1,j}, \vec{n}_{(i-1,j),(i,j+1)} \right) l_{(i-1,j)(i,j+1)} +$$
$$\left(\vec{W}_{i-1,j-1}, \vec{n}_{(i,j-1),(i-1,j)} \right) l_{(i,j-1)(i-1,j)} +$$
$$\left. \left(\vec{W}_{i,j-1}, \vec{n}_{(i+1,j),(i,j-1)} \right) l_{(i+1,j)(i,j-1)} \right] \Big/ (2 \, VN_{i,j}),$$

where $(\, , \,)$ denotes the usual dot product of the two vectors.

Reminded that the invariant definition of the operator div is

$$\text{div} \vec{w} = \lim_{V \to 0} \frac{\oint_{\partial V} (\vec{w}, \vec{n}) \, dV}{V}, \tag{3.9}$$

we can see that the expression for DIV can be considered in the form which is the analog of 3.9 if we take the quadrangle $\diamondsuit_{i,j}$ as V in 3.9. Therefore,

$$VN_{i,j} = \frac{\diamondsuit_{i,j}}{2}.$$

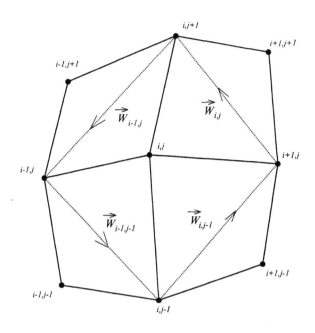

Figure 3.15: *Contour for operator* DIV.

The last formula can also be explained if we take into account that for $VN_{i,j}$ we must have

$$\sum_i VN_{i,j} \approx \text{volume of the domain.}$$

In fact,

$$\sum_i \Diamond_{i,j} \approx 2 * \text{volume of the domain}$$

because each $\Diamond_{i,j}$ contains four triangles, and for the example of the rectangular grid, $\Diamond_{i,j}$ is twice the size of $\Omega_{i,j}$.

It is possible to show that for the general regular grid, we have a zero-order truncation error for this choice of $VN_{i,j}$. That is,

$$\psi_{\text{DIV}} = O(1).$$

It is also possible to prove that it is not possible to find some formula for $VN_{i,j}$ which will give us the first-order truncation error for a general grid.

For a smooth grid, it is possible to prove that this choice of VN_{ij} gives us the first-order truncation error. Also, any choice of VN_{ij} where

$$VN_{ij} = \Diamond_{ij} + O(h^2),$$

will give us the first-order truncation error on a smooth grid. For example, in some applications the following formula is used:

$$V N_{ij} = 0.25 \left(\Omega_{ij} + \Omega_{i-1,j} + \Omega_{i-1,j-1} + \Omega_{i,j-1} \right).$$

Let us now consider operator $\mathcal{B}^h$ on the boundary. For example, for the bottom boundary $i = 2, \ldots, M - 1$ and $j = 1$ when

$$(\mathcal{B}^h \vec{W})_{i,1} = \tag{3.10}$$
$$\left[\left(\frac{y_{i,2} - y_{i+1,1}}{2} W X_{i,1} - \frac{y_{i,2} - y_{i-1,1}}{2} W X_{i-1,1} \right) - \left(\frac{x_{i,2} - x_{i+1,1}}{2} W Y_{i,1} - \frac{x_{i,2} - x_{i-1,1}}{2} W Y_{i-1,1} \right) \right] / l_{i,1}.$$

Similar to the case of DIV, we can rewrite this formula as follows:

$$(\mathcal{B}^h \vec{W})_{i,1} =$$
$$\frac{\left(\vec{W}_{i,1}, \vec{n}_{(i,2),(i+1,1)} \right) l_{(i,2)(i+1,1)} + \left(\vec{W}_{i-1,1}, \vec{n}_{(i-1,1),(i,2)} \right) l_{(i-1,1)(i,2)}}{2 \, l_{i,1}}.$$

From this formula we can see that $\mathcal{B}^h_{i,1}$ is the approximation of some normal component of vector $\vec{W}$ because it looks like some average of the components which are normal to $l_{(i,2)(i+1,1)}$ and $l_{(i-1,1)(i,2)}$. This formula has to be exact at least for the constant vectors, that is, $\mathcal{B}^h_{i,1}$ must approximate the normal component of vector $\vec{W} = (WX, WY), WX, WY = const$, with respect to some direction. If we substitute the constant vector $\vec{W} = (WX, WY)$ into the formula for $\mathcal{B}^h_{i,1}$ (3.10), we get

$$(\mathcal{B}^h \vec{W})_{i,1} =$$
$$\frac{y_{i-1,1} - y_{i+1,1}}{2 \, l_{i,1}} W X - \frac{x_{i-1,1} - x_{i+1,1}}{2 \, l_{i,1}} W Y.$$

From the theoretical considerations, we know that this expression must approximate the expression $-(\vec{W}, \vec{n})$, and from the previous formula it is easy to see that there is only one possibility where the expression in the right-hand side can be a minus normal component of vector $\vec{W}$. That is,

$$l_{i,1} = 0.5 \, l_{(i-1,1)(i+1,1)}.$$

Therefore $(\mathcal{B}^h \vec{W})_{i,1}$ approximates the expression $-(\vec{W}, \vec{n}_{(i-1,1)(i+1,1)})$ (see Figure 3.16).

If we consider our problem more carefully, we find that this is the natural approximation for $-(\vec{W}, \vec{n})$ in node $(i, 1)$ because we do not have a unique direction in the node.

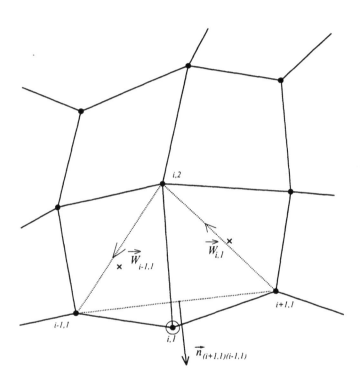

Figure 3.16: *Approximations on the boundary.*

Again, if we assume that the grid is smooth and the curve which represents a boundary is also smooth, then we can prove that we have the first-order truncation error for operator B^h on the boundary.

Let us note that for a smooth grid, we can use another formula for $l_{i,1}$. For example,

$$l_{i,1} = 0.5 \left(l\xi_{i-1,1} + l\xi_{i,1} \right).$$

Similar to the case of the bottom boundary, we find the expressions for the length of the node on the other boundaries and get the following:

$$l_{i,1} = 0.5 \, l_{(i-1,1)(i+1,1)}, \quad i = 2, \ldots, M-1,$$
$$l_{M,j} = 0.5 \, l_{(M,j-1)(M,j+1)} \quad j = 2, \ldots, N-1,$$
$$l_{i,N} = 0.5 \, l_{(i-1,N)(i+1,N)} \quad i = 2, \ldots, M-1,$$
$$l_{1,j} = 0.5 \, l_{(1,j-1)(1,j+1)} \quad j = 2, \ldots, N-1.$$

Similar considerations for the corner nodes

$$(1,1); \ (M,1); \ (M,N); \ (1,N),$$

give us the following formulas:

$$l_{1,1} = 0.5 \, l_{(1,2)(2,1)}, \, l_{M,1} = 0.5 \, l_{(M-1,1)(M,2)},$$
$$l_{M,N} = 0.5 \, l_{(M,N-1)(M-1,N)}, \, l_{1,N} = 0.5 \, l_{(2,N)(1,N-1)}.$$

The graphical illustration of the approximation in the corner nodes $(1,1)$ are presented in Figure 3.17. Thus, the expression $(B^h \, \vec{W})_{1,1}$ is the approximation of $-\left(\vec{W}, \vec{n}_{(2,1)(1,2)} \right)$.

Therefore, we have constructed the approximation for the operator divergence in internal nodes, and the approximation for the normal component of the vector on the boundary.

Now following the 1-D considerations, we must construct approximations for operators $\mathcal{K}$ and $\mathcal{D}$.

In 2-D K is symmetric and a positive definite matrix:

$$K = \begin{pmatrix} KXX & KXY \\ KXY & KYY \end{pmatrix} \tag{3.11}$$

and

$$K \vec{W} = \begin{pmatrix} KXX \, WX + KXY \, WY \\ KXY \, WX + KYY \, WY \end{pmatrix}.$$

If we assume that the components of matrix K are given by their values in the cells (similar to the components of vector $\vec{W}$) then we can define the discrete operator $\mathcal{K}^h$ as follows:

$$(\mathcal{K}^h \, \vec{W})_{i,j} = \begin{pmatrix} KXX_{i,j} \, WX_{i,j} + KXY_{i,j} \, WY_{i,j} \\ KXY_{i,j} \, WX_{i,j} + KYY_{i,j} \, WY_{i,j} \end{pmatrix}.$$

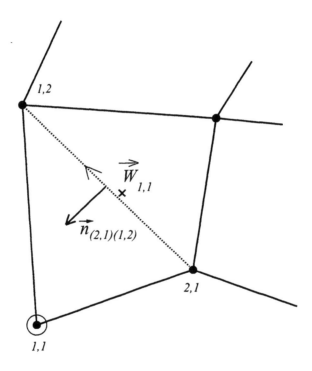

Figure 3.17: *Approximations in the corner.*

Let us define operator $\mathcal{D}$. If we assume that function α is given by its value in the nodes on the boundary, then

$$(\mathcal{D}^h u)_{i,j} = \alpha_{i,j} u_{i,j},$$

where nodes (i,j) belong to the boundary.

3.1.1 The Finite-Difference Scheme for the Dirichlet Boundary Problem

Using the constructed operators DIV and GRAD, the finite-difference scheme can be written in the following form:

$$(-\text{DIV} \, \mathcal{K}^h \, \text{GRAD} \, u)_{i,j} = \varphi_{i,j}; \quad \begin{array}{l} i = 2, \ldots, M-1; \\ j = 2, \ldots, N-1, \end{array}$$

where $\varphi_{i,j} = f(x_{i,j}, y_{i,j})$.

Approximations for the Dirichlet boundary conditions can be shown as:

$$
\begin{aligned}
u_{i,1} &= \psi_{i,1}, & i = 1, \ldots, M, \\
u_{M,j} &= \psi_{M,j}, & j = 1, \ldots, N, \\
u_{i,N} &= \psi_{i,N}, & i = 1, \ldots, M, \\
u_{1,j} &= \psi_{1,j}, & j = 1, \ldots, N.
\end{aligned}
$$

From the Dirichlet boundary conditions we know the value of function U on the boundary, and therefore, the unknowns are

$$u_{i,j}, \, i = 2, \ldots, M-1; \, j = 2, \ldots, N-1.$$

To write the matrix form of the finite-difference equations, we must move the known values to the right-hand side. Similar to the 1-D case, the matrix related to the finite-difference scheme will be symmetric and positive if we multiply both sides of the difference equations by $VN_{i,j}$.

For example, for the internal nodes $i = 3, \ldots, M-2; \, j = 3, \ldots, N-2$, we get

$$(\text{DIV} \, \mathcal{K} \, \text{GRAD} \, u)_{ij} =$$

$$
\left\{ \left[\frac{y_{i-1,j} - y_{i,j-1}}{2} * \right. \right.
$$

$$(KXX_{i-1,j-1} \left((u_{i,j} - u_{i-1,j-1}) (y_{i-1,j} - y_{i,j-1}) - \right.$$

$$(u_{i-1,j} - u_{i,j-1})(y_{i,j} - y_{i-1,j-1}))/(2\,\Omega_{i-1,j-1}) -$$

$$KXY_{i-1,j-1} \left((u_{i,j} - u_{i-1,j-1}) (x_{i-1,j} - x_{i,j-1}) - \right.$$

$$(u_{i-1,j} - u_{i,j-1})(x_{i,j} - x_{i-1,j-1}))/(2\,\Omega_{i-1,j-1})) +$$

$$\frac{y_{i,j+1} - y_{i+1,j}}{2} *$$

$$(KXX_{i,j} \, ((u_{i+1,j+1} - u_{ij}) \, (y_{i,j+1} - y_{i+1,j}) -$$
$$(u_{i,j+1} - u_{i+1,j}) \, (y_{i+1,j+1} - y_{i,j})) / \, (2 \, \Omega_{ij}) -$$
$$KXY_{i,j} \, ((u_{i+1,j+1} - u_{ij}) \, (y_{i,j+1} - y_{i+1,j}) -$$
$$(u_{i,j+1} - u_{i+1,j}) \, (y_{i+1,j+1} - y_{i,j})) / \, (2 \, \Omega_{ij})) +$$
$$\frac{y_{i+1,j} - y_{i,j-1}}{2} *$$
$$(KXX_{i,j-1} \, ((u_{i+1,j} - u_{i,j-1}) \, (y_{i,j} - y_{i+1,j-1}) -$$
$$(u_{i,j} - u_{i+1,j-1}) \, (y_{i+1,j} - y_{i,j-1})) / \, (2 \, \Omega_{i,j-1}) -$$
$$KXY_{i,j-1} \, ((u_{i+1,j} - u_{i,j-1}) \, (x_{i,j} - x_{i+1,j-1}) -$$
$$(u_{i,j} - u_{i+1,j-1}) \, (x_{i+1,j} - x_{i,j-1})) / \, (2 \, \Omega_{i,j-1})) +$$
$$\frac{y_{i,j+1} - y_{i-1,j}}{2} *$$
$$(KXX_{i-1,j} \, ((u_{i,j+1} - u_{i-1,j}) \, (y_{i-1,j+1} - y_{i,j}) -$$
$$(u_{i-1,j+1} - u_{i,j}) \, (y_{i,j+1} - y_{i-1,j})) / \, (2 \, \Omega_{i-1,j}) -$$
$$KXY_{i-1,j} \, ((u_{i,j+1} - u_{i-1,j}) \, (x_{i-1,j+1} - x_{i,j}) -$$
$$(u_{i-1,j+1} - u_{i,j}) \, (x_{i,j+1} - x_{i-1,j})) / \, (2 \, \Omega_{i-1,j}))] -$$
$$\left[\frac{x_{i-1,j} - x_{i,j-1}}{2} * \right.$$
$$(KXY_{i-1,j-1} \, ((u_{i,j} - u_{i-1,j-1}) \, (y_{i-1,j} - y_{i,j-1}) -$$
$$(u_{i-1,j} - u_{i,j-1}) \, (y_{i,j} - y_{i,j-1})) / \, (2 \, \Omega_{i-1,j-1}) -$$
$$KYY_{i-1,j-1} \, ((u_{i,j} - u_{i-1,j-1}) \, (x_{i-1,j} - x_{i,j-1}) -$$
$$(u_{i-1,j} - u_{i,j-1}) \, (x_{i,j} - x_{i-1,j-1})) / \, (2 \, \Omega_{i-1,j-1})) +$$
$$\frac{x_{i,j+1} - x_{i+1,j}}{2} *$$
$$(KXY_{i,j} \, ((u_{i+1,j+1} - u_{ij}) \, (y_{i,j+1} - y_{i+1,j}) -$$
$$(u_{i,j+1} - u_{i+1,j}) \, (y_{i+1,j+1} - y_{i,j})) / \, (2 \, \Omega_{ij}) -$$
$$KYY_{i,j} \, ((u_{i+1,j+1} - u_{ij}) \, (y_{i,j+1} - y_{i+1,j}) -$$
$$(u_{i,j+1} - u_{i+1,j}) \, (y_{i+1,j+1} - y_{i,j})) / \, (2 \, \Omega_{ij})) +$$
$$\frac{x_{i+1,j} - x_{i,j-1}}{2} *$$
$$(KXY_{i,j-1} \, ((u_{i+1,j} - u_{i,j-1}) \, (y_{i,j} - y_{i+1,j-1}) -$$
$$(u_{i,j} - u_{i+1,j-1}) \, (y_{i+1,j} - y_{i,j-1})) / \, (2 \, \Omega_{i,j-1}) -$$
$$KYY_{i,j-1} \, ((u_{i+1,j} - u_{i,j-1}) \, (x_{i,j} - x_{i+1,j-1}) -$$
$$(u_{i,j} - u_{i+1,j-1}) \, (x_{i+1,j} - x_{i,j-1})) / \, (2 \, \Omega_{i,j-1})) +$$
$$\frac{x_{i,j+1} - x_{i-1,j}}{2} *$$
$$(KXY_{i-1,j} \, ((u_{i,j+1} - u_{i-1,j}) \, (y_{i-1,j+1} - y_{i,j}) -$$
$$(u_{i-1,j+1} - u_{i,j}) \, (y_{i,j+1} - y_{i-1,j})) / \, (2 \, \Omega_{i-1,j}) -$$

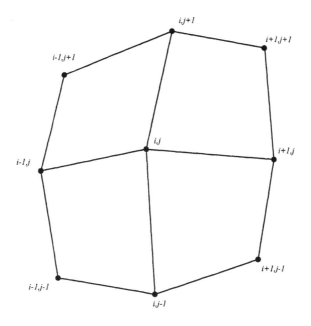

Figure 3.18: *Stencil for operator* DIV GRAD.

$$KYY_{i-1,j} \left((u_{i,j+1} - u_{i-1,j})(x_{i-1,j+1} - x_{i,j}) - \right.$$
$$\left. (u_{i-1,j+1} - u_{i,j})(x_{i,j+1} - x_{i-1,j})\right)/(2\,\Omega_{i-1,j}))]\}/V\,N_{i,j}\,.$$

The stencil for this operator contains nine points and is presented in Figure 3.18. It is clear that this explicit form of difference equations does not give us any new information about the properties of finite-difference schemes. Therefore, we will not write the explicit form for the difference equations on the boundary. We just present the stencils on the bottom boundary and the corner cell.

Readers can refer to the subroutine which computes the coefficients of finite-difference equations in the Fortran directory on the attached floppy disk (see the Fortran Guide chapter for details).

3.1.2 The Finite-Difference Scheme for the Robin Boundary Problem

For Robin boundary conditions, the equations for internal points are the same as the equations for the Dirichlet boundary conditions, but now, because values of the difference function on the boundary are also unknown, we do not need to move them into the right-hand side.

The general Robin boundary condition is

$$(\vec{n}, K \operatorname{grad} u) + \alpha\, u = \gamma .$$

Then, because

$$B \vec{W} = -(\vec{W}, \vec{n}) = K \operatorname{grad} u$$

on the boundary and we have already obtained the approximation for this operator, there is no problem in finding the approximation of the first term in the Robin boundary condition. Also, because we use nodal discretisation, we do not have a problem in finding the approximation of the second term in the left-hand side.

Let us demonstrate how we approximate the Robin boundary conditions on the bottom boundary $(i = 2, \ldots, M - 1)$ and in the left bottom corner $(i = 1; j = 1)$.

For $i = 2, \ldots, M - 1$, we have

$$
\begin{aligned}
- \Bigg\{ \Bigg[& \frac{y_{i,2} - y_{i+1,1}}{2} * \\
& (KXX_{i,1}\,((u_{i+1,2} - u_{i1})\,(y_{i,2} - y_{i+1,1}) - \\
& (u_{i,2} - u_{i+1,1})\,(y_{i+1,2} - y_{i,1}))/\,(2\,\Omega_{i,1}) - \\
& KXY_{i,1}\,((u_{i+1,2} - u_{i,1})\,(x_{i,2} - x_{i+1,1}) - \\
& (u_{i,2} - u_{i+1,1})\,(x_{i+1,2} - x_{i,1}))/\,(2\,\Omega_{i,1})) - \\
& \frac{y_{i,2} - y_{i-1,1}}{2} * \\
& (KXX_{i-1,1}\,((u_{i,2} - u_{i-1,1})\,(y_{i-1,2} - y_{i,1}) - \\
& (u_{i-1,2} - u_{i,1})\,(y_{i,2} - y_{i-1,1}))/\,(2\,\Omega_{i-1,1}) - \\
& KXY_{i-1,1}\,((u_{i,2} - u_{i-1,1})\,(x_{i-1,2} - x_{i,1}) - \\
& (u_{i-1,2} - u_{i,1})\,(x_{i,2} - x_{i-1,1}))/\,(2\,\Omega_{i-1,1}))\Bigg] - \\
\Bigg[& \frac{x_{i,2} - x_{i+1,1}}{2} * \\
& (KXY_{i,1}\,((u_{i+1,2} - u_{i1})\,(y_{i,2} - y_{i+1,1}) - \\
& (u_{i,2} - u_{i+1,1})\,(y_{i+1,2} - y_{i,1}))/\,(2\,\Omega_{i,1}) - \\
& KYY_{i,1}\,((u_{i+1,2} - u_{i,1})\,(x_{i,2} - x_{i+1,1}) - \\
& (u_{i,2} - u_{i+1,1})\,(x_{i+1,2} - x_{i,1}))/\,(2\,\Omega_{i,1})) - \\
& \frac{x_{i,2} - x_{i-1,1}}{2} * \\
& (KXY_{i-1,1}\,((u_{i,2} - u_{i-1,1})\,(y_{i-1,2} - y_{i,1}) - \\
& (u_{i-1,2} - u_{i,1})\,(y_{i,2} - y_{i-1,1}))/\,(2\,\Omega_{i-1,1}) - \\
& KYY_{i-1,1}\,((u_{i,2} - u_{i-1,1})\,(x_{i-1,2} - x_{i,1}) - \\
& (u_{i-1,2} - u_{i,1})\,(x_{i,2} - x_{i-1,1}))/\,(2\,\Omega_{i-1,1}))\Bigg]\Bigg\}\,/l_{i,1} + \\
& + \alpha_{i,1}\,u_{i,1} = \gamma_{i,1} .
\end{aligned}
\tag{3.12}
$$

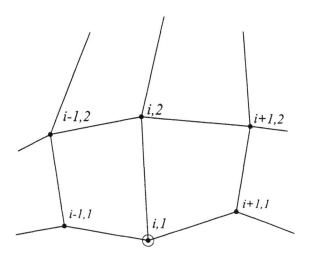

Figure 3.19: *N-C Discretisation, stencil for Robin boundary conditions, bottom boundary.*

For approximation of the Robin boundary conditions, we get the 6-node stencil as presented in Figure 3.19.

For the corner node $(1, 1)$ we get

$$
\left\{ \left[\frac{y_{1,2} - y_{2,1}}{2} * \right. \right.
$$

$$
KXX_{1,1} \left((u_{2,2} - u_{1,1})(y_{1,2} - y_{2,1}) - \right.
$$
$$
(u_{1,2} - u_{2,1})(y_{2,2} - y_{1,1}))/(2\,\Omega_{1,1}) -
$$
$$
KXY_{1,1} \left((u_{2,2} - u_{1,1})(x_{1,2} - x_{2,1}) - \right.
$$
$$
(u_{1,2} - u_{2,1})(x_{2,2} - x_{1,1}))/(2\,\Omega_{1,1}))] -
$$
$$
\left[\frac{x_{1,2} - x_{2,1}}{2} * \right.
$$
$$
(KXY_{1,1} \left((u_{2,2} - u_{1,1})(y_{1,2} - y_{2,1}) - \right.
$$
$$
(u_{1,2} - u_{2,1})(y_{2,2} - y_{1,1}))/(2\,\Omega_{1,1}) -
$$
$$
KYY_{1,1} \left((u_{2,2} - u_{1,1})(x_{1,2} - x_{2,1}) - \right.
$$
$$
(u_{1,2} - u_{2,1})(x_{2,2} - x_{1,1}))/(2\,\Omega_{1,1}))] \right\} / l_{1,1} +
$$
$$
\alpha_{1,1}\,u_{1,1} = \gamma_{1,1}
$$

$$(3.13)$$

The stencil for this approximation is presented in Figure 3.20. For other boundary and corner nodes, the equations look similar.

Similar to the 1-D case, to obtain a symmetric and positive definite matrix problem, we must multiply the equation in internal nodes by volume $VN_{i,j}$ and the equations on the boundary by $l_{i,j}$.

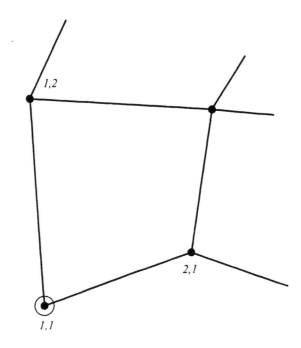

Figure 3.20: *N-C discretisation, stencil for Robin boundary conditions, corner (1,1).*

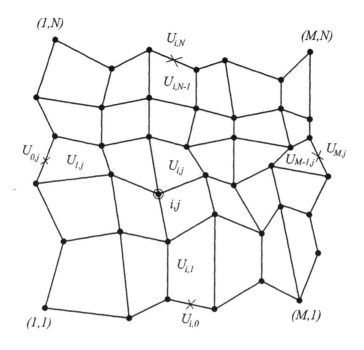

Figure 3.21: *C-N discretisation in 2-D.*

It is important to note that these properties of matrix problems are guaranteed by the method of construction of finite-difference schemes.

The reader can find all formulas in the Fortran directory on the attached floppy disk.

3.2 Cell-Valued Discretisation of Scalar Functions and Nodal Discretisation of Vector Functions

For this type of discretisation, we must introduce some values of the scalar function on the boundary. These values will be associated with sides of cells which form the boundary (see Figure 3.21).

For this type of discretisation, similar to the 1-D case, it is natural to take the finite-difference analog of operator B as prime. That is, we must construct operator DIV with domain $\mathcal{HN}$ and range of values HC, and also approximation of normal component of vector $(\vec{W}, \vec{n})$, $\vec{W} \in \mathcal{HN}$, on the boundary.

To construct operator DIV, we will use its invariant definition:

$$\operatorname{div} \vec{w} = \lim_{V \to 0} \frac{\oint_{\partial V} (\vec{w}, \vec{n}) \, dS}{V}.$$

In the discrete case, we will take cell $\Omega_{i,j}$ as V in the previous formula. We first divide the contour integral over the boundary of cell $\Omega_{i,j}$ into four integrals over the sides:

$$\oint_{\partial\Omega_{i,j}} (\vec{w}, \vec{n})\, dS =$$

$$\int_{l\xi_{i,j}} (\vec{w}, \vec{n})\, dS + \int_{l\eta_{i+1,j}} (\vec{w}, \vec{n})\, dS +$$

$$\int_{l\xi_{i,j+1}} (\vec{w}, \vec{n})\, dS + \int_{l\eta_{i,j}} (\vec{w}, \vec{n})\, dS .$$

Let us consider the process of approximation of the first integral in the right-hand side. Because the side of the cell is a straight line, the vector of $\vec{n}$ is the constant vector on each side. For example, for side $l\xi i, j$, we get

$$\vec{n}_{(i,j)(i+1,j)} = \begin{pmatrix} \dfrac{y_{i+1,j} - y_{i,j}}{l\xi_{i,j}} \\ \dfrac{-(x_{i+1,j} - x_{i,j})}{l\xi_{i,j}} \end{pmatrix}$$

and the trapezium formula gives us the following approximation for the integral over the side $l\xi_{i,j}$:

$$\oint_{\partial l\xi_{i,j}} (\vec{w}, \vec{n})\, dS \approx$$

$$\left(\frac{WX_{i,j} + WX_{i+1,j}}{2} \frac{y_{i+1,j} - y_{i,j}}{l\xi_{i,j}} - \right.$$

$$\left. \frac{WY_{i,j} + WY_{i+1,j}}{2} \frac{(x_{i+1,j} - x_{i,j})}{l\xi_{i,j}} \right) l\xi_{i,j} =$$

$$\frac{WX_{i,j} + WX_{i+1,j}}{2} (y_{i+1,j} - y_{i,j}) - \frac{WY_{i,j} + WY_{i+1,j}}{2} (x_{i+1,j} - x_{i,j}).$$

For the integrals over the other sides, we can get similar formulas. Finally, we obtain the following formula for operator $\mathrm{DIV} : \mathcal{HN} \to HC$:

$$(\mathrm{DIV}\,\vec{W})_{i,j} =$$

$$\left[\left(\frac{WX_{i,j} + WX_{i+1,j}}{2} (y_{i+1,j} - y_{i,j}) - \right. \right.$$

$$\left. \frac{WY_{i,j} + WY_{i+1,j}}{2} (x_{i+1,j} - x_{i,j}) \right) +$$

$$\left(\frac{WX_{i+1,j} + WX_{i+1,j+1}}{2} (y_{i+1,j+1} - y_{i+1,j}) - \right.$$

$$\left. \frac{WY_{i+1,j} + WY_{i+1,j+1}}{2} (x_{i+1,j+1} - x_{i+1,j}) \right) +$$

$$\left(\frac{WX_{i+1,j+1} + WX_{i,j+1}}{2}\left(y_{i,j+1} - y_{i+1,j+1}\right)-\right.$$

$$\left.\frac{WY_{i+1,j+1} + WY_{i,j+1}}{2}\left(x_{i,j+1} - x_{i+1,j+1}\right)\right) +$$

$$\left(\frac{WX_{i,j+1} + WX_{i,j}}{2}\left(y_{i,j} - y_{i,j+1}\right)-\right.$$

$$\left.\left.\frac{WY_{i,j+1} + WY_{i,j}}{2}\left(x_{i,j} - x_{i,j+1}\right)\right)\right] \Big/ \Omega_{i,j}\,.$$

After some regrouping, the expression for operator DIV can be presented in the following form:

$$(\text{DIV}\,\vec{W})_{i,j} = \tag{3.14}$$
$$((WX_{i+1,j+1} - WX_{ij})(y_{i,j+1} - y_{i+1,j})-$$
$$(WX_{i,j+1} - WX_{i+1,j})(y_{i+1,j+1} - y_{i,j}))/(2\,\Omega_{ij}) -$$
$$((WY_{i+1,j+1} - WY_{ij})(x_{i,j+1} - x_{i+1,j})-$$
$$(WY_{i,j+1} - WY_{i+1,j})(x_{i+1,j+1} - x_{i,j}))/(2\,\Omega_{ij})\,.$$

Now if we take into account formulas 3.1 and 3.2 for operators D_x and D_y, we can write

$$(\text{DIV}\,\vec{W})_{i,j} = (D_x\,WX)_{i,j} + (D_y\,WY)_{i,j}\,.$$

To complete the construction of the finite-difference analog of operator $\mathcal{B}^h$, we must construct the approximation of expression $(\vec{W}, \vec{n})$. The finite-difference analog of this expression must be defined at the same place as the values of the scalar function on the boundary, that is, on the boundary sides. Let us consider the bottom boundary $j = 1; i = 1, \ldots, N - 1$. We will use the following definition:

$$(\mathcal{B}^h\,\vec{W})_{i,1} = \tag{3.15}$$
$$-\left(\frac{WX_{i,1} + WX_{i+1,1}}{2}\frac{y_{i+1,1} - y_{i,1}}{l\xi_{i,1}} -\right.$$
$$\left.\frac{WY_{i,1} + WY_{i+1,1}}{2}\frac{x_{i+1,1} - x_{i,1}}{l\xi_{i,1}}\right)\,.$$

Let us note that for the case of the cell-valued discretisation of scalar functions and nodal discretisation for vector functions, we have the natural approximation for the normal component of the vector and we do not need to consider questions relating to the definition of the normal direction in the nodes as we did for the case of nodal discretisation of scalar functions.

The next step is the construction of the inner products in spaces of discrete functions. For the space HC, we have

$$(u, v)_{HC} =$$
$$\sum_{i,j=1}^{M-1,N-1} u_{ij}\, v_{ij}\, \Omega_{ij} +$$
$$\sum_{i=1}^{M-1} u_{i,0}\, v_{i,0}\, l\xi_{i,1} + \sum_{j=1}^{N-1} u_{M,j}\, v_{M,j}\, l\eta_{M,j} +$$
$$\sum_{i=1}^{M-1} u_{i,N}\, v_{i,N}\, l\xi_{i,N} + \sum_{j=1}^{N-1} u_{0,j}\, v_{0,j}\, l\eta_{1,j}\, .$$

For the space of vector functions $\mathcal{HN}$, we define the inner product as follows:

$$(\vec{A}, \vec{B})_{\mathcal{HN}} = \sum_{i,j=1}^{M,N} \left(AX_{ij}\, BX_{ij} + AY_{ij}\, BY_{ij} \right) VN_{ij}\, .$$

Volume VN_{ij} will be defined later when we will consider the approximation properties of operator GRAD.

Having the expressions for the inner products, we can now construct the derived operator $C^h = -\text{GRAD}$. To do this, we will transform the expression for $(B^h \vec{W}, u)_{HC}$ to $(\vec{W}, C^h u)_{\mathcal{HN}}$. Technically, this procedure looks similar to other cases where we constructed the derived operator. That is, we must do some regrouping to collect coefficients near the components of vector $\vec{W}$. Here we present only the final results. We will use notations GX, GY for the components of vector GRAD u, that is,

$$\text{GRAD}\, u = \begin{pmatrix} GX \\ GY \end{pmatrix}.$$

For the internal nodes $i = 2, \ldots, M - 1;\ j = 2, \ldots, N - 1$, we get

$$GX_{ij} = \left(\frac{y_{i,j+1} - y_{i+1,j}}{2}\, u_{ij} + \frac{y_{i-1,j} - y_{i,j+1}}{2}\, u_{i-1,j} + \right.$$
$$\left. \frac{y_{i,j-1} - y_{i-1,j}}{2}\, u_{i-1,j-1} + \frac{y_{i+1,j} - y_{i,j-1}}{2}\, u_{i,j-1} \right) / VN_{ij}\, ,$$

$$(3.16)$$

$$GY_{ij} = -\left(\frac{x_{i,j+1} - x_{i+1,j}}{2}\, u_{ij} + \frac{x_{i-1,j} - x_{i,j+1}}{2}\, u_{i-1,j} + \right.$$
$$\left. \frac{x_{i,j-1} - x_{i-1,j}}{2}\, u_{i-1,j-1} + \frac{x_{i+1,j} - x_{i,j-1}}{2}\, u_{i,j-1} \right) / VN_{ij}\, .$$

For the bottom boundary $j = 1$; $i = 2, \ldots, N - 1$, we get

$$GX_{i,1} = \left(\frac{y_{i,2} - y_{i+1,1}}{2} u_{i,1} + \frac{y_{i-1,1} - y_{i,2}}{2} u_{i-1,1} + \right.$$

$$\left. \frac{y_{i,1} - y_{i-1,1}}{2} u_{i-1,0} + \frac{y_{i+1,1} - y_{i,1}}{2} u_{i,0} \right) / V N_{i,1} ,$$

$$(3.17)$$

$$GY_{i,1} = - \left(\frac{x_{i,2} - x_{i+1,1}}{2} u_{i,1} + \frac{x_{i-1,1} - x_{i,2}}{2} u_{i-1,1} + \right.$$

$$\left. \frac{x_{i,1} - x_{i-1,1}}{2} u_{i-1,0} + \frac{x_{i+1,1} - x_{i,1}}{2} u_{i,0} \right) / V N_{i,1} .$$

Similar equations can be derived for other boundaries.

It is easy to see that if we introduce fictitious nodes for $j = 1, \ldots, N$:

$$(x_{0,j} = x_{1,j}, y_{0,j} = y_{1,j}), (x_{M+1,j} = x_{M,j}, y_{M+1,j} = y_{M,j}),$$

and for $i = 1, \ldots, M$:

$$(x_{i,0} = x_{i,1}, y_{i,0} = y_{i,1}), (x_{i,N+1} = x_{i,N}, y_{i,N+1} = y_{i,N}),$$

then the formula on the boundary 3.17 can be written in the same form as the formula for the internal nodes 3.16.

Let us now consider the corner node $i = 1$; $j = 1$. For this node, we get

$$GX_{1,1} =$$

$$\left(\frac{y_{1,2} - y_{2,1}}{2} u_{1,1} + \frac{y_{1,1} - y_{1,2}}{2} u_{0,1} + \frac{y_{2,1} - y_{1,1}}{2} u_{1,0} \right) / V N_{1,1} ,$$

$$(3.18)$$

$$GY_{1,1} =$$

$$- \left(\frac{x_{1,2} - x_{2,1}}{2} u_{1,1} + \frac{x_{1,1} - x_{1,2}}{2} u_{0,1} + \frac{x_{2,1} - x_{1,1}}{2} u_{1,0} \right) / V N_{1,1} .$$

These formulas can also be written in the same form as for internal nodes using fictitious nodes and the assumption that values $u_{0,0}$, $u_{0,N}$, $u_{M,N}$, $u_{M,0}$ do not exist and, therefore, are not contained in the formula.

Using fictitious nodes is very useful for programming because we do not need to program different formulas for internal nodes, boundaries, and corners. We use this technique when writing the code for this finite-difference scheme.

Now we can choose the formula for the volume $V N_{ij}$. For internal nodes $i = 2, \ldots, M - 1$; $j = 2, \ldots, N - 1$, we can see from formula 3.16 that the numerators in the left-hand side in the expressions for GX_{ij} and GY_{ij} can be interpreted as the finite-difference analogs of integrals

$$\oint_{\partial \Diamond_{ij}} u \, dy; \quad - \oint_{\partial \Diamond_{ij}} u \, dx ,$$

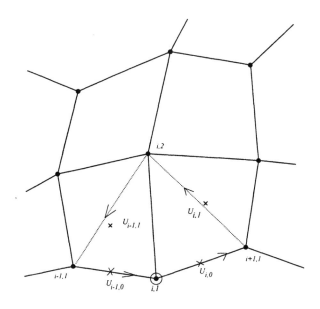

Figure 3.22: *Contour for* GRAD *on the boundary.*

where $\Diamond_{ij}$ is presented in Figure 3.15.

If we take into account Green's formula

$$\frac{\partial u}{\partial x} = \lim_{S \to 0} \frac{\oint_{\partial S} u \, dy}{S},$$

$$\frac{\partial u}{\partial y} = -\lim_{S \to 0} \frac{\oint_{\partial S} u \, dx}{S},$$

and compare it with the formula for components of the gradient, we can conclude that, similar to the case of cell-valued discretisation of vector functions, we must take

$$VN_{ij} = \frac{\Diamond_{ij}}{2}.$$

Again, similar to the case of cell-valued discretisation of vector functions, we can show that

$$\psi_{\text{GRAD}} = O(1)$$

for a general regular grid, and for a smooth grid, we get the first-order truncation error.

For the bottom boundary $j = 1$; $i = 2, \ldots, N - 1$, the interpretation in the framework of the approximation Green formula can be done if we consider the contour given in Figure 3.22. Therefore, volume $VN_{i,1}$ for $i = 2, \ldots, N - 1$ must be equal to one half the volume of the quadrangle

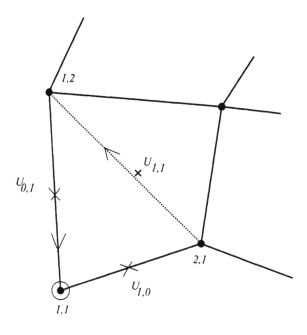

Figure 3.23: *Contour for* GRAD *at the corner.*

with vertices $(i, 1)$, $(i+1, 1)$, $(i, 2)$, $(i-1, 1)$, in which the boundary is the contour of the integration. For uniformity, we will denote this quadrangle as $\Diamond_{i,1}$.

Approximation properties of operator GRAD on the boundary are the same as those in interior nodes.

Let us now consider the corner node $(1, 1)$. The graphical illustration for the formula for GRAD is presented in Figure 3.23. Therefore, we must assume that $VN_{i,j}$ at the corner node $(1, 1)$ is equal to one half the volume of triangle $(1, 1)$, $(2, 1)$, $(1, 2)$:

$$VN_{1,1} =$$
$$\frac{(y_{1,2} - y_{2,1}) \frac{x_{1,2} + x_{2,1}}{2} + (y_{1,1} - y_{1,2}) \frac{x_{1,1} + x_{1,2}}{2} + (y_{2,1} - y_{1,1}) \frac{x_{2,1} + x_{1,1}}{2}}{2}.$$

Again, the approximation properties of operator GRAD at the corner node are the same as those in the interior nodes and on the boundary.

Following the support-operator method, we must construct approximations for operators $\mathcal{K}$ and $\mathcal{D}$. If we use nodal discretisation for the components of matrix K (see 3.11), we get

$$(\mathcal{K}^h \vec{W})_{i,j} = \begin{pmatrix} KXX_{i,j}\, WX_{i,j} + KXY_{i,j}\, WY_{i,j} \\ KXY_{i,j}\, WX_{i,j} + KYY_{i,j}\, WY_{i,j} \end{pmatrix},$$

where components $WX_{i,j}$ and $WY_{i,j}$ are given in the nodes as well as in the results of the application of operator $\mathcal{K}^h$.

We can now approximate operator $\mathcal{D}$, which is equal to zero in the internal nodes. On the boundary it is

$$\mathcal{D}\,u = \alpha\,u.$$

In the discrete case, for the bottom boundary we will use the following formula for the right-hand side:

$$\alpha \left(\frac{x_{i,1} + x_{i+1,1}}{2}, \frac{y_{i,1} + y_{i+1,1}}{2} \right) u_{i,0}.$$

This value will be defined in the same place as $u_{i,0}$, that is, on the boundary sides. On the other boundary sides, the discrete analog of operator $\mathcal{D}$ looks similar.

3.2.1 The Finite-Difference Scheme for the Dirichlet Boundary Problem

Using the constructed operators DIV and GRAD, the finite-difference scheme can be written in the following form:

$$(-\text{DIV}\,\mathcal{K}^h\,\text{GRAD}\,u)_{i,j} = \varphi_{i,j}; \quad \begin{array}{l} i = 1, \ldots, M - 1; \\ j = 1, \ldots, N - 1, \end{array}$$

where

$$\varphi_{i,j} = f(xc_{i,j}, yc_{i,j})$$

and

$$xc_{i,j} = 0.25\,(x_{ij} + x_{i+1,j} + x_{i+1,j+1} + x_{i,j+1}),$$
$$yc_{i,j} = 0.25\,(y_{ij} + y_{i+1,j} + y_{i+1,j+1} + y_{i,j+1})$$

are coordinates of the geometrical center of the cell i, j.

The approximation for the Dirichlet boundary conditions is shown as

$$u_{i,0} = \psi \left(\frac{x_{i,1} + x_{i+1,1}}{2}, \frac{y_{i,1} + y_{i+1,1}}{2} \right), i = 1, \ldots, M - 1,$$

$$u_{M,j} = \psi \left(\frac{x_{M,j} + x_{M,j+1}}{2}, \frac{y_{M,j} + y_{M,j+1}}{2} \right), j = 1, \ldots, N - 1,$$

$$\tag{3.19}$$

$$u_{i,N} = \psi \left(\frac{x_{i,N} + x_{i+1,N}}{2}, \frac{y_{i,N} + y_{i+1,N}}{2} \right) i = 1, \ldots, M - 1,$$

$$u_{0,j} = \psi \left(\frac{x_{1,j} + x_{1,j+1}}{2}, \frac{y_{1,j} + y_{1,j+1}}{2} \right), j = 1, \ldots, N - 1.$$

Since we know the value of function U on the boundary from the Dirichlet boundary conditions, then the unknowns are

$$U_{i,j}, \quad i = 1, \ldots, M - 1; \quad j = 1, \ldots, N - 1.$$

To write the matrix form of the finite-difference equations, we must move known values to the right-hand side. Similar to the 1-D case, the matrix related to the finite-difference scheme will be symmetric and positive if we multiply both sides of the difference equations by $\Omega_{i,j}$.

To give insight to what the finite-difference scheme looks like, we present the explicit form of the difference equations for cells $i = 2, \ldots, M - 2; j = 2, \ldots, N - 2$:

$$(\mathrm{DIV}\, \mathcal{K}^h\, \mathrm{GRAD}\, U)_{ij} =$$

$$\frac{y_{i,j+1} - y_{i+1,j}}{2\,\Omega_{ij}} *$$

$\{KXX_{i+1,j+1}\, [(y_{i+1,j+2} - y_{i+2,j+1})\, u_{i+1,j+1}+$

$(y_{i,j+1} - y_{i+1,j+2})\, u_{i,j+1} +$

$(y_{i+1,j} - y_{i,j+1})\, u_{ij} + (y_{i+2,j+1} - y_{i+1,j})\, u_{i+1,j})\, /(2\, V N_{i+1,j+1})] -$

$KXY_{i+1,j+1}\, [(x_{i+1,j+2} - x_{i+2,j+1})\, u_{i+1,j+1}+$

$(x_{i,j+1} - x_{i+1,j+2})\, u_{i,j+1} + (x_{i+1,j} - x_{i,j+1})\, u_{ij} +$

$(x_{i+2,j+1} - x_{i+1,j})\, u_{i+1,j})\, /(2\, V N_{i+1,j+1})] -$

$KXX_{i,j}\, [(y_{i,j+1} - y_{i+1,j})\, u_{i,j} + (y_{i-1,j} - y_{i,j+1})\, u_{i-1,j}+$

$(y_{i,j-1} - y_{i-1,j})\, u_{i-1,j-1} + (y_{i+1,j} - y_{i,j-1})\, u_{i,j-1})\, /(2\, V N_{i,j})] +$

$KXY_{i,j}\, [(x_{i,j+1} - x_{i+1,j})\, u_{i,j} + (x_{i-1,j} - x_{i,j+1})\, u_{i-1,j}+$

$(x_{i,j-1} - x_{i-1,j})\, u_{i-1,j-1} + (x_{i+1,j} - x_{i,j-1})\, u_{i,j-1})\, /(2\, V N_{i,j})]\} -$

$$\frac{y_{i+1,j+1} - y_{i,j}}{2\,\Omega_{ij}} *$$

$\{KXX_{i,j+1}\, [(y_{i,j+2} - y_{i+1,j+1})\, u_{i,j+1} + (y_{i-1,j+1} - y_{i,j+2})\, u_{i-1,j+1}+$

$(y_{i,j} - y_{i-1,j+1})\, u_{i-1,j} + (y_{i+1,j+1} - y_{i,j})\, u_{i,j})\, /(2\, V N_{i,j+1})] -$

$KXY_{i,j+1}\, [(x_{i,j+2} - x_{i+1,j+1})\, u_{i,j+1} + (x_{i-1,j+1} - x_{i,j+2})\, u_{i-1,j+1}+$

$(x_{i,j} - x_{i-1,j+1})\, u_{i-1,j} + (x_{i+1,j+1} - x_{i,j})\, u_{ij})\, /(2\, V N_{i,j+1})] -$

$KXX_{i+1,j}\, [(y_{i+1,j+1} - y_{i+2,j})\, u_{i+1,j} + (y_{i,j} - y_{i+1,j+1})\, u_{i,j}+$

$(y_{i+1,j-1} - y_{i,j})\, u_{i,j-1} + (y_{i+2,j} - y_{i+1,j-1})\, u_{i+1,j-1})\, /(2\, V N_{i+1,j})] +$

$KXY_{i+1,j}\, [(x_{i+1,j+1} - x_{i+2,j})\, u_{i+1,j} + (x_{i,j} - x_{i+1,j+1})\, u_{i,j}+$

$(x_{i+1,j-1} - x_{i,j})\, u_{i,j-1} +$

$(x_{i+2,j} - x_{i+1,j-1})\, u_{i+1,j-1})\, /(2\, V N_{i+1,j})]\} -$

$$\frac{x_{i,j+1} - x_{i+1,j}}{2\,\Omega_{ij}} *$$

$\{KXY_{i+1,j+1}\, [(y_{i+1,j+2} - y_{i+2,j+1})\, u_{i+1,j+1}+$

$$(y_{i,j+1} - y_{i+1,j+2}) u_{i,j+1} +$$

$$(y_{i+1,j} - y_{i,j+1}) u_{ij} + (y_{i+2,j+1} - y_{i+1,j}) u_{i+1,j}) / (2\,V N_{i+1,j+1})] -$$

$$KYY_{i+1,j+1} [(x_{i+1,j+2} - x_{i+2,j+1}) u_{i+1,j+1} +$$

$$(x_{i,j+1} - x_{i+1,j+2}) u_{i,j+1} +$$

$$(x_{i+1,j} - x_{i,j+1}) u_{ij} + (x_{i+2,j+1} - x_{i+1,j}) u_{i+1,j}) / (2\,V N_{i+1,j+1})] -$$

$$KXY_{i,j} [(y_{i,j+1} - y_{i+1,j}) u_{i,j} + (y_{i-1,j} - y_{i,j+1}) u_{i-1,j} +$$

$$(y_{i,j-1} - y_{i-1,j}) u_{i-1,j-1} + (y_{i+1,j} - y_{i,j-1}) u_{i,j-1}) / (2\,V N_{i,j})] +$$

$$KYY_{i,j} [(x_{i,j+1} - x_{i+1,j}) u_{i,j} + (x_{i-1,j} - x_{i,j+1}) u_{i-1,j} +$$

$$(x_{i,j-1} - x_{i-1,j}) u_{i-1,j-1} + (x_{i+1,j} - x_{i,j-1}) u_{i,j-1}) / (2\,V N_{i,j})]\} +$$

$$\frac{x_{i+1,j+1} - x_{i,j}}{2\,\Omega_{ij}} *$$

$$\{KXY_{i,j+1} [(y_{i,j+2} - y_{i+1,j+1}) u_{i,j+1} + (y_{i-1,j+1} - y_{i,j+2}) u_{i-1,j+1} +$$

$$(y_{i,j} - y_{i-1,j+1}) u_{i-1,j} + (y_{i+1,j+1} - y_{i,j}) u_{i,j}) / (2\,V N_{i,j+1})] -$$

$$KYY_{i,j+1} [(x_{i,j+2} - x_{i+1,j+1}) u_{i,j+1} + (x_{i-1,j+1} - x_{i,j+2}) u_{i-1,j+1} +$$

$$(x_{i,j} - x_{i-1,j+1}) u_{i-1,j} + (x_{i+1,j+1} - x_{i,j}) u_{ij}) / (2\,V N_{i,j+1})] -$$

$$KXY_{i+1,j} [(y_{i+1,j+1} - y_{i+2,j}) u_{i+1,j} + (y_{i,j} - y_{i+1,j+1}) u_{i,j} +$$

$$(y_{i+1,j-1} - y_{i,j}) u_{i,j-1} + (y_{i+2,j} - y_{i+1,j-1}) u_{i+1,j-1}) / (2\,V N_{i+1,j})] +$$

$$KYY_{i+1,j} [(x_{i+1,j+1} - x_{i+2,j}) u_{i+1,j} + (x_{i,j} - x_{i+1,j+1}) u_{i,j} +$$

$$(x_{i+1,j-1} - x_{i,j}) u_{i,j-1} +$$

$$(x_{i+2,j} - x_{i+1,j-1}) u_{i+1,j-1}) / (2\,V N_{i+1,j})]\}.$$

The stencil for operator DIV $\mathcal{K}^h$ GRAD is presented in Figure 3.24.

It is clear that this explicit form of difference equations does not give us any new information about the properties of the finite-difference scheme. Therefore, we will not write the explicit form for the difference equations on the boundary. Instead, we will present stencils for operator DIV $\mathcal{K}^h$ GRAD in the cells which are adjacent to the bottom boundary and in the corner cell $(1,1)$. For cells that are adjacent to the bottom boundary, we have a stencil that contains 6 cells and three values on the boundary sides (see Figure 3.25). For corner $(1,1)$, we get a stencil that is presented in Figure 3.26.

Readers can find subroutines which compute the coefficients of finite-difference equations in the Fortran directory on the attached floppy disk (see the Fortran Guide chapter for details).

3.2.2   The Finite-Difference Scheme for the Robin Boundary Problem

For Robin boundary conditions, equations for internal points are the same as the equations for the Dirichlet boundary conditions. Because the

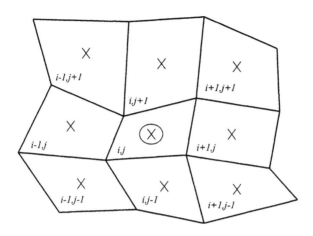

Figure 3.24: *C-N discretisation, stencil for operator* DIV GRAD.

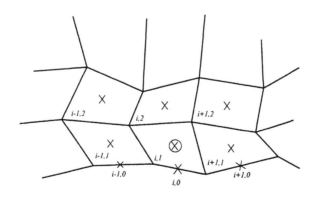

Figure 3.25: *Stencil for operator* DIV GRAD.

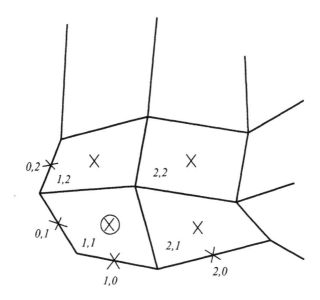

Figure 3.26: *Stencil for operator* DIV GRAD.

values of the difference function on the boundary are also unknown, we do not need to move it into the right-hand side.

The general Robin boundary condition is

$$(\vec{n}, K \operatorname{grad} u) + \alpha \, u = \gamma .$$

For cell-valued discretisation of scalar functions and nodal discretisation of vector functions, we will obtain the approximation of these boundary conditions on the boundary sides where we have the definition of the operator

$$\mathcal{B} \vec{W} = -(\vec{W}, \vec{n}),$$

where this approximation is given by formula 3.15. Thus, there is no problem in finding the approximation of the first term in the Robin boundary condition. Because of special discretisation for scalar functions on the boundary (side related), we do not have a problem with finding the approximation of the second term in the left-hand side.

Let us demonstrate how we approximate the Robin boundary conditions on the bottom boundary ($j = 1; i = 1, \ldots, M - 1$).

For $i = 1, \ldots, M - 1$, we have

$$\left[\left(K X X_{i,1} \left(\frac{y_{i,2} - y_{i+1,1}}{2} \, u_{i,1} + \frac{y_{i-1,1} - y_{i,2}}{2} \, u_{i-1,1} + \right. \right. \right.$$

$$\left. \frac{y_{i,1} - y_{i-1,1}}{2} u_{i-1,0} + \frac{y_{i+1,1} - y_{i,1}}{2} u_{i,0} \right) / V N_{i,1}$$

$$- KXY_{i,1} \left(\frac{x_{i,2} - x_{i+1,1}}{2} u_{i,1} + \frac{x_{i-1,1} - x_{i,2}}{2} u_{i-1,1} + \right.$$

$$\left. \frac{x_{i,1} - x_{i-1,1}}{2} u_{i-1,0} + \frac{x_{i+1,1} - x_{i,1}}{2} u_{i,0} \right) / V N_{i,1} +$$

$$\left(KXX_{i+1,1} \left(\frac{y_{i+1,2} - y_{i+2,1}}{2} u_{i+1,1} + \frac{y_{i,1} - y_{i+1,2}}{2} u_{i,1} + \right. \right.$$

$$\left. \frac{y_{i+1,1} - y_{i,1}}{2} u_{i,0} + \frac{y_{i+2,1} - y_{i+1,1}}{2} u_{i+1,0} \right) / V N_{i+1,1}$$

$$- KXY_{i+1,1} \left(\frac{x_{i+1,2} - x_{i+2,1}}{2} u_{i+1,1} + \frac{x_{i,1} - x_{i+1,2}}{2} u_{i,1} + \right.$$

$$\left. \left. \frac{x_{i+1,1} - x_{i,1}}{2} u_{i,0} + \frac{x_{i+2,1} - x_{i+1,1}}{2} u_{i+1,0} \right) / V N_{i+1,1} \right] \frac{y_{i+1,1} - y_{i,1}}{2 \, l\xi_{i,1}}$$

$$- \left[\left(KXY_{i,1} \left(\frac{y_{i,2} - y_{i+1,1}}{2} u_{i,1} + \frac{y_{i-1,1} - y_{i,2}}{2} u_{i-1,1} + \right. \right. \right.$$

$$\left. \frac{y_{i,1} - y_{i-1,1}}{2} u_{i-1,0} + \frac{y_{i+1,1} - y_{i,1}}{2} u_{i,0} \right) / V N_{i,1}$$

$$- KYY_{i,1} \left(\frac{x_{i,2} - x_{i+1,1}}{2} u_{i,1} + \frac{x_{i-1,1} - x_{i,2}}{2} u_{i-1,1} + \right.$$

$$\left. \frac{x_{i,1} - x_{i-1,1}}{2} u_{i-1,0} + \frac{x_{i+1,1} - x_{i,1}}{2} u_{i,0} \right) / V N_{i,1} +$$

$$\left(KXY_{i+1,1} \left(\frac{y_{i+1,2} - y_{i+2,1}}{2} u_{i+1,1} + \frac{y_{i,1} - y_{i+1,2}}{2} u_{i,1} + \right. \right.$$

$$\left. \frac{y_{i+1,1} - y_{i,1}}{2} u_{i,0} + \frac{y_{i+2,1} - y_{i+1,1}}{2} u_{i+1,0} \right) / V N_{i+1,1}$$

$$- KYY_{i+1,1} \left(\frac{x_{i+1,2} - x_{i+2,1}}{2} u_{i+1,1} + \frac{x_{i,1} - x_{i+1,2}}{2} u_{i,1} + \right.$$

$$\left. \left. \frac{x_{i+1,1} - x_{i,1}}{2} u_{i,0} + \frac{x_{i+2,1} - x_{i+1,1}}{2} u_{i+1,0} \right) / V N_{i+1,1} \right] \frac{x_{i+1,1} - x_{i,1}}{2 \, l\xi_{i,1}}$$

$$+ \alpha \left(\frac{x_{i,1} + x_{i+1,1}}{2}, \frac{y_{i,1} + y_{i+1,1}}{2} \right) u_{i,0} =$$

$$\gamma \left(\frac{x_{i,1} + x_{i+1,1}}{2}, \frac{y_{i,1} + y_{i+1,1}}{2} \right).$$

3.3   The Numerical Solution of Test Problems

3.3.1   The Dirichlet Boundary Value Problem

In this section we present the results of computation for both types of discretisation in case of the Dirichlet boundary conditions.

To test the finite-difference algorithms, we chose a problem that is sufficient to exercise all terms in the algorithms. First, K is chosen to be a rotation of a diagonal matrix:

$$K = R D R^T,$$

where

$$R = \begin{pmatrix} +\cos(\theta) & -\sin(\theta) \\ +\sin(\theta) & +\cos(\theta) \end{pmatrix}, \quad D = \begin{pmatrix} d_1 & 0 \\ 0 & d_2 \end{pmatrix},$$

and

$$\begin{aligned} \theta &= \frac{3\pi}{12}, \\ d_1 &= 1 + 2x^2 + y^2 + y^5, \\ d_2 &= 1 + x^2 + 2y^2 + x^3. \end{aligned}$$

The solution is chosen to be

$$u(x, y) = \sin(\pi x)\sin(\pi y),$$

which means that for a given domain, the Dirichlet boundary conditions coincide with this solution. The right-hand side of the Poisson equation is

$$f = -\operatorname{div} K \operatorname{grad} u,$$

where matrix K and function u have been given in previous formulas.

To check the convergence properties of the constructed finite-difference schemes on a smooth grid, we use a grid in the domain presented in Figure 3.27. The top boundary of this domain is the graph of the following function:

$$Y(x) = 1 + 0.5\sin(2\pi x).$$

To construct the grid, we use the following formulas:

$$x_{i,j} = x_i = \frac{i - 1}{M - 1}.$$

To determine $y_{i,j}$ we use the following procedure: for each i we divide the length of the segment of the straight line between the top and bottom boundaries on $M - 1$ equal intervals of size:

$$hy_i = \frac{Y(x_i)}{N - 1},$$

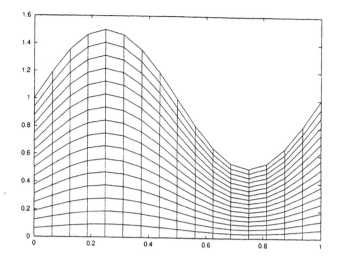

Figure 3.27: *Grid in domain with curvilinear boundaries.*

and then take
$$y_{i,j} = (j-1)\, hy_i\,.$$

The resulting grid for $M = N = 17$ is presented in Figure 3.27. The results of the computation for nodal discretisation for scalar functions and cell-valued discretisation for vector functions are presented in Table 3.10. This table is arranged similar to the 1-D case. In the first column, we present the number of nodes $(N = M)$; in the second and third columns, we present the max and L_2 norms respectively; in the fourth and fifth columns, we present the numerically computed constant

$$\frac{\|\, u^h - p_h\, u\,\|_{max}}{h^2}, \quad \frac{\|\, u^h - p_h\, u\,\|_{L_2}}{h^2}\,.$$

In the last two columns, we present a ratio between the error for the number of nodes given in this row and the error for the number of nodes given in the previous row. Then, similar to the 1-D case when $h \to 0$, this ratio must converge to 4 for the second-order method. The results presented in this table demonstrate the second-order convergence of the finite-difference scheme.

The results for cell-valued discretisation for scalar functions and nodal discretisation for vector functions are presented in Table 3.11. This table also demonstrates the second-order convergence rate.

To compare a convergence rate on smooth and non-smooth grids, we present the results of the test problem in the unit square on the "random" grid, which is presented in Figure 3.28.

M	max norm	L_2 norm	Const max	Const L_2	R max	R L_2
17	0.809E-02	0.295E-02	2.071	0.755	-	-
33	0.218E-02	0.719E-03	2.232	0.736	3.711	4.102
65	0.546E-03	0.178E-03	2.236	0.729	3.992	4..039

Table 3.10: *Convergence analysis, Dirichlet boundary conditions, 2-D-N-C discretisation, smooth-grid.*

M	max norm	L_2 norm	Const max	Const L_2	R max	R L_2
17	0.157E-01	0.496E-02	4.019	1.269	-	-
33	0.403E-02	0.129E-02	4.126	1.320	3.895	3.844
65	0.103E-02	0.328E-03	3.959	1.260	3.950	3.932

Table 3.11: *Convergence analysis, Dirichlet boundary conditions, 2-D-C-N discretisation, smooth-grid.*

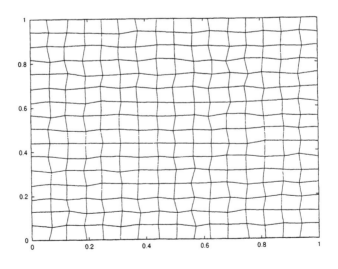

Figure 3.28: *Random grid.*

M	max-norm	L_2-norm	q-max	$q - L_2$
17	0.128E-01	0.634E-02	2.009	2.004
33	0.318E-02	0.158E-02	1.819	1.942
65	0.901E-03	0.411E-03	-	-

Table 3.12: *Convergence analysis, Dirichlet boundary conditions, 2-D-N-C discretisation, random-grid.*

The coordinates are computed by formulas

$$x_{i,j} = \xi_i - 0.125\,h + 0.25\,h\,R_x, \tag{3.20}$$

$$y_{i,j} = \eta_j - 0.125\,h + 0.25\,h\,R_y, \tag{3.21}$$

where R_x, R_y are random numbers from the interval $(0,1)$.

For this type of non-uniform grid, there is no reason to expect a second-order convergence rate. In fact, we can only expect a first-order accuracy. Another difficulty related to the random grid is that error behavior for this type of grid is not regular when we diminish h. This means that to see asymptotic behavior of the error, h must be very small.

To demonstrate the behavior of the error for a real grid, we will use another type of table. To explain this type of table, let us consider the expression for the error. Let us assume that error E has the following form:

$$E_h = C\,h^q + O(h^{q+1}).$$

Then E for $h/2$ will have the following form:

$$E_{h/2} = C\,\frac{h^q}{2^q} + O(h^{q+1}).$$

This means that

$$q \approx \log_2 \frac{E_h}{E_{h/2}}. \tag{3.22}$$

The tables for random grids are arranged as follows. The first three columns are the same as for the case of a smooth grid. In the fourth column, we put the approximate order of convergence in max norm which is computed by formula 3.22. In the last column, we put a similar value for L_2 norm.

The results for the nodal discretisation for scalar functions and cell-valued discretisation for vector functions are presented in Table 3.12.

This table demonstrates that for this type of grid and for this particular solution, we have approximately a second-order convergence rate.

The results for cell-valued discretisation for scalar functions and nodal discretisation for vector functions are presented in Table 3.13. In this case,

M	max-norm	L_2-norm	q-max	$q - L_2$
17	0.169E-01	0.552E-02	1.371	1.716
33	0.653E-02	0.168E-02	0.664	1.066
65	0.412E-02	0.802E-03	0.738	0.383
129	0.247E-02	0.615E-03	-	-

Table 3.13: *Convergence analysis, Dirichlet boundary conditions, 2-D-C-N discretisation, random-grid.*

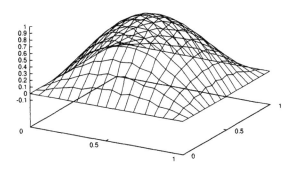

Figure 3.29: *Exact solution.*

since the behavior of the error is more irregular than for the previous case, we also present the results for $M = N = 129$.

From this table we can conclude that we have approximately a first-order convergence rate.

The exact solution is presented in Figure 3.29. The approximate solution for $M = N = 17$ for the case of N-C discretisation is presented in Figure 3.30 and the approximate solution for the case of C-N discretisation is presented in Figure 3.31.

Again we must emphasize that the random grid is a very irregular grid and to obtain a first-order convergence rate for this type of grid is not so easy. These results also confirm the robustness of our finite-difference schemes.

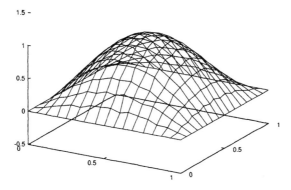

Figure 3.30: *Approximate solution, N-C discretisation, $N = M = 17$.*

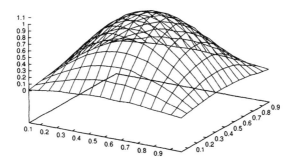

Figure 3.31: *Approximate solution, C-N discretisation, $N = M = 17$.*

3.3.2   The Robin Boundary Value Problem

In this section, we present the results of the computations for both types of discretisation for Robin boundary conditions.

The matrix K and the exact solution are the same as for the case of Dirichlet boundary conditions. Function α in the left-hand side of the Robin boundary condition is equal to 1, and function γ in the right-hand side is computed using the expression for matrix K, the exact solution, and α.

To check the convergence properties, we use the same domain and grid as in the example for the Dirichlet problem, the grid presented in Figure 3.27.

First, we must make some comments about the Robin boundary conditions. It is clear that in the "corners" of the curvilinear domain presented in Figure 3.27, the normal vector to the boundary is not defined and the traditional statement of the Robin boundary conditions in these points does not make sense. The physical problem still makes sense and this is the problem with discontinuous boundary conditions. From the physical point of view, it is evident that the problem with discontinuous boundary conditions has a unique solution. Therefore, from the computational point of view, it is important to make the correct discretisation, which in some sense will correspond to the physical problem.

If we use nodal discretisation for scalar functions and cell-valued discretisation for vector functions, our computational boundary conditions must be formulated in nodes (see 3.12 and 3.13). Then in the corner $(1, 1)$, we have two problems. The first is that the normal in this node is not defined and the second is that we do not know what values of α and γ we must use, i.e., are the values from the boundary where $i = 1$ or where $j = 1$?

At this point, we will not discuss what the right decision is. We will just emphasize that this type of discretisation is not consistent with the problem for discontinuous boundary conditions and there is no reason to expect good accuracy.

On the contrary, for cell-valued discretisation for scalar functions and nodal discretisation for vector functions, we have the approximation for Robin boundary conditions on the boundary sides, where normal is defined and there is no problem with the definition of α and γ. Therefore, when we use the C-N type, we do not have singularity in the statement of the discrete problem and it corresponds to the statement of the original physical problem, which also does not have singularity in the statement. Thus, we can expect that for the C-N type of discretisation, we will obtain better results than for the N-C type of discretisation.

The results of the computation for N-C discretisation are presented in Table 3.14. In this table, we present the approximate value for the convergence rate. From this table, we can conclude that the convergence rate is near one. If we take into account singularities in the corner nodes, then the convergence rate looks very good.

M	max-norm	L_2-norm	q-max	$q - L_2$
17	0.858E-00	0.515E-00	1.535	1.201
33	0.296E-00	0.224E-00	1.209	1.025
65	0.128E-00	0.110E-00	-	-

Table 3.14: *Convergence analysis, Robin boundary conditions, 2-D-N-C discretisation, smooth-grid.*

M	max norm	L_2 norm	Const max	Const L_2	R max	R L_2
17	0.123E-00	0.106E-00	31.48	27.13	-	-
33	0.343E-01	0.263E-01	35.12	26.93	3.586	4.030
65	0.981E-02	0.653E-02	40.18	26.74	3.496	4.027

Table 3.15: *Convergence analysis, Robin boundary conditions, 2-D-C-N discretisation, smooth-grid.*

The results for C-N discretisation are presented in Table 3.15. Because the grid is smooth and we do not have singularities in the discrete problem, we can expect that the finite-difference scheme will have a second-order convergence rate. From this table, we can conclude that the convergence rate is near two.

4 Conclusion

Let us summarize the primary results of this chapter.

- The detailed description of the support-operators method is given with an application to the construction of a finite-difference scheme on a logically rectangular grid for second-order elliptic equations with the general matrix of coefficients and the general boundary conditions.

- The constructed finite-difference scheme mimics the main operators' properties of the original differential problem. That is, the discrete operators DIV and GRAD satisfy the relation GRAD $= -$DIV*, and the discrete analog of the general elliptic operator is self-adjoint and positive. The discrete operator related to the Robin boundary problem is also self-adjoint and positive. These properties of the discrete problem make it possible to use effective iteration methods for solving the system of finite-difference equations [111], [50], [89].

- The finite-difference scheme for the two main types of discretisations are constructed. The cases of nodal discretisations of scalar functions and cell-valued discretisation of vector functions, and vice versa, are considered. Then the user can choose the type of discretisation which is more appropriate for his problem, where the solution of the elliptic equations can be only one stage of the computational algorithm.

- It was demonstrated by numerical examples that for a smooth grid, constructed finite-difference schemes have a second order convergence rate. The robustness of constructed finite-difference schemes on rough, strongly non-smooth grids was also demonstrated.

- Constructed finite-difference operators DIV and GRAD (and also finite-difference approximations for finding the normal component of a vector) can be used for the approximation of other differential equations which contain differential operators div and grad. In the next chapters, we will use these discrete operators for constructing finite-difference schemes for the heat equation and equations of fluid dynamics.

Chapter 4

The Heat Equation

1 Introduction

In this section, for the example of the 1-D heat equation, we consider new notions and ideas in regard to the theory of finite-difference schemes, which are related to the heat equation.

Let us consider the one-dimensional heat equation

$$\frac{\partial u}{\partial t} = \operatorname{div} K \operatorname{grad} u + f = \frac{\partial}{\partial x}\left(K\frac{\partial u}{\partial x}\right) + f \quad , 0 \leq x \leq 1,\ t > 0,$$

where the unknown function $u = u(x,t)$ is the function of x and time t, $K(x,t) > 0$, the conductivity coefficient in 1-D is the scalar function. At the initial moment of time, when $t = 0$, the value of u is given as

$$u(x,0) = T^0(x).$$

On the boundary, we have the same boundary conditions as for the elliptic problem, Dirichlet, Neumann, or Robin. There is only one difference: all functions in this boundary condition can also be functions of time.

In this section, for simplicity, we will consider Dirichlet boundary conditions when temperature $u(x,t)$ is given on the boundary as

$$u(0,t) = \mu_0(t), \quad u(1,t) = \mu_1(t).$$

1.1 The Conservation Law

As we already know, there is only one conservation law for the heat equation, it is the law of heat conservation,

$$\frac{dQ}{dt} = -\oint_S (\vec{w}, \vec{n})\, dS + \int_V f\, dV,$$

149

where

$$Q = \int_V u\, dV\,, \quad \vec{w} = -K\, \mathrm{grad}\, u\,,$$

are the total amount of heat and heat flux.

From a formal viewpoint, this relation can be obtained as follows:

$$\frac{dQ}{dt} = \frac{d}{dt} \int_V u\, dV = -\int_V \mathrm{div}\, \vec{w}\, dV + \int_V f\, dV = -\oint_S (\vec{w}, \vec{n})\, dS + \int_V f\, dV\,.$$

Then we want to preserve the property

$$\int_V \mathrm{div}\, \vec{w}\, dV = \oint_S (\vec{w}, \vec{n})\, dS$$

in the discrete case.

All considerations that we have shown about the operator $-\mathrm{div}\, K\, \mathrm{grad}$ in the chapter relating to the elliptic equation, are also true for the heat equation.

Therefore, for the approximation of the differential operator in the right-hand side of the heat equation, we can use the same difference operators that we have constructed in previous chapters. We can also use all the approximations for the boundary conditions that we have already constructed for the elliptic problems.

The main subject of this chapter is what new features are brought in by the presence of the time derivative.

1.2   Time Discretisation

We introduce a grid in time as follows:

$$t_0 = 0, \quad t_{n+1} = t_n + \tau_n, \quad n = 0, 1, \dots,$$

where τ_n is called the *time step*. When $\tau_n \neq const$, it is called the *variable time step* and when $\tau_n = const$, we will use the notation Δt for the time step. Methods with a variable time step are usually used when the coefficients of the differential equations have some singularity as the functions of time.

To denote the difference analog of function $u(x, t)$, we will use subscripts for discretisation in space (discretisation in space is absolutely the same as for elliptic equations) in the same way as we did for function $u(x)$. We will use superscripts for discretisation in time. Hence, we get

$$u_i^n \approx u(x_i, t_n).$$

Again, we will use integer indices for nodal and cell-valued discretisation in space.

To approximate a time derivative $\partial u/\partial t$, we can use the formula

$$\left(\frac{\partial u}{\partial t}\right)\bigg|_{(x_i, t_n)} \approx \frac{u_i^{n+1} - u_i^n}{\tau_n}.$$

This approximation has a first-order truncation error for $t = t_n$, because

$$\frac{u(x_i, t_{n+1}) - u(x_i, t_n)}{\tau_n} = \frac{\partial u}{\partial t}\bigg|_{(x_i, t_n)} + \frac{\tau_n}{2}\frac{\partial^2 u}{\partial t^2}\bigg|_{(x_i, t_n)} + 0\left((\tau_n)^2\right).$$

It is also easy to see that for $t = t_{n+1/2} = 0.5\,(t_n + t_{n+1})$, we have a second-order truncation error

$$\frac{u(x_i, t_{n+1}) - u(x_i, t_n)}{\tau_n} = \frac{\partial u}{\partial t}\bigg|_{(x_i, t_{n+1/2})} + 0\left((\tau_n)^2\right).$$

1.3 Explicit and Implicit Finite-Difference Schemes

The simplest finite-difference scheme for Dirichlet boundary conditions and for nodal space discretisation on a uniform grid looks as follows: (here we used $K = 1$ for simplicity)

$$\frac{u_i^{n+1} - u_i^n}{\Delta t} = \frac{u_{i+1}^n - 2\,u_i^n + u_{i-1}^n}{h^2} + \varphi_i^n, \tag{1.1}$$

where φ_i^n is some approximation for function f, and in the simplest case $\varphi_i^n = f(x_i, t^n)$. The approximation for the initial and boundary conditions looks as follows:

$$u_i^0 = T^0(x_i), \quad i = 1, \ldots, M, \tag{1.2}$$

$$u_1^n = \mu_0(t_n), \quad u_M^n = \mu_1(t_n). \tag{1.3}$$

This scheme is called *explicit* because one can compute values on the new $n + 1$ time level by the *explicit* formula, which follows from 1.1:

$$u_i^{n+1} = \frac{\Delta t}{h^2}\,u_{i+1}^n\left(1 - \frac{2\,\Delta t}{h^2}\right)u_i^n + \frac{\Delta t}{h^2}\,u_{i-1}^n + \Delta t\,\varphi_i^n \tag{1.4}$$

and values for $n = 0$ are given by the initial conditions.

Another possibility is to take the approximation of the Laplacian on the new $n + 1$ time level

$$\frac{u_i^{n+1} - u_i^n}{\Delta t} = \frac{u_{i+1}^{n+1} - 2\,u_i^{n+1} + u_{i-1}^{n+1}}{h^2} + \varphi_i^{n+1} \tag{1.5}$$

and use the following approximation for the initial and boundary conditions

$$u_i^0 = T^0(x_i), \quad i = 1, \ldots, M, \tag{1.6}$$

$$u_1^n = \mu_0(t_n), \quad u_M^n = \mu_1(t_n). \tag{1.7}$$

The finite-difference scheme 1.5 is called *implicit*, or more exactly, *fully implicit* because to find a value of temperature on the new time level, we must solve the linear system of equations

$$-\frac{\Delta t}{h^2}\, u_{i+1}^n \left(1 + \frac{2\,\Delta t}{h^2}\right) u_i^{n+1} - \frac{\Delta t}{h^2}\, u_{i-1}^{n+1} = u_i^n + \Delta t\, \varphi_i^{n+1}, \tag{1.8}$$

$$u_1^{n+1} = \mu_0(t_{n+1}), \quad u_M^{n+1} = \mu_1(t_{n+1}), \tag{1.9}$$

where u_i^n are known values of temperature from the previous time step, and for $n = 0$, the values are given by the initial conditions.

In general, to approximate the right-hand side, we can take some combination of Laplacian on the n and $n + 1$ time levels and take the function f in a related moment of time:

$$\frac{u_i^{n+1} - u_i^n}{\Delta t} = \tag{1.10}$$

$$(1 - \sigma)\,\frac{u_{i+1}^n - 2\,u_i^n + u_{i-1}^n}{h^2} + \sigma\,\frac{u_{i+1}^{n+1} - 2\,u_i^{n+1} + u_{i-1}^{n+1}}{h^2} +$$

$$f(x_i, (1 - \sigma)\, t_n + \sigma\, t_{n+1}).$$

Then $\sigma = 0$ corresponds to the explicit scheme, and $\sigma = 1$ corresponds to the fully implicit scheme. For $0 < \sigma < 1$, the finite difference scheme 1.10 is implicit.

The finite-difference which uses only the values of the function from the two time levels n and $n + 1$ is called *the two-level finite-difference scheme*.

Now let us investigate the *residual or error of approximation* of heat equations by the discrete equation 1.10 *on the solution of the differential problem*. To do this, we must find the equation for error for which there is a difference between the solution of discrete equations and the projections of the solution of the differential equation

$$z_i^n = u_i^n - (\mathcal{P}_h^n\, u)_i^n = u_i^n - u(x_i, t^n).$$

To obtain the formula for the residual, we express u_i^n from the previous equation as follows:

$$u_i^n = u(x_i, t^n) + z_i^n$$

and substitute this formula for the difference equation

$$\frac{z_i^{n+1} - z_i^n}{\Delta t} = \tag{1.11}$$

$$(1 - \sigma)\,\frac{z_{i+1}^n - 2\,z_i^n + z_{i-1}^n}{h^2} + \sigma\,\frac{z_{i+1}^{n+1} - 2\,z_i^{n+1} + z_{i-1}^{n+1}}{h^2} + \psi_i^n,$$

where ψ_i^n is the residual:

$$\psi_i^n = \sigma \, \mathrm{DIV\,GRAD}(\mathcal{P}_h^{n+1}\,u) + (1-\sigma)\,\mathrm{DIV\,GRAD}(\mathcal{P}_h^n\,u) - \left[\frac{(\mathcal{P}_h^{n+1}\,u)_i - (\mathcal{P}_h^n\,u)_i}{\Delta t}\right] + f(x_i,(1-\sigma)\,t_n + \sigma\,t_{n+1}).$$

Now if we take into account that

$$0 = \left(\frac{\partial u}{\partial t} - \frac{\partial^2 u}{\partial x^2} - f\right)\Bigg|_{x_i, t_{n+1/2}},$$

we can rewrite the expression for the residual as follows:

$$\psi_i^n =$$
$$[\sigma \, \mathrm{DIV\,GRAD}(\mathcal{P}_h^{n+1}\,u) + (1-\sigma)\,\mathrm{DIV\,GRAD}(\mathcal{P}_h^n\,u) - \frac{\partial^2 u}{\partial x^2}\Bigg|_{x_i, t_{n+1/2}}] +$$
$$\left[\frac{(\mathcal{P}_h^{n+1}\,u)_i - (\mathcal{P}_h^n\,u)_i}{\Delta t} - \frac{\partial u}{\partial t}\Bigg|_{x_i, t_{n+1/2}}\right] +$$
$$[f(x_i,(1-\sigma)\,t_n + \sigma\,t_{n+1}) - f(x_i, t_{n+1/2})].$$

That is, we can separate the truncation error related to the space and time discretisation, and the residual is the sum of the truncation error for Laplacian and for the first time derivative and for function f. Let us make the Taylor expansion in point $(x_i, t_{n+1/2})$. Then for the time derivative, we get

$$\frac{(\mathcal{P}_h^{n+1}\,u)_i - (\mathcal{P}_h^n\,u)_i}{\Delta t} = \frac{\partial u}{\partial t}\Bigg|_{x_i, t_{n+1/2}} + O\left((\Delta t)^2\right).$$

For Laplacian on time level t_n, we get

$$\mathrm{DIV\,GRAD}(\mathcal{P}_h^n\,u) =$$
$$\frac{\partial^2 u}{\partial x^2}\Bigg|_{x_i, t_{n+1/2}} - \frac{\Delta t}{2}\frac{\partial^3 u}{\partial x^2 \partial t}\Bigg|_{x_i, t_{n+1/2}} + O\left(h^2 + (\Delta t)^2\right).$$

For Laplacian on time level t_{n+1}, we get

$$\mathrm{DIV\,GRAD}(\mathcal{P}_h^{n+1}\,u) =$$
$$\frac{\partial^2 u}{\partial x^2}\Bigg|_{x_i, t_{n+1/2}} + \frac{\Delta t}{2}\frac{\partial^3 u}{\partial x^2 \partial t}\Bigg|_{x_i, t_{n+1/2}} + O\left(h^2 + (\Delta t)^2\right).$$

Finally, we get following formula for the residual:

$$\psi_i^n = (2\sigma - 1)\frac{\Delta t}{2}\frac{\partial^3 u}{\partial x^2 \partial t}\Bigg|_{x_i, t_{n+1/2}} + (\sigma - 0.5)\,\Delta t\,\frac{\partial f}{\partial t} + O\left(h^2 + (\Delta t)^2\right).$$

From this formula we can see that for $\sigma \neq 0.5$, we get the second-order approximation in space and the first-order approximation in time

$$\psi_i^n = O\left(h^2 + \Delta t\right).$$

For $\sigma = 0.5$, we get the second-order approximation for space and time variables:

$$\psi_i^n = O\left(h^2 + (\Delta t)^2\right).$$

From a practical point of view (if we consider only the value of the truncation error), we can conclude that it makes no sense to choose a very small space step. Space and time steps must be coordinated with each other to give a similar contribution to the truncation error.

The real restriction on the ratio of time and space steps is implied by the *stability of the finite-difference scheme*.

1.4   Stability of the Finite-Difference Scheme for the Heat Equation

The stability of a finite-difference scheme is an internal property of the finite-difference scheme which expresses the continuous dependence of the solution on input data. Let us consider the finite-difference schemes 1.1, 1.2, 1.3 and let $\{u_i^n\}$ and $\{\tilde{u}_i^n\}$ be the solutions which correspond to the initial and boundary conditions given by functions T^0, μ_0, μ_1 and $\tilde{T}^0, \tilde{\mu}_0, \tilde{\mu}_1$, respectively.

Then, the finite-difference scheme is stable if

$$\| u^n - \tilde{u}^n \|_{H_1} \leq M_1 \, \| T^0 - \tilde{T}^0 \|_{H_1} +$$
$$M_2 \max_{k=1,\dots,n} \| \varphi - \tilde{\varphi} \|_{H_1} +$$
$$M_3 \max_{k=1,\dots,n} |\mu_0 - \tilde{\mu}_0| + M_4 \max_{k=1,\dots,n} |\mu_1 - \tilde{\mu}_1| \, ,$$

where the $M_1, M_2, M_3, M_4 > 0$ constants are not dependent on h and Δt, and $\| \bullet \|_{H_1}$, are some norm.

That is, small variations in the input data stipulate a small variation in the solution.

It is useful to distinguish stability with respect to the initial data and stability with respect to the right-hand side.

A finite-difference scheme is called stable with respect to initial data if

$$\| u^n - \tilde{u}^n \|_{H_1} \leq M_1 \, \| T^0 - \tilde{T}^0 \|_{H_1} \, . \tag{1.12}$$

A finite-difference scheme is called stable with respect to the right-hand side if

$$\| u^n - \tilde{u}^n \|_{H_1} \leq M_2 \max_{k=1,\dots,n} \| \varphi - \tilde{\varphi} \|_{H_1} \, . \tag{1.13}$$

There is no special consideration for stability with respect to boundary values because it can be considered as being stable with respect to the right-hand side. For example, for the Dirichlet boundary condition, known boundary values are used to modify the right-hand side of finite-difference equations near the boundary.

The general theory of stability of finite-difference schemes is described in [107] (see also [46], [100]). It is proved that for the linear problem, finite-difference schemes which are stable with respect to the initial data are also stable with respect to the right-hand side. Therefore, we will consider only stability with respect to the initial data.

It is important to note that for the case of linear problems, stability conditions with respect to initial data can be written as follows:

$$\| u^n \|_{H_1} \le M_1 \| T^0 \|_{H_1} . \tag{1.14}$$

To see this, let us construct the equation for

$$(\Delta u)^n_i = u^n_i - \tilde{u}^n_i . \tag{1.15}$$

It is easy to see that for the linear problem, Δu satisfies the similar equation as u^n_i with the initial condition

$$(\Delta u)^0_i = T^0_i - \tilde{T}^0_i . \tag{1.16}$$

From 1.14 and 1.16, we obtain

$$\| (\Delta u)^n \|_{H_1} \le M_1 \| (\Delta u)^0 \|_{H_1} = M_1 \| T^0 - \tilde{T}^0 \|_{H_1} ,$$

which is the same as 1.12.

Neumann's Method for the Investigation of Stability

The method of *von Neumann or the method of harmonics* [100], [46] is one of the most widely used practical methods for the investigation of stability of the finite-difference scheme for non-stationary problems. More exactly, the method of harmonics makes it possible to establish the necessary conditions for stability. Directly, this method can be applied only to the finite-difference scheme with a constant coefficient. In our case, it means the constant space step, or uniform grid. There are two main steps when you use this method. At first, the original finite-difference problem is considered in whole space, and secondly, the right-hand side is taken to zero. Then instead of the original problem 1.10, 1.6, 1.7, we consider the following problem:

$$\frac{u^{n+1}_i - u^n_i}{\Delta t} = \tag{1.17}$$

$$(1 - \sigma) \frac{u^n_{i+1} - 2\,u^n_i + u^n_{i-1}}{h^2} + \sigma \frac{u^{n+1}_{i+1} - 2\,u^{n+1}_i + u^{n+1}_{i-1}}{h^2} ,$$

$$u_i^0 = T^0(x_i), \tag{1.18}$$

where index i is now varied from $-\infty, +\infty$ and

$$x_i = i\,h, \quad -\infty < x_i < +\infty.$$

Now we can try to find a particular solution for problems 1.17 and 1.18 in the following form:

$$u_k^n = q^n\, e^{ik\varphi}, \tag{1.19}$$

where i is the unit imaginary number, q is the complex number, which must be defined, and φ is the arbitrary real number, and we investigate the stability of this particular solution. The particular solution 1.19 will be stable if

$$|q^n\, e^{ik\varphi}| < |q^0\, e^{ik\varphi}| = |e^{ik\varphi}| = 1.$$

Now, because the module of the product of the two complex numbers is equal to the product of its modules and because $|e^{ik\varphi}| = 1$, the left-hand side in the previous inequality is

$$|q^n\, e^{ik\varphi}| = |q^n| = |q|^n.$$

Finally, the stability condition is

$$|q|^n < 1$$

or

$$|q| < 1. \tag{1.20}$$

Therefore, the necessary condition for stability is inequality (1.20), which must be valid for any φ, because function T^0, which determined by the initial condition, can contain any harmonic. Even if the exact function T^0 contains only one harmonic, then the round-off error from the computational point of view can contain any harmonic. Also, because it is possible to consider each time level as the initial condition for following the time levels, even if function T^0 contains some of the harmonics, the values of some time levels can contain any harmonic.

Let us now consider what stability condition we obtain for 1.17. To find q, let us substitute the expression 1.19 into equation 1.17 (just for this consideration, to avoid misunderstanding, we will change the space index i in 1.17 to k, then i is the unit imaginary number):

$$\frac{q^{n+1}\, e^{ik\varphi} - q^n\, e^{ik\varphi}}{\Delta t} =$$

$$(1 - \sigma)\,\frac{q^n\, e^{i(k+1)\varphi} - 2\,q^n\, e^{ik\varphi} + q^n\, e^{i(k-1)\varphi}}{h^2} +$$

$$\sigma\,\frac{q^{n+1}\, e^{i(k+1)\varphi} - 2\,q^{n+1}\, e^{ik\varphi} + q^{n+1}\, e^{i(k-1)\varphi}}{h^2}.$$

After cancellation by $q^n e^{ik\varphi}$, we get

$$\frac{q-1}{\Delta t} =$$

$$(1-\sigma)\frac{e^{i\varphi}-2+e^{-i\varphi}}{h^2} +$$

$$\sigma q \frac{e^{i\varphi}-2+e^{-i\varphi}}{h^2} =$$

$$[(1-\sigma)+q\sigma]\frac{e^{i\varphi}-2+e^{-i\varphi}}{h^2}.$$

Now if we take into account that

$$\frac{e^{i\varphi}-2+e^{-i\varphi}}{4} = -\left(\frac{e^{i\frac{\varphi}{2}}-e^{-i\frac{\varphi}{2}}}{2i}\right)^2 = -\sin^2\frac{\varphi}{2},$$

we get

$$q(\varphi) = \frac{1-4\frac{\Delta t}{h^2}(1-\sigma)\sin^2\frac{\varphi}{2}}{1+4\frac{\Delta t}{h^2}\sigma\sin^2\frac{\varphi}{2}}.$$

Because q is a real number stability condition, $|q| < 1$ is equivalent to the inequality

$$-1 \le q \le 1.$$

Right inequality is always valid for any φ and σ. Left inequality will be valid if

$$\sigma \ge \frac{1}{2} - \frac{h^2}{4\Delta t \sin^2\frac{\varphi}{2}}.$$

The last inequality will be valid for any φ if

$$\sigma \ge \frac{1}{2} - \frac{h^2}{4\Delta t}. \tag{1.21}$$

It is shown in the general theory of stability for finite-difference schemes [107], that condition 1.21 not only is necessary but also a sufficient condition for the stability of finite-difference schemes.

Now let us consider some particular cases. For the fully implicit finite-difference scheme, when $\sigma = 1$, it is easy to see that condition 1.21 will be valid for any h and Δt. This type of finite-difference scheme is called *absolutely stable*. If we consider the explicit scheme $\sigma = 0$, then from 1.21, we can conclude that the following condition must be satisfied:

$$\frac{\Delta t}{h^2} \le 0.5. \tag{1.22}$$

Therefore, for the explicit scheme, the necessary condition is satisfied only when the space and time step satisfy some special condition 1.22. This type of finite-difference scheme is called *conditionally stable*.

Finally, we must make some comments regarding the use of harmonic methods. If the necessary condition is not satisfied, there is no point in expecting stability for *any* reasonable choice of norms, and if the necessary conditions are satisfied, then there is some hope in finding an appropriate norm for which the finite-difference scheme will be stable.

From a practical point of view, it is more important to check the necessary condition of stability. This and the constructive nature of the von Neumann method are the reasons for its wide use.

For finite-difference equations with variable coefficients (when coefficients of the original differential equations depend on x, or when we use a non-uniform grid in space), the method of harmonics is used in combination with the method of *frozen coefficients* [100], [46].

The method of frozen coefficients makes it possible to use the method of harmonics for finite-difference schemes with "continuous", but not constant, coefficients, as well as for problems in bounded domains when the boundary conditions are given on the boundary of the domain. This method can also be used for the study of non-linear problems.

Let us, for example, consider a finite-difference scheme for the heat equation on a non-uniform grid:

$$\frac{u_i^{n+1} - u_i^n}{\Delta t} = \frac{\dfrac{u_{i+1}^n - u_i^n}{x_{i+1} - x_i} - \dfrac{u_i^n - u_{i-1}^n}{x_i - x_{i-1}}}{0.5\left(x_{i+1} - x_{i-1}\right)},$$

with initial conditions

$$u_i^0 = T_i^0 \tag{1.23}$$

and boundary conditions

$$u_0^n = \mu_0(t^n), \quad u_M^n = \mu_1. \tag{1.24}$$

We can write this equation as follows:

$$\frac{u_i^{n+1} - u_i^n}{\Delta t} = L_i\, u_{i+1}^n + C_i\, u_i^n + R_i\, u_{i-1}^n, \tag{1.25}$$

where

$$L_i = \frac{1}{\left(x_{i+1} - x_i\right) 0.5\left(x_{i+1} - x_{i-1}\right)},$$

$$R_i = \frac{1}{\left(x_i - x_{i-1}\right) 0.5\left(x_{i+1} - x_{i-1}\right)}, \tag{1.26}$$

$$C_i = -\left(L_i + R_i\right).$$

Now we take some interior node i_0 and denote the coefficients in this node as follows:

$$\tilde{L} = L_{i_0}, \quad \tilde{R} = R_{i_0}, \quad \tilde{C} = C_{i_0}.$$

Finally, instead of the original problem, determined by equations 1.25, 1.23, and 1.24, we consider the following problem in the unbounded domain:

$$\frac{u_i^{n+1} - u_i^n}{\Delta t} = \tilde{L}\, u_{i+1}^n + \tilde{C}\, u_i^n + \tilde{R}\, u_{i-1}^n \qquad (1.27)$$

with the initial conditions from 1.23. Now we can formulate the *principle of frozen coefficients*. For stability of the original problem it is necessary that the Cauchy problem 1.27 with the constant coefficients should satisfy the necessary von Neumann stability condition. Some heuristics basis for the method of frozen coefficients can be found in [46]. In general, the argument is as follows. When we refine the grid, the coefficients of the finite-difference scheme in the neighborhood of node i_0, for any given number of space steps, changes less and less and differs less and less from the values in node i_0. Moreover, the number of grid points from the boundary to the point given by coordinate x_{i_0} goes to infinity. It gives us hope that for the fine grid, the perturbations induced in the solution for the original problem at the moment $t = \tilde{t}$ and in the neighborhood of point x_{i_0}, develops for some short time period and is approximately the same as for problem 1.27. This type of consideration does not depend on the number of space variables or on the form of finite-difference equations, as well as on the assumption of linearity of the original equations.

Taking into account the influence of the approximation of boundary conditions on stability of the finite-difference scheme, instead of the problem in all space, we can consider the boundary problem with boundary conditions only on one end of the original segment, and take the frozen coefficients at this boundary. For example,

$$\frac{u_i^{n+1} - u_i^n}{\Delta t} = L_1\, u_{i+1}^n + C_1\, u_i^n + R_1\, u_{i-1}^n, \quad i = 1, \ldots,$$
$$u_1^n = \mu_1(t^n)$$

and the initial conditions from 1.23. Certainly this type of analysis makes more sense for those more complicated than Dirichlet boundary conditions. Examples for using this method of frozen coefficients can be found in [46].

1.5 The Positivity Preserving Methods

In the example of the heat equation, we can also introduce a new notion, the notion of the *positivity preserving* finite-difference scheme.

In the differential case, the heat equation satisfies the so-called *maximum principle*. The statement of the maximum principle for the initial-boundary value problem is: the maximum and minimum of the solution for the heat equation can be reached either at the initial moment or at the boundary of the domain (see, for example [133]). There is a special class of finite-difference schemes which satisfies the discrete analog of this principle

[109]. We will not consider this question, however, because for a 2-D case finite-difference schemes on non-uniform grids do not usually satisfy this principle.

Here we will demonstrate a more simple notion of the positivity preserving finite-difference scheme and what type of restrictions may be required. If we consider temperature as the *absolute temperature*, then from the physical point of view, any time moment temperature must be non-negative. Let us consider the finite-difference scheme 1.10, where we will use $f = 0$ for simplicity. Suppose that the grid contains only three nodes $i = 1, 2, 3$, and on the boundary, we have zero Dirichlet boundary conditions where $u_1^n = u_3^n = 0$. Then for node $i = 2$, we get

$$\frac{u_2^{n+1} - u_2^n}{\Delta t} = \tag{1.28}$$
$$(1 - \sigma) \frac{-2 u_2^n}{h^2} + \sigma \frac{-2 u_2^{n+1}}{h^2}.$$

Assuming that the temperature in the initial moment is positive, then it must be positive for any time moment. From 1.28, we get

$$u_2^{n+1} = \frac{1 - 2(1 - \sigma)\frac{2\Delta t}{h^2}}{1 + \sigma \frac{2\Delta t}{h^2}} u_2^n.$$

Then, the denominator is always positive and temperature on the new time level will be positive if the numerator is positive. That is,

$$1 - (1 - \sigma)\frac{2\Delta t}{h^2} > 0$$

or

$$\sigma > 1 - \frac{h^2}{2\Delta t}. \tag{1.29}$$

If we compare this inequality with the inequality of 1.21, which follows from the von Neumann condition, then we can see that the expression in the right-hand side in the positivity condition is twice as big. This means that for a given h and Δt, we must choose a σ twice as big as the value of σ that is necessary for stability.

It is interesting to note that for the explicit scheme, the positivity condition gives us the same restriction on the ratio of space and time steps. As for the fully implicit scheme, the positivity condition is satisfied for any h and Δt, which is necessary for the condition for stability.

1.6   The Method of Lines

There is another approach to time integration of partial differential equations called the *method of lines*. In this method only approximation of spatial operators are made. Then, instead of the original partial differential equation, we have the system of ordinary differential equations, or ODE, with respect to functions $u_i(t)$. To make the integration of this system in time, we can use some standard packages for ODE [56], [75].

2 Finite-Difference Schemes for the Heat Equation in 1-D

2.1 Introduction

In this section we will construct finite-difference schemes for the 1-D heat equation

$$\frac{\partial u}{\partial t} = \text{div}\, K\, \text{grad} u + f = \frac{\partial}{\partial x}\left(K\frac{\partial u}{\partial x}\right) + f \ , 0 \le x \le 1,\ t > 0,$$

where the unknown function $u = u(x,t)$ is a function of x and time t, $K(x,t) > 0$, the conductivity coefficient in 1-D is the scalar function, and $f = f(x,t)$ is the density of heat sources or heat sinks. At the initial moment of time, when $t = 0$, the value of u is given as

$$u(x,0) = T^0(x).$$

On the boundary, we have the same boundary conditions as for the elliptic problem, Dirichlet, Neumann, or Robin. There is only one difference: all functions in the boundary conditions can also be functions of time.

One of the main advantages of the support-operators method is that for the approximation of spatial differential operators, we can use the finite-difference operators that we constructed in the previous section for elliptic problems.

Therefore, in this section, we will concentrate our attention on new questions that arise with the presence of the time variable.

For all types of discretisation, we can write the finite-difference scheme for Dirichlet boundary conditions in following form:

$$\frac{u^{n+1} - u^n}{\Delta t} = \text{DIV}\, K\, \text{GRAD} u^\sigma$$

or

$$[E - \Delta t\, \sigma\, \text{DIV}\, K\, \text{GRAD}]\, u^{n+1} = u^n + \Delta t\,(1 - \sigma)\, \text{DIV}\, K\, \text{GRAD} u^n,$$

and the operator in the left-hand side is positive definite and self-adjoint. Therefore, if we use the finite-difference operators that we constructed in the previous chapter, we automatically obtain the implicit finite-difference scheme for heat equations, for which in order to find a temperature on the new time level, we need to solve the system of linear equations with a positive definite and symmetric matrix. Moreover, the matrix which corresponds to the operator

$$E - \Delta t\,\sigma\,\mathrm{DIV}\,K\,\mathrm{GRAD}, \qquad (2.1)$$

is essentially diagonally dominant because the discrete Laplacian is multiplied by the small Δt. This fact and the availability of a good initial guess (values taken from the previous time step), gives us the ability to use effective iterative methods for the solution to the system of linear equations.

For the case of Robin boundary conditions, the situation is similar if we change the operators DIV and GRAD to the operators $\mathcal{B}$ and $\mathcal{C}$. Then to find the temperature on the new time level t^{n+1}, we obtain the system of linear equations

$$\mathcal{A}\,u^{n+1} = F^n,$$

where the internal nodes operator $\mathcal{A}$ is the same as 2.1, and on the boundary it is the same as for the elliptic problem. It is clear that operator $\mathcal{A}$ will also be positive definite and self-adjoint.

These properties of operator $\mathcal{A}$ guarantee that the finite-difference scheme is conditionally stable for $\sigma > 0$. This fact can be proved using the *energy method* of investigation of stability (see, for example, [80], [107]). Derivation of the exact condition of stability requires the investigation of spectral properties of operator $\mathcal{A}$, which depend on distribution of grid nodes and coefficient K.

2.2   Nodal Discretisation for Scalar Functions and Cell-Valued Discretisation for Vector Functions

We will consider in detail only the case of cell-valued discretisation for scalar functions and nodal discretisation for vector functions. Let us note that in the Fortran supplement to this book, the reader can find programs for all cases.

Let us also introduce some notations. We denote the following:

$$u_i^\sigma = (1 - \sigma)\,u_i^n + \sigma\,u_i^{n+1}$$

and similarly,

$$t^\sigma = (1 - \sigma)\,t^n + \sigma\,t^{n+1}$$

and

$$f_i^\sigma = f(x_i, t^\sigma).$$

Then using the finite-difference operators from the previous chapter, we have the following finite-difference scheme for the Dirichlet problem:

$$\frac{u_i^{n+1} - u_i^n}{\Delta t} = \frac{K_i^{0.5}\left(\frac{u_{i+1}^\sigma - u_i^\sigma}{x_{i+1} - x_i}\right) - K_{i-1}^{0.5}\left(\frac{u_i^\sigma - u_{i-1}^\sigma}{x_i - x_{i-1}}\right)}{0.5\,(x_{i+1} - x_{i-1})} + f_i^\sigma. \qquad (2.2)$$

The choice of weight for K, when K is function only of coordinates, is not so important. We have chosen weight 0.5, because then truncation error for $\sigma = 0.5$ is $O(\Delta t^2)$. The approximation of Dirichlet boundary conditions is as follows:

$$u_1^n = \mu_1(t_n) \quad u_M^n = \mu_2(t_n). \tag{2.3}$$

At the initial moment of time, when $t = 0$, the value of u is given as

$$u_i^0 = T^0(x_i).$$

If $\sigma = 0$, then the finite-difference scheme is explicit and values on the new time level t^{n+1} can be found by the explicit formula for $i = 2, \ldots, M-1$:

$$u_i^{n+1} = u_i^n + \Delta t \, \frac{K_i^{0.5}\left(\frac{u_{i+1}^n - u_i^n}{x_{i+1} - x_i}\right) - K_{i-1}^{0.5}\left(\frac{u_i^n - u_{i-1}^n}{x_i - x_{i-1}}\right)}{0.5\left(x_{i+1} - x_{i-1}\right)} + f_i^n, \tag{2.4}$$

wherein the values at initial moment u_i^0 are given by the initial conditions, and the values on the boundary for any moment are given by the boundary conditions in 2.3.

For the implicit finite-difference scheme $\sigma \neq 0$, we obtain the system of linear equations:

$$\left[-\frac{\sigma \Delta t \, K_i^{0.5}}{x_{i+1} - x_i}\right] u_{i+1}^{n+1} +$$

$$\left[0.5\left(x_{i+1} - x_{i-1}\right) + \sigma \Delta t \left(\frac{K_i^{0.5}}{x_{i+1} - x_i} + \frac{K_{i-1}^{0.5}}{x_i - x_{i-1}}\right)\right] u_i^{n+1} +$$

$$\left[-\frac{\sigma \Delta t \, K_{i-1}^{0.5}}{x_i - x_{i-1}}\right] u_{i-1}^{n+1} = \tag{2.5}$$

$$\left[0.5\left(x_{i+1} - x_{i-1}\right)\right] f_i^{n+1} +$$

$$(1 - \sigma) \, \Delta t \left\{ K_i^{0.5} \frac{u_{i+1}^n - u_i^n}{x_{i+1} - x_i} - K_{i-1}^{0.5} \frac{u_i^n - u_{i-1}^n}{x_i - x_{i-1}} \right\}.$$

As it follows from the general considerations that we presented in the beginning of this section, the matrix of the system of linear equations is symmetric and positive definite. This means that it is possible to use effective iterative methods for the solution of this system.

Let us now consider the case of Robin boundary conditions:

$$-(K \frac{du}{dx}) + \alpha_a(t) \, u = \psi_a(t), \quad x = a,$$

$$\tag{2.6}$$

$$(K \frac{du}{dx}) + \alpha_b(t) \, u = \psi_b(t), \quad x = b.$$

For Robin boundary conditions we will consider only implicit finite-difference schemes. The approximations for these conditions in the case of the heat equation look similar to the case for the elliptic equation:

$$- K_1^{0.5} \frac{(u_2^h)^\sigma - (u_1^h)^\sigma}{x_2 - x_1} + \alpha_a^{0.5} (u_1^h)^\sigma = \psi_a^{0.5},$$

$$(2.7)$$

$$K_{M-1}^{0.5} \frac{(u_M^h)^\sigma - (u_{M-1}^h)^\sigma}{x_M - x_{M-1}} + \alpha_b^{0.5} (u_M^h)^\sigma = \psi_b^{0.5}.$$

The truncation error for these approximations is the first order in space.

For the case of the Robin boundary conditions in addition to equation 2.5, in the internal nodes, we get from 2.7 the following equations on the boundary:

$$\left[-\frac{\sigma K_1^{0.5}}{x_2 - x_1} \right] u_2^{n+1} + \left[\frac{\sigma K_1^{0.5}}{x_2 - x_1} + \alpha_a^{0.5} \right] u_1^{n+1} =$$

$$\psi_a^{0.5} - (1 - \sigma) \alpha_a^{0.5} u_1^n + (1 - \sigma) K_1^{0.5} \frac{u_2^n - u_1^n}{x_2 - x_1},$$

$$(2.8)$$

$$\left[-\frac{\sigma K_{M-1}^{0.5}}{x_M - x_{M-1}} \right] u_{M-1}^{n+1} + \left[\frac{\sigma K_{M-1}^{0.5}}{x_M - x_{M-1}} + \alpha_b^{0.5} \right] u_M^{n+1} =$$

$$\psi_b^{0.5} - (1 - \sigma) \alpha_b^{0.5} u_M^n + (1 - \sigma) K_{M-1}^{0.5} \frac{u_M^n - u_{M-1}^n}{x_M - x_{M-1}}.$$

Second-Order Approximation for Robin Boundary Conditions

To obtain the second-order approximation for Robin boundary conditions for heat equations, we can use the following formal considerations. From a formal point of view, the change from a stationary elliptic equation to a non-stationary heat equation can be considered as a substitution of f by $f - du/dt$. We can use this idea for deriving an improved approximation for the Robin boundary condition, similar to what was done for the elliptic problem (see formula 2.22 in previous chapter). Let us remember that for the elliptic problem, the second-order approximation in space for Robin boundary conditions is

$$- K_1 \frac{u_2^h - u_1^h}{x_2 - x_1} + \alpha_a u_1^h = \psi_a + \frac{x_2 - x_1}{2} f(a),$$

where $f(a)$ is the right-hand side in the Poisson equation $-d^2u/dx^2 = f$. For heat equations, we get

$$- K_1^{0.5} \frac{(u_2^h)^\sigma - (u_1^h)^\sigma}{x_2 - x_1} + \alpha_a^{0.5} (u_1^h)^\sigma =$$

$$\psi_a^{0.5} + \frac{x_2 - x_1}{2} \left(f^{0.5}(a) - \frac{(u_1^h)^{n+1} - (u_1^h)^n}{\Delta t} \right).$$

If we compare this equation with equation 2.7, we can see that only the coefficient near u_1^{n+1} is changed. This means that for the related system of linear equations, we change only the diagonal by adding the positive expression $(x_2 - x_1)/(2\,\Delta t)$. Therefore, the new matrix will also be positive definite and symmetric.

2.2.1 Stability of the Finite-Difference Scheme and Results of the Solution for Test Problems

Let us now investigate the stability of the explicit finite-difference scheme starting with the simplest case where $K(x,t) = 1$ and the grid in space is uniform with step h. This is a case of difference equations with constant coefficients when we can directly use the necessary Neumann condition:

$$\Delta t \leq \frac{h^2}{2}.$$

Let us consider the case where $M = 17$ and consequently $h = 1/16 = 0.0625$. Therefore, the Neumann condition for the time step is

$$\Delta t \leq 0.001953125.$$

To demonstrate the stability properties of the finite-difference scheme, we will use the test problem where $K(x,t) = 1$, $f(x,t) = 0$, and the case of zero Dirichlet boundary conditions and initial conditions:

$$u(x, 0) = \sin(2\pi x).$$

The exact solution to this problem is

$$u(x, t) = e^{-4\pi^2 t} \sin(2\pi x).$$

In Figure 2.1, we present the numerical solution for

$$t = 0.045, 0.9, 0.135, 0.18, 0.225, 0.270, 0.315$$

for the case where $\Delta t = 0.0019$. That is, it satisfies the necessary Neumann condition and the finite-difference scheme is stable. We can see this in Figure 2.1.

In Figure 2.2, we present the numerical solution for the same moments of time as for the previous case, for $\Delta t = 0.0021$, where the Neumann condition is not satisfied. In this figure, we can see the development of instability.

The solution for the first three time moments is stable, but then a rapid growth of instability begins. This means that during this time some unstable harmonics appear.

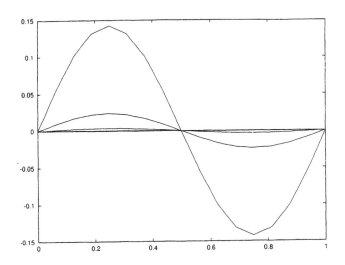

Figure 2.1: *Uniform grid, $K = 1$, $\Delta t = 0.0019$.*

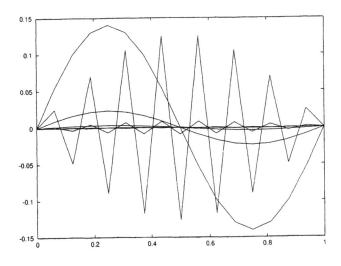

Figure 2.2: *Uniform grid, $K = 1$, $\Delta t = 0.0021$.*

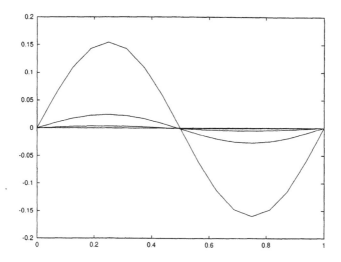

Figure 2.3: *Uniform grid, $K = x^4 + 1$, $\Delta t = 0.0012$.*

To demonstrate how the presence of variable coefficients can affect the stability condition, we consider the case where $K(x,t) = K(x) = x^4 + 1$, and the exact solution, as in the previous case, and function $f(x,t)$ correspond to this exact solution. Now to investigate stability, we must use the method of frozen coefficients. A rough estimation will give us the following condition:

$$\Delta t \leq \frac{h^2}{2 \max\limits_{0 \leq x \leq 1} K(x)} = \frac{h^2}{4}\,.$$

For our grid, the stability condition is

$$\Delta t \leq 0.009765625\,.$$

In Figures 2.3 and 2.4, we present examples of stable and unstable behavior for the solution. Numerical experiments demonstrate that the actual threshold of stability is a little bit bigger than that which is given by a rough theoretical estimation and equal to 0.0012.

Correspondingly, in the theoretical consideration, instability begins to go up near $x = 1$ where the coefficient $K(x)$ is large.

Let us now demonstrate how the stability condition changes on a non-uniform grid. To show this, we are solving the previous problem on a smooth grid:

$$x_i = \xi_i^2,\ \xi = \frac{i-1}{M-1},\ i = 1, \ldots, M.$$

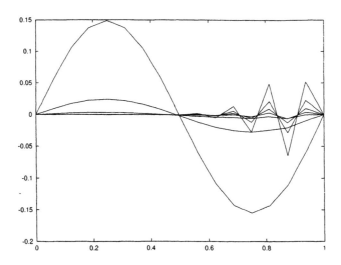

Figure 2.4: *Uniform grid, $K = x^4 + 1$, $\Delta t = 0.0014$.*

In this case, we have finite-difference equations with variable coefficients, but the coefficients change not only because $K(x)$ is variable, but also because the mesh size is variable. In some sense, we can take into account the variable mesh size by changing the discrete coefficient K_i to $\tilde{K}_i = K_i/(x_{i+1} - x_i)$. A similar rough estimation by the method of frozen coefficients gives us the following stability condition:

$$\Delta t \leq \frac{\left(\min_i (x_{i+1} - x_i)\right)^2}{2 \max_{0 \leq x \leq 1} K(x)}.$$

For $M = 17$
$$\min_i (x_{i+1} - x_i) = 3.90625 \cdot 10^{-3},$$

and we get the stability condition

$$\Delta t \leq 3.81146 \cdot 10^{-6}.$$

Numerical experiments show that the real threshold is $4.410 \cdot 10^{-5}$. The difference can be explained as follows. At first, the mesh size is small (near $x = 0$) where the coefficient $K(x) \approx 1$ and vise versa. In addition, near $x = 0$, the mesh size changes very rapidly and it is necessary to use a more accurate estimation which takes this fact into account.

In the next two figures, 2.5 and 2.6, we demonstrate stable and unstable behavior of the solution for Δt near the stability threshold.

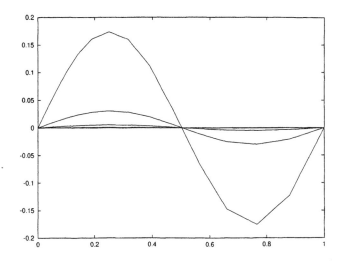

Figure 2.5: *Non-uniform grid,* $K = x^4 + 1,$ $\Delta t = 4.410 \cdot 10^{-5}.$

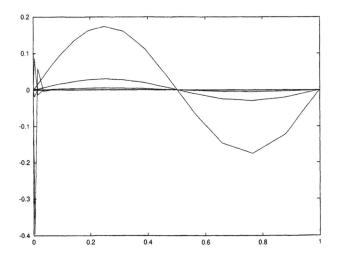

Figure 2.6: *Non-uniform grid,* $K = x^4 + 1,$ $\Delta t = 4.4155 \cdot 10^{-5}.$

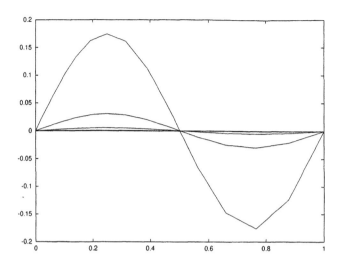

Figure 2.7: *Implicit scheme, non-uniform grid,* $K = x^4 + 1$, $\Delta t = 1.0 \cdot 10^{-4}$.

In accordance with theoretical considerations, instability begins near $x = 0$ where the mesh size is small.

To show the difference between explicit and implicit schemes with respect to the choice of the time step Δt, in the next picture we present the results for the computation of the same problem by the fully implicit finite-difference scheme $\sigma = 1$, with $\Delta t = 1.0 \cdot 10^{-4}$, which is approximately twice as large as the stability threshold time step for the explicit scheme.

The accuracy for the explicit scheme with $\Delta t = 4.410 \cdot 10^{-5}$ and the implicit scheme with $\Delta t = 1.0 \cdot 10^{-4}$ is approximately the same, for example, for $t = 0.09$, max norm of error is $0.21 \cdot 10^{-2}$ for the explicit scheme and $0.23 \cdot 10^{-2}$ for the implicit scheme.

To make a decision as to what type of scheme (either explicit or implicit) is better to use, each particular case must be considered separately. In making this decision, the following must be considered: contributions of terms with Δt and h into the truncation error; for a real computation, not only asymptotic behavior is important but also the constant near Δt and h; in general, each time step for the implicit scheme takes more time than for the explicit scheme; because we need to solve the system of linear equations, in each particular case, we need to take into account what iterative method we use; how the efficiency of this method depends on the initial guess; is the solution of our problem smooth in time so that the values from the previous time step give us a good initial guess; and so on.

In addition, for solving a real problem, we must consider some questions relating to the computer on which the algorithm will be implemented. For example, what is the possibility of using parallel computations, what is the memory capacity, and so on.

If coefficients of the differential equation strongly depend on time, it makes sense to use the *variable time step*, which for each time level, satisfies the Neumann condition.

Robin Boundary Conditions

As a test problem for the finite-difference scheme for Robin boundary conditions, we use the following problem:

$$\frac{du}{dt} = \frac{d}{dx}\left(K\frac{\partial u}{\partial x}\right) + f \quad , 0 \le x \le 1, \ t > 0,$$

where $K(x,t) = x^4 + 1$ and function $f(x,t)$ is given as follows:

$$f(x,t) = e^{-4\pi^2 t}\left(4\pi^2 x^4 \sin(2\pi x) + 4x^3 2\pi \cos(2\pi x)\right),$$

and at the initial moment, the temperature is given as follows:

$$u(x,0) = \sin(2\pi x),$$

$$-\frac{du}{dx} + u = -2\pi e^{-4\pi t}, \quad x = 0,$$

$$\frac{du}{dx} + u = 4\pi e^{-4\pi t}, \quad x = 1.$$

These problems have the same exact solution as the problem with Dirichlet boundary conditions which we considered before:

$$u(x,t) = e^{-4\pi^2 t}\sin(2\pi x).$$

Here we present the results of the computation for the first- and second-order approximations of the Robin boundary condition for the same time moments as for the case of Dirichlet boundary conditions. The results presented for the case of a non-uniform grid are given by transformation $x = \xi^2$. In Figure 2.8, we present the results for the first-order approximation of Robin boundary conditions for $M = 17$ and $\Delta t = 1.0 \cdot 10^{-5}$. In Figure 2.9, we present the results for the second-order approximation of Robin boundary conditions for $M = 17$ and $\Delta t = 1.0 \cdot 10^{-5}$. If we compare the accuracy of the finite-difference scheme with the first- and second-order approximations of boundary conditions for moment $t = 0.045$ and the same time step $\Delta t = 1.0 \cdot 10^{-5}$, we get the following results: for the second-order approximation, the max norm of error is equal to $0.31 \cdot 10^{-1}$, and for the first-order approximation, the max norm of error is equal to 0.12. That is, the second-order scheme is four times more accurate than the first-order scheme.

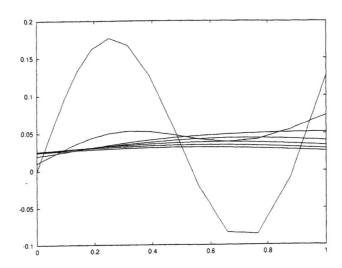

Figure 2.8: *First-order approximation of Robin boundary conditions, non-uniform grid, $K = x^4 + 1$, $\Delta t = 1.0 \cdot 10^{-5}$.*

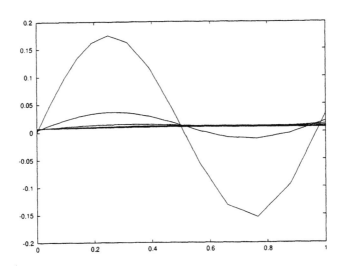

Figure 2.9: *Second-order approximation of Robin boundary conditions, non-uniform grid, $K = x^4 + 1$, $\Delta t = 1.0 \cdot 10^{-5}$.*

2.3   Cell-Valued Discretisation of Scalar Functions and Nodal Discretisation of Vector Functions

For the case of cell-valued discretisation of scalar functions and nodal discretisation of vector functions, all considerations are similar. Only the stability conditions can be a little bit different because of the different discrete operators. Let us consider the explicit form of the finite-difference scheme for Dirichlet boundary conditions. For the sake of brevity, we will consider only the explicit scheme and the case where coefficient K depends only on the coordinate and does not depend on time. Let us note that the finite-difference scheme for the general case can be obtained in a similar manner as the case of nodal discretisation of scalar functions and, in the Fortran supplement, we implement the general case $K(x, t)$. Therefore, the explicit finite-difference scheme is

$$\frac{u^{n+1} - u^n}{\Delta t} = \frac{K_{i+1}\frac{u_{i+1}^n - u_i^n}{VN_{i+1}} - K_i\frac{u_i^n - u_{i-1}^n}{VN_i}}{VC_i} + \varphi_i^n, \quad i = 1, \cdots, M - 1 \quad (2.9)$$

and, on the boundary, we have

$$u_0^n = \mu_0(t_n), \quad u_M^n = \mu_1(t_n).$$

Let us recall that the approximation of the partial operator is exactly the same as for elliptic equations.

For Robin boundary conditions, we get

$$-K_1 \frac{u_1^{n+1} - u_0^{n+1}}{VN_1} + \alpha_a u_0^{n+1} = \psi_a^{n+1},$$

$$\frac{u^{n+1} - u^n}{\Delta t} = \frac{K_{i+1}\frac{u_{i+1}^n - u_i^n}{VN_{i+1}} - K_i\frac{u_i^n - u_{i-1}^n}{VN_i}}{VC_i} + \varphi_i^n, \quad i = 1, \cdots, M - 1$$

$$+K_M \frac{u_M^{n+1} - u_{M-1}^{n+1}}{VN_M} + \alpha_b u_M^{n+1} = \psi_b^{n+1}.$$

Let us present some numerical results for Dirichlet boundary conditions. The test problem is the same as for the case of nodal discretisation of scalar functions. First, let us note that in numerical experiments we have found that the stability condition for cell-valued discretisation is approximately four times more rigid than for nodal discretisation. This can be explained as follows. In the finite-difference scheme 2.9, in cell $i = 1$, the coefficient near u_0 is equal to $1/(VC_1 VN_1)$ and $VN_1 = 0.5 VC_1$. Volume VC_1 is small for our grid, which is given by the transformation $x = \xi^2$, and the actual mesh size decreasing near $x = 0$ is very rapid. In the case of nodal discretisation, the finite-difference scheme in node $i = 2$, the coefficient near u_1, which is the boundary value, is equal to $1/(VC_1 VN_2)$ and $VN_2 = 0.5 (VC_1 + VC_2)$. That is, in the case of cell discretisation, the coefficient is much smaller, and

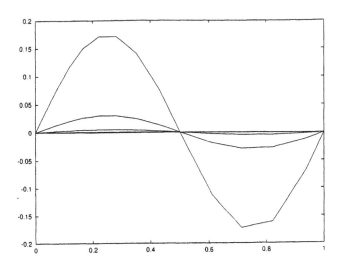

Figure 2.10: *Explicit scheme, C-N, Dirichlet BC, non-uniform grid, $K = x^4 + 1$, $\Delta t = 1.15 \cdot 10^{-5}$.*

consequently, the restriction on the time step is more rigid. The results of the computation for Dirichlet boundary conditions are presented in Figure 2.10.

The comparison of accuracy of the finite-difference schemes for nodal and cell-valued discretisation gives us the following results. For $t = 0.045$ and $\Delta t = 1.14 \cdot 10^{-5}$, nodal discretisation gives the max error $0.77 \cdot 10^{-2}$ and cell-valued discretisation gives the max error $0.91 \cdot 10^{-2}$. That is, in this case, nodal discretisation gives us more accurate results.

Now let us consider some numerical results for the case of Robin boundary conditions. As we know from the previous chapter, the finite-difference scheme has a second-order truncation error including the approximations of boundary conditions. Therefore, the finite-difference scheme for the heat equation has the same property. The statement for the test problem is the same as for the case of nodal discretisation of scalar functions. The results of the computations are presented in Figure 2.11. As we know from the previous section, for $t = 0.045$ and $\Delta t = 1.0 \cdot 10^{-5}$, the second-order approximation of boundary conditions for nodal discretisation gives us three times more accurate results than the first-order approximation of boundary conditions. The finite-difference scheme with cell-valued discretisation gives us accuracy $0.2 \cdot 10^{-1}$, that is, 1.5 times more accurate than the second-order scheme with nodal discretisation. The graphical comparison of the three schemes is presented in Figure 2.12.

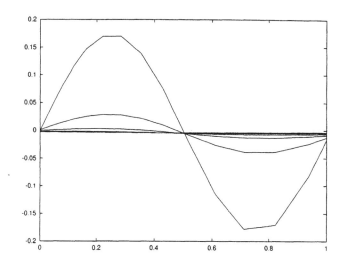

Figure 2.11: *Explicit scheme, C-N, Robin BC, non-uniform grid, $K = x^4 + 1$, $\Delta t = 4.0 \cdot 10^{-5}$.*

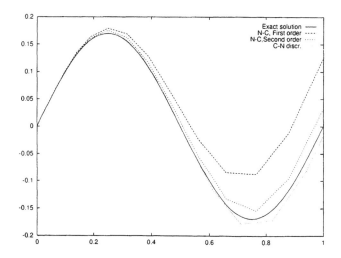

Figure 2.12: *Comparison of finite-difference schemes for Robin boundary conditions, non-uniform grid, $K = x^4 + 1$, $\Delta t = 1.0 \cdot 10^{-5}$.*

Implementations of explicit and implicit finite-difference schemes for different types of boundary conditions are presented as Fortran code on the attached floppy disk (for details, see the Fortran Guide chapter).

2.4   Conservation Laws and the Iteration Process

Here we want to consider one important question relating to the implementation of implicit finite-difference schemes.

Suppose a conservative implicit finite-difference scheme is constructed by some method, and correspondingly, the system of linear equations is solved by an iteration method. The result of the iteration process is some discrete function which satisfies the equations related to the finite-difference scheme with some accuracy. Consequently, even if the finite-difference scheme is conservative, the solution obtained by the iterative method may not satisfy the conservation law. Then the question of importance is how much the magnitude of the resulting energy imbalances depends on the accuracy of the iteration process, and how much it depends on the parameters of the finite-difference scheme itself (e.g., on the time and space steps). It is particularly important to know how to assess the imbalance in multidimensional problems.

Let us consider the example of the non-conservative iteration process. For the purpose of this demonstration, we consider the Neumann problem for the heat equation

$$\frac{\partial u}{\partial t} = \frac{\partial^2 u}{\partial x^2}, \quad 0 < x < 1,$$

$$\left.\frac{\partial u}{\partial x}\right|_{x=0} = \gamma_1(t), \quad \left.\frac{\partial u}{\partial x}\right|_{x=1} = \gamma_2(t),$$

$$u|_{t=0} = u_0(x)$$

where γ_1, γ_2 and u_0 are given functions.

The amount of heat Q contained in the system at the instant of time t is given by

$$Q(t) = \int_0^1 u(x, t)\, dt.$$

From the statement of this problem, we get

$$\frac{dQ}{dt} = \gamma_2(t) - \gamma_1(t). \tag{2.10}$$

This equation expresses the fact that the change in the amount of heat in the entire system occurs because of the heat inflow through the boundary of the region.

The conservative implicit finite-difference scheme, which has truncation error $O(\Delta t + h^2)$, is

$$\frac{u_2^{n+1} - u_1^{n+1}}{h} - \frac{h}{2}\frac{u_1^{n+1} - u_1^n}{\Delta t} = \gamma_1(t^{n+1}),$$

$$\frac{u_i^{n+1} - u_i^n}{\Delta t} = \frac{u_{i+1}^{n+1} - 2u_i^{n+1} + u_{i-1}^{n+1}}{h^2}, \quad i = 2, 3, \ldots, M-1,$$

$$\frac{u_M^{n+1} - u_{M-1}^{n+1}}{h} + \frac{h}{2}\frac{u_M^{n+1} - u_M^n}{\Delta t} = \gamma_2(t^{n+1}).$$

The discrete analog for the amount of heat Q has the following form:

$$Q = \sum_{i=1}^{M-1} h\frac{u_{i+1} - u_i}{2} = \frac{h}{2}u_1 + \sum_{i=2}^{M-1} h\,u_i + \frac{h}{2}u_M.$$

Using the equations for the finite-difference scheme, we get

$$\frac{Q^{n+1} - Q^n}{\Delta t} = \gamma_2(t^{n+1}) - \gamma_1(t^{n+1}).$$

This equation is the discrete analog of 2.10, and expresses the fact that the finite-difference scheme is conservative and no additional "discrete" inputs or outflows of heat are present.

Now let us consider what happens if we solve the system of difference equations by the usual Gauss-Seidel method. Formulas for the Gauss-Seidel method are

$$\frac{u_2^{(s)} - u_1^{(s+1)}}{h} - \frac{h}{2}\frac{u_1^{(s+1)} - u_1^n}{\Delta t} = \gamma_1(t^{n+1}),$$

$$\frac{u_i^{(s+1)} - u_i^n}{\Delta t} = \frac{u_{i+1}^{(s)} - 2u_i^{(s+1)} + u_{i-1}^{(s+1)}}{h^2}, \quad i = 2, 3, \ldots, M-1,$$

$$\frac{u_M^{(s+1)} - u_{M-1}^{(s+1)}}{h} + \frac{h}{2}\frac{u_M^{(s+1)} - u_M^n}{\Delta t} = \gamma_2(t^{n+1}),$$

where (s) is the iteration number.

We can write these equations in the form of analogs of the finite-difference scheme:

$$\frac{u_2^{(s+1)} - u_1^{(s+1)}}{h} - \frac{h}{2}\frac{u_1^{(s+1)} - u_1^n}{\Delta t} = \gamma_1(t^{n+1}) + \left(\frac{u_2^{(s+1)} - u_2(s)}{h}\right), \quad (2.11)$$

$$\frac{u_i^{(s+1)} - u_i^n}{\Delta t} = \frac{u_{i+1}^{(s+1)} - 2u_i^{(s+1)} + u_{i-1}^{(s+1)}}{h^2} + \left(\frac{u_{i+1}^{(s)} - u_{i+1}^{(s+1)}}{h^2}\right), \quad (2.12)$$

$$\frac{u_M^{(s+1)} - u_{M-1}^{(s+1)}}{h} + \frac{h}{2}\frac{u_M^{(s+1)} - u_M^n}{\Delta t} = \gamma_2(t^{n+1}), \quad (2.13)$$

where, in the second equation, $i = 2, 3, \ldots, M - 1$. After carrying out
a certain number of iterations necessary to satisfy the chosen criterion of
convergence, the process is terminated, and $u_i^{(s+1)}$ is taken as the value of
the temperature on the next time layer t^{n+1}. Thus, $u_i^{n+1} \equiv u_i^{(s+1)}$, obtained
by the iteration process, satisfies equations 2.11, 2.12, and 2.13, which are
the same as the original finite-difference schemes with the exception that the
right-hand sides contain terms that are in addition to the original equations
and can also be regarded as heat sources and heat sinks.

Let the condition of the termination of the iteration process have the
form

$$\max_i \left| u_i^{(s+1)} - u_i^{(s)} \right| < \varepsilon,$$

where ε is a given small number. Let us estimate the imbalance in the
amount of heat caused by using the Gauss-Seidel method with this termination condition. From equations 2.11, 2.12, 2.13 and the definition of Q
in discrete case, we get

$$\frac{Q^{n+1} - Q^n}{\Delta t} = \gamma_2^{n+1} - \gamma_1^{n+1} + \sum_{i=2}^{M} \frac{u_i^{(s)} - u_i^{(s+1)}}{h^2} \, h.$$

Therefore, the imbalance in the amount of heat is

$$\Delta Q = \Delta t \sum_{i=2}^{M} \frac{u_i^{(s)} - u_i^{(s+1)}}{h^2} \, h.$$

Now using the termination condition, we can estimate this imbalance as
follows:

$$|\Delta Q| \leq \frac{\varepsilon}{h^2} \left(M - 1 \right) h \, \Delta t = \frac{\varepsilon \, \Delta t}{h^2}.$$

The expression on the right-hand side gives the upper limit of the imbalance
occurring at the one time step.

This estimation must be taken into account when using the Seidel
method for the solution of difference equations. For example, in order
for the law of change in the amount of heat for the difference scheme for an
instant of time t^k to be satisfied with accuracy h^2, i.e., with the accuracy
of the finite-difference scheme, the quantity, ε, should be chosen from the
condition

$$\sum_{p=1}^{k} \frac{\varepsilon \, \Delta t}{h^2} \sim h^2$$

or

$$\varepsilon \sim \frac{h^4}{\Delta t \, (k - 1)}.$$

In the article [138] some numerical results are presented that confirm the
theoretical considerations which are made here.

There is some approach to solving the system of difference equations that eliminates the problems connected with the non-conservative properties of the iteration process. Let us first introduce fluxes

$$W_i^{n+1} = \frac{u_{i+1}^{n+1} - u_i^{n+1}}{h}. \tag{2.14}$$

These fluxes are given in cells and the previous relationship is valid for any n, which means that flux on some time level is determined by the temperature from the same time level.

Using the definition of flux, we can rewrite the original finite-difference scheme as follows:

$$\frac{u_1^{n+1} - u_1^n}{\Delta t} = \frac{W_1^{n+1} - \gamma_1^{n+1}}{0.5\,h}, \tag{2.15}$$

$$\frac{u_i^{n+1} - u_i^n}{\Delta t} = \frac{W_i^{n+1} - W_{i-1}^{n+1}}{h}, \quad i = 2, 3, \dots, M-1, \tag{2.16}$$

$$\frac{u_M^{n+1} - u_M^n}{\Delta t} = \frac{\gamma_2^{n+1} - W_{M-1}^{n+1}}{0.5\,h}. \tag{2.17}$$

Now if we express the temperature u_i^{n+1} from these equations and substitute it into the definition of fluxes (2.14), we get the following equations for fluxes:

$$\frac{W_1^{n+1} - W_1^n}{\Delta t} = \frac{\dfrac{W_2^{n+1} - W_1^{n+1}}{h} - \dfrac{W_1^{n+1} - \gamma_1^{n+1}}{0.5\,h}}{h},$$

$$\frac{W_i^{n+1} - W_i^n}{\Delta t} = \frac{W_{i+1}^{n+1} - 2\,W_i^{n+1} + W_{i-1}^{n+1}}{h^2}, \quad i = 2, 3, \dots M-2,$$

$$\frac{W_{M-1}^{n+1} - W_{M-1}^n}{\Delta t} = \frac{\dfrac{\gamma_2^{n+1} - W_{M-1}^{n+1}}{0.5\,h} - \dfrac{W_{M-1}^{n+1} - W_{M-2}^{n+1}}{h}}{h}.$$

This is the system of linear equations with respect to fluxes W_i^{n+1}, because fluxes W_i^n are known from the previous time step, and in the initial moment, can be computed from the initial conditions.

Assuming we solved this system of equations by some iterative method, we can then find the temperature u_i^{n+1} on the new time level by using the explicit formulas 2.15, 2.16, and 2.17. It is important to note that this process will preserve the conservative properties of the original finite-difference scheme for *any fluxes* W_i^{n+1}, and in particular, *does not* depend on the accuracy of the iterative process for the solution of the system for fluxes. Of course, accuracy of the finite-difference scheme is strongly dependent on the accuracy of the iterative solution of equations for fluxes.

It is important to understand that the conservative properties and the accuracy of the discrete solution are different properties for finite-difference algorithms.

3   Finite-Difference Schemes for the Heat Equation in 2-D

Similar to the 1-D case, we can use all the discrete operators for the approximations of partial differential operators and boundary conditions that we obtained in the previous chapter in Section 3 when we constructed the finite-difference schemes for 2-D elliptic equations. Then using the approximation for partial operators from the previous chapter, we can write the finite-difference schemes for heat equations for the case of Dirichlet boundary conditions in operator form

$$\frac{u^{n+1} - u^n}{\Delta t} = \text{DIV } K \text{ GRAD } u^\sigma .$$

For the case of Robin boundary conditions, we only need to use operator $\mathcal{B}$, which includes the approximation of normal flux instead of operator DIV. Therefore we will not consider here the explicit form of discrete equations for the heat equation, because all the approximations for spatial operators are the same and the approximation for the time derivative is trivial.

Most of the qualitative effects relating to the presence of the time derivative are similar to the case of the 1-D equation. Therefore, we will explain some new points which are related to 2-D.

In the appendix relating to the Fortran code directory, the reader can find a description of all the subroutines which implement finite-difference schemes for the heat equation. There are four packages which correspond to different types of discretisations and different types of boundary conditions. In each case, the user can choose between implicit and explicit finite-difference schemes.

3.1   Stability Conditions in 2-D

Let us consider the simplest case of a uniform grid in a unit square and nodal discretisation for scalar functions and cell-valued discretisation for vector functions. Then the explicit finite-difference scheme will look as follows:

$$\frac{u_{i,j}^{n+1} - u_{i,j}^n}{\Delta t} =$$
$$\frac{u_{i+1,j}^n - 2\,u_{i,j}^n + u_{i-1,j}^n}{h^2} + \frac{u_{i,j+1}^n - 2\,u_{i,j}^n + u_{i,j-1}^n}{h^2} .$$

To find the necessary condition for stability, we can use a procedure that is similar to 1-D. That is, we will investigate the stability of particular solution

$$u_{k,l}^n = q^n\, e^{i\,(k\varphi + l\psi)}.$$

After substitution of this solution for finite-difference equations and cancellation by

$$u_{k,l}^n = q^n \, e^{i(k\varphi + l\psi)} \,,$$

we get

$$\frac{q-1}{\Delta t} = \frac{e^{i\varphi} - 2 + e^{-i\varphi}}{h^2} + \frac{e^{i\psi} - 2 + e^{-i\psi}}{h^2} \,.$$

And finally, using trigonometric identities, we get

$$q = 1 - \frac{4\,\Delta t}{h^2} \left(\sin^2 \frac{\varphi}{2} + \sin^2 \frac{\psi}{2} \right) \,.$$

Then the condition $|q| \leq 1$ gives us the following necessary condition:

$$\frac{\Delta t}{h^2} \leq 0.25 \,,$$

which means that for a square grid in 2-D, the restriction for the time step is twice as stringent than it is for the 1-D case. This must be taken into account in practical computations. All considerations that were done regarding the influence of variable coefficients and non-uniform grids on the stability condition, are held in 2-D for tensor product grids. A general, logically rectangular grid situation is more complicated, and the role of h in a stability condition is played by the ratio of some effective linear size and area of the cell. From a practical point of view, we can use the following condition:

$$\Delta t \leq 0.25 \left(\min \frac{VC_{i,j}}{L_{i,j}} \right)^2 \,,$$

where $L_{i,j}$ is some effective line size which characterizes the cell (i,j), and can be used as an example for the length of the diagonal.

Let us make some general comments about the stability condition for the case of general matrix K and general, logically rectangular grid. It is clear that it is not easy to obtain the exact conditions in this case because the properties of the grid are changed with the coordinates, and the coefficients can be varied with the coordinates very quickly. In addition, for a given node (i,j), the nearest nodes can be distributed very non-uniformly. This means that from a practical point of view, it is easy to make some rough estimations and then find a reasonable time step Δt by using numerical experiments. This usually takes less computer and human time than finding a precise estimation.

3.2 Symmetry Preserving Finite-Difference Schemes and Iteration Methods

Here we will demonstrate with a simple example some very important properties of two-dimensional finite-difference schemes and iteration methods for its solution.

Suppose that the initial and boundary conditions for the original differential problem are given so that the differential problem has a solution which depends only on one space variable or maybe has some special type of symmetry. Then for many applications, it is important to construct the discrete algorithm which will preserve these properties for the solution. This is important for a problem where we investigate the behavior of some small perturbation of a symmetric solution.

The violation of symmetry can happen for two reasons. First, the finite-difference scheme may not preserve these properties, and second, the violation can happen during the iterative solution for difference equations. The finite-difference scheme in Cartesian coordinates usually preserves symmetry of the solution, and in particular, all finite-difference schemes presented in this book do.

Here we will consider the violation of symmetry relating to the iteration methods. More exactly, we will show what happens if the original differential problem and finite-difference scheme has a 1-D solution. We will solve the system of linear equations by using the Gauss-Seidel method.

Let us describe the test problem. We will consider a 2-D heat equation in a unit square,

$$\frac{\partial u}{\partial t} = \frac{\partial^2 u}{\partial x^2} + \frac{\partial^2 u}{\partial y^2} + f(x), \tag{3.1}$$

with the following initial and boundary conditions:

$$u(x, y, 0) = u_0(x), \tag{3.2}$$

$$u(0, y, t) = \mu_1(t), \ u(1, y, t) = \mu_2(t), \tag{3.3}$$

$$\frac{\partial u}{\partial y}\bigg|_{y=0} = 0, \ \frac{\partial u}{\partial y}\bigg|_{y=1} = 0. \tag{3.4}$$

This problem has a solution which depends only on x, that is, $u(x, y, t) = u(x, t)$ and satisfies the following 1-D problem:

$$\frac{\partial u}{\partial t} = \frac{\partial^2 u}{\partial x^2} + f(x) \tag{3.5}$$

with the following initial and boundary conditions:

$$u(x, 0) = u_0(x), \tag{3.6}$$

$$u(0, t) = \mu_1(t), \ u(1, t) = \mu_2(t). \tag{3.7}$$

The analog of the solution $u(x, t)$ will be the grid function u which depends on time and only one index i, that is, u_i^n.

Let us check that our finite difference scheme for problems 3.1, 3.2, 3.3, and 3.4 also has a 1-D solution.

For these boundary and initial conditions, the implicit finite-difference scheme is

$$\frac{u_{i,j}^{n+1} - u_{i,j}^{n}}{\Delta t} =$$
$$\frac{u_{i+1,j}^{n+1} - 2\,u_{i,j}^{n+1} + u_{i-1,j}^{n+1}}{h^2} + \frac{u_{i,j+1}^{n+1} - 2\,u_{i,j}^{n+1} + u_{i,j-1}^{n+1}}{h^2} + f_i , \quad (3.8)$$

$$u_{1,j}^{n} = \mu_1(t^n), \ u_{M,j}^{n} = \mu_2(t^n) , \tag{3.9}$$

$$-\frac{u_{i,2}^{n+1} - u_{i,1}^{n+1}}{h} = 0, \quad \frac{u_{i,N}^{n+1} - u_{i,N-1}^{n+1}}{h} , = 0 \tag{3.10}$$

$$u_{i,j}^{0} = u_0(x_i). \tag{3.11}$$

Let us try to find the solution to these difference equations in the form $u_{i,j}^n = u_i^n$. First, let us note that the Neumann boundary condition for $j = 1$ and $j = N$ are satisfied automatically and the second term in the right-hand side of 3.8 is equal to zero. After substitution of such a solution to our finite-difference scheme, we get

$$\frac{u_i^{n+1} - u_i^n}{\Delta t} = \tag{3.12}$$
$$\frac{u_{i+1}^{n+1} - 2\,u_i^{n+1} + u_{i-1}^{n+1}}{h^2} + f_i ,$$

$$u_1^n = \mu_1(t^n), \ u_M^n = \mu_2(t^n) , \tag{3.13}$$

$$u_i^0 = u_0(x_i). \tag{3.14}$$

This is the exact 1-D finite-difference scheme for the heat equation. This means that our 2-D finite-difference scheme allows a 1-D solution for the appropriate initial and boundary conditions similar to the continuous case.

Now let's see what happens if we solve the difference equations related to the 2-D finite difference scheme by using the Gauss-Seidel method. Let us note that for the initial moment we have a 1-D solution from the initial conditions. We take this solution as the initial guess in the Gauss-Seidel iteration for computing the solution for the new time level. Formulas for the Gauss-Seidel method for solutions of fully implicit finite-difference schemes are

$$\frac{u_{i,j}^{(s+1)} - u_{i,j}^n}{\Delta t} = \tag{3.15}$$
$$\frac{u_{i+1,j}^{(s)} - 2\,u_{i,j}^{(s+1)} + u_{i-1,j}^{(s+1)}}{h^2} + \frac{u_{i,j+1}^{(s)} - 2\,u_{i,j}^{(s+1)} + u_{i,j-1}^{(s+1)}}{h^2} + f_i ,$$

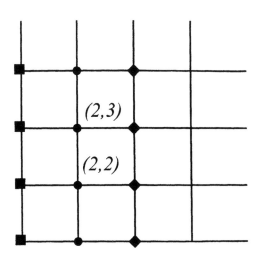

Figure 3.13: *Initial guess.*

$$u_{1,j}^{(s+1)} = \mu_1(t^n), \quad u_{M,j}^{(s+1)} = \mu_2(t^n) , \tag{3.16}$$

$$-\frac{u_{i,2}^{(s)} - u_{i,1}^{(s+1)}}{h} = 0, \tag{3.17}$$

$$\frac{u_{i,N}^{(s+1)} - u_{i,N-1}^{(s+1)}}{h} = 0 . \tag{3.18}$$

Let us consider the first iteration. First, from the boundary conditions on the bottom boundary 3.17, we find $u_{i,1}^{(1)}$, and because the initial guess does not depend on j, values of $u_{i,1}^{(1)}$ will not be dependent on j either.

It is evident that values which are related to the first iteration in the internal nodes, will not be dependent only on index i. Considering the stencil for node $(2, 2)$ and $(2, 3)$, we show that these two values are different. The initial guess is shown in Figure 3.13, where the same symbols are denoted by the same values of the function. The stencil for nodes $(2, 2)$ and $(2, 3)$ are shown in Figure 3.14. It is clear that when we compute the value in node $(2, 3)$ on the new iteration, we use a new value in node $(2, 2)$ which is different from the initial guess, and because the other values in the stencil are the same as for node $(2, 2)$, the values in nodes $(2, 2)$ and $(2, 3)$ in the new iteration will be different.

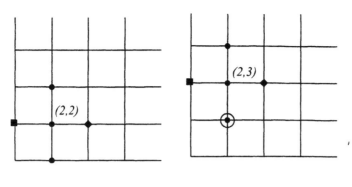

Stencil for node (2,2).   Stencil for node (2,3).

Figure 3.14: *Stencils.*

Now if we rewrite the formulas for the Seidel process in a form that is similar to the form of the original finite-difference scheme, we get

$$
\frac{u_{i,j}^{(s+1)} - u_{i,j}^{n}}{\Delta t} = \tag{3.19}
$$
$$
\frac{u_{i+1,j}^{(s+1)} - 2\,u_{i,j}^{(s+1)} + u_{i-1,j}^{(s+1)}}{h^2} + \frac{u_{i,j+1}^{(s+1)} - 2\,u_{i,j}^{(s+1)} + u_{i,j-1}^{(s+1)}}{h^2} + f_i +
$$
$$
\frac{u_{i+1,j}^{(s)} - u_{i+1,j}^{(s+1)}}{h^2} + \frac{u_{i,j+1}^{(s)} - u_{i,j+1}^{(s+1)}}{h^2} .
$$

That is, we have some additional terms in the right-hand side. If the termination condition for the Seidel process is

$$
\max_{i,j} |u_{i,j}^{(s+1)} - u_{i,j}^{(s)}| < \varepsilon ,
$$

then these additional terms have the order

$$
\frac{\varepsilon}{h^2} .
$$

Then, if we want to withstand a one-dimensional solution with accuracy h^2, we must take $\varepsilon \sim h^4$.

Let us describe one possible approach to the solution of the finite-difference equations which will preserve the symmetry of the solution. To do this we will apply *the block or line Gauss-Seidel method*, to finite-difference equations, where in each step, we will determine all values for a given j;

$$\frac{u_{i,j}^{(s+1)} - u_{i,j}^n}{\Delta t} = \tag{3.20}$$

$$\frac{u_{i+1,j}^{(s+1)} - 2\,u_{i,j}^{(s+1)} + u_{i-1,j}^{(s+1)}}{h^2} + \frac{u_{i,j+1}^{(s)} - 2\,u_{i,j}^{(s)} + u_{i,j-1}^{(s)}}{h^2} + f_i\,.$$

Then for each line $j = const$, we will have the same three-point equations that we can solve by some direct method. Finally, for each j, we obtain the same solution which depends only on i.

It is interesting to note that the explicit scheme also preserves the symmetry.

3.3   Numerical Examples

In this section we consider some results of the numerical computation for the 2-D heat equation.

To test the finite-difference algorithms for the case of Dirichlet boundary conditions, we chose a problem that is sufficient to exercise all terms in the algorithm. First, K is chosen to be a rotation of a diagonal matrix:

$$K = R\,D\,R^T\,,$$

where

$$R = \begin{pmatrix} +\cos(\theta) & -\sin(\theta) \\ +\sin(\theta) & +\cos(\theta) \end{pmatrix}, \quad D = \begin{pmatrix} d_1 & 0 \\ 0 & d_2 \end{pmatrix},$$

and

$$\begin{aligned} \theta &= \frac{3\,\pi}{12}, \\ d_1 &= 1 + 2\,x^2 + y^2 + y^5, \\ d_2 &= 1 + x^2 + 2\,y^2 + x^3. \end{aligned}$$

This is the exact same matrix K used in the test problem for the elliptic equation. The solution is chosen to be

$$u(x, y, t) = e^{-2\pi^2 t}\,\sin(\pi\,x)\,\sin(\pi\,y),$$

which means that for a given domain, the Dirichlet boundary conditions coincide with this solution and the right-hand side is

$$f = -\operatorname{div} K \operatorname{grad} u + \frac{\partial u}{\partial t},$$

where matrix K and function u are given by previous formulas.

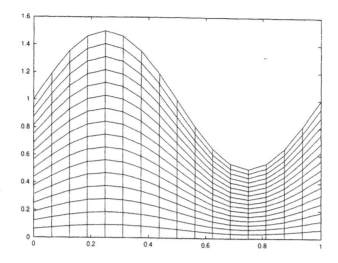

Figure 3.15: *Grid in domain with curvilinear boundaries.*

To check the convergence properties of the constructed finite-difference schemes, we use the same grid as we used for the elliptical equation and that is shown in Figure 3.15

To compare the exact and approximate solutions in each time moment, we introduce the same norms for the errors as for the case of elliptic equations. That is, the max norm and the discrete analog of L_2 norm.

In all computations, we take the time step Δt small enough so that we can check for accuracy in space.

The results of the computations for nodal discretisation for scalar functions and cell-valued discretisation for vector functions are presented in Table 3.1 for time moment $t = 0.01$. This table is arranged similar to that for the 1-D case. In the first column, we present the number of nodes $(N = M)$. In the second and third columns, we present the max and L_2 norm respectively, and in the fourth and fifth columns, we present the numerically computed constant

$$\frac{\parallel u^h - p_h\,u \parallel_{max}}{h^2}, \quad \frac{\parallel u^h - p_h\,u \parallel_{L_2}}{h^2}.$$

In the last two columns, we present a ratio between the error for the number of nodes given in this row and the error for the number of nodes given in the previous row, then, similar to the 1-D case when $h \rightarrow 0$, this ratio must converge to 4 for the second-order method. The results presented in the table demonstrate the second-order convergence for the finite-difference scheme.

M	max norm	L_2 norm	Const max	Const L_2	R max	R L_2
17	0.687E-02	0.225E-02	1.740	0.576	-	-
33	0.186E-02	0.572E-03	1.904	0.585	3.693	3.933

Table 3.1: *Convergence analysis, heat equation, Dirichlet boundary conditions, 2-D-N-C discretisation, smooth-grid.*

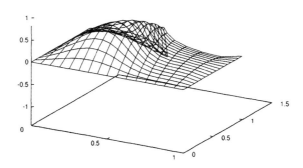

Figure 3.16: *Exact solution for $t = 0.01$.*

In Figures 3.16 and 3.17, we present the exact and approximate solution for $t = 0.01$.

In the next table, we present the results for the case of Robin boundary conditions. Because we do not have an improvement in the truncation error on the boundary, we expect first-order accuracy. Therefore, constants in the table are the ratio of errors for h and the ratio of errors for h and $h/2$ must be close to 2. This table actually shows that for this particular problem, we have a convergence rate a little bit higher than the first order. In Figure 3.18, we present the approximate solution. It is easy to see that the error is large near the corners where the vector of normal to the boundary can not be correctly defined (see the discussion about Robin boundary conditions for the case of nodal discretisation of scalar functions in Section 3.3.2 in the previous chapter).

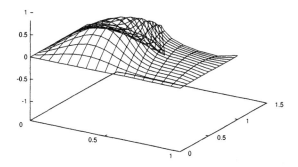

Figure 3.17: *Approximate solution for $t = 0.01$ and $M = N = 17$, Dirichlet boundary conditions.*

M	max norm	L_2 norm	Const max	Const L_2	R max	R L_2
17	0.581E-00	0.211E-00	9.296	3.376	-	-
33	0.199E-00	0.823E-01	6.368	2.633	2.919	2.563

Table 3.2: *Convergence analysis, heat equation, Robin boundary conditions, 2-D-N-C discretisation, smooth-grid.*

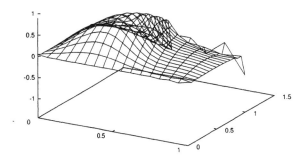

Figure 3.18: *Approximate solution for $t = 0.01$, and $M = N = 17$, Robin boundary conditions, 2-D-N-C discretisation.*

M	max norm	L_2 norm	Const max	Const L_2	R max	R L_2
17	0.129E-01	0.367E-02	3.302	0.939	-	-
33	0.314E-02	0.942E-03	3.215	0.946	4.108	3.895

Table 3.3: *Convergence analysis, heat equation, Dirichlet boundary conditions, 2-D-C-N discretisation, smooth-grid.*

Let us now consider the case of cell-valued discretisation of scalar functions and nodal discretisation of vector functions. The test problems are the same as for the previous case. In Table 3.3, we present the convergence analysis for the case of Dirichlet boundary conditions. This table demonstrates the second-order convergence rate. The convergence analysis for the case of Robin boundary conditions is presented in Table 3.4. This also demonstrates the second-order convergence rate. It is interesting to note that the accuracy for Robin boundary conditions and C-N discretisation is approximately seven times better than for the case of N-C discretisation. This can be explained by the fact that for the C-N discretisation case, we do not have any singularity in the approximation of boundary conditions. In Figure 3.19, we present the approximate solution for Robin boundary conditions, which is visually much better than that for the approximate solution for N-C discretisation presented in Figure 3.18.

M	max norm	L_2 norm	Const max	Const L_2	R max	R L_2
17	0.846E-01	0.431E-01	21.65	11.03	-	-
33	0.241E-01	0.109E-01	24.67	11.16	3.510	3.954

Table 3.4: *Convergence analysis, heat equation, Robin boundary conditions, 2-D-C-N discretisation, smooth-grid.*

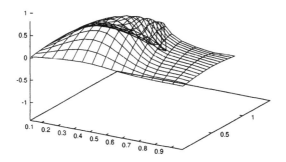

Figure 3.19: *Approximate solution for t = 0.01, and M = N = 17, Robin boundary conditions, 2-D-C-N discretisation.*

Chapter 5

Lagrangian Gas Dynamics

In this chapter we describe the application of the support-operators method for construction of finite-difference schemes for gas dynamics equations in the Lagrangian form.

The support-operators method itself is the method that gives us the ability to construct a system of discrete operators (analogs of differential operators) which satisfy some integral identities related to conservation laws. Therefore, we will describe the technique which is based on using differential equations for the description of gas dynamics flows. In fact, we will use the Lagrangian form of gas dynamics equations when independent space variables refer to the coordinate system fixed in gas and undergo all the motions and distortion of the gas, in order that the particles of the gas are permanently identified by their Lagrangian variables.

Fluid flows are frequently characterized by internal discontinuities, such as shock, for which special conditions are required. These conditions are provided by the well-known Rankine-Hugoniot equations, but their application in practice is encumbered by difficulties where the surfaces on which the conditions are to be applied are in motion through the gas and their motion is not known in advance, but is determined by the equations. Therefore, we want to construct our finite-difference algorithm by some modification of the finite-difference scheme which describes smooth flows and where we can use the appropriate discrete operators. The most widespread method for a unified description of gas dynamical flows is the "pseudoviscosity" method. In this method, we avoid a detailed consideration of shock waves and consider the flow of gases possessing some viscosity which sometimes is quite unrelated to the physical viscosity. By introducing this viscosity into the differential equations of gas dynamics, we approximately describe the shock waves as a smooth shock transition.

The introduction of viscosity into the differential equation can be done in different ways, but in this book, we consider the simplest way where we add some artificial "pressure" to real pressure. The expression for artificial pressure can usually be formulated as the discrete analog of some differential operator. For example, "linear artificial viscosity" is proportional to div $\vec{W}$, where $\vec{W}$ is the velocity vector. To approximate these differential operators, we use the method of support-operators, which gives us the "right" approximation for the viscous term. For example, using the support-operator method, we satisfy the condition where artificial viscosity causes the increase of internal energy on shock waves.

There is no doubt that the nature of equations for gas dynamics is much more rich than it can be derived from the formal properties of differential operators. This is the reason why we describe in the introduction the basic notion related to gas dynamics equations, such as the description of the approach for continuum media, some elements of thermodynamics, integral and differential forms of gas dynamics equations, linear approximation, that is, acoustic equations, the general definition of hyperbolic equations, the notion of characteristics of differential equations, Riemann's invariants, the domain of dependence, and the range of influence and so on. A special section is devoted to the description of discontinuous solutions of gas dynamics equations and conditions on discontinuity. The effect of artificial viscosity on the structure of the solution is considered on a 1-D case. This consideration is very useful for introducing artificial viscosity and understanding how artificial viscosity works.

All considerations in the first section are very short and present the basic notion in a form that may be useful for future sections. Readers who want to know more about equations of gas dynamics in general, and in particular about computational fluid dynamics, will have to read systematic books such as [1], [10], [19], [32], [33], [52], [62], [78], [95], [96], [97], [100], [101], [102], [110], [114], [122], [148]. The reader who has basic knowledge in fundamentals of gas dynamics and is primarily interested in the numerical methods can skip most part of the first section of this chapter.

Equations of gas dynamics in Lagrangian form are formulated in terms of the first-order invariant differential operators div and grad. It is also important to note that the differential term in artificial viscosity can be formulated using the same operators. In the second section of this chapter, we analyze the properties of the differential operators that provide the conservation laws of mass, momentum, and full energy. It is shown that these properties are: divergence properties of operators grad and div and property div $= -$grad. It is usual for the Lagrangian methods that the velocity vector is measured in nodes, and that pressure, internal energy, and density are associated with cells. In previous chapters, we have already constructed discrete operators DIV and GRAD for this type of discretisation and constructed operators to satisfy conditions which provide conservation laws in a discrete case. Therefore, we can use these discrete operators for construction of finite-difference schemes for equations of gas dynamics. Another confirmation of the capabilities of the support-operators method is that we can use the approximations that were created for one type of equation to construct the finite-difference schemes for other types of equations.

To make the main ideas clearer in the special section, we considered a finite-difference algorithm for the 1-D case. In this section, we consider an application of the support-operators algorithm for deriving the finite-difference scheme and describe the computational algorithm for the case of implicit and explicit finite-difference schemes, and consider the treatment of impermeability and free boundary conditions. Also for the 1-D case, we explain typical restrictions for the ratio of space and time steps which are imposed by stability conditions. Finally, we describe how we introduce artificial viscosity.

The main contents of this chapter describe the 2-D finite-difference algorithm. Here we describe in detail all the questions related to discretisation of spatial differential operators and consider the computational algorithm for explicit and implicit finite-difference schemes. The implicit finite-difference scheme is a system of non-linear algebraic equations. We describe one possible algorithm for the solution of this system that does not violate the conservation properties of the original finite-difference scheme. We also describe in detail how to implement impermeability and free boundary conditions.

1 Fundamentals of Gas Dynamics

Introduction to Continuum Motion

The analysis of fluid motion assumes that the body of the fluid under consideration forms a physical continuum. A physical continuum is a medium filled with continuous matter such that every part of the medium, however small, is itself a continuum and is entirely filled with the matter. Since matter is composed of molecules, the continuum hypothesis implies that every small volume will contain a large number of molecules. For example, $1\ cm^3$ of air contains 2.687×10^{19} molecules under normal conditions (Avogadro number). Thus, in a cube 0.002 cm on a side, there are 2.687×10^{10}, which is a large number. We are not interested in the properties of each molecule at point P but rather in the average over a large number of molecules in the neighborhood of point P. Mathematically, the association of averaged values of properties at point P also gives rise to a continuum of points and numbers. In the summary, the continuum hypothesis implies the following postulate: *Matter is continuously distributed throughout the region under consideration with a large number of molecules, even in macroscopically small volumes.*

Though the single postulate of continuum mechanics satisfactorily describes fluid motion, it is imperative to consider some statistical aspects of molecular motions. These considerations distinguish different continua by their physical properties. The most common fluids are either gases or liquids. In gases, the molecules are far apart having an average separation between the molecules approximately 3.5×10^{-7} cm. The cohesive forces between the molecules are weak. The molecules randomly collide and exchange their momentum, heat, and other properties and thus give rise to viscosity, thermal conductivity, etc. These effects, though molecular in origin, are considered to be the physical properties of the continuum itself. In liquids, the separation between the molecules is much smaller and the cohesive forces between a molecule and its neighbors are quite strong. Again, the averaged molecular properties resulting from these cohesive forces are taken as the properties of the medium. While air and water are treated through the same continuum hypothesis, the effects of their motions are different due to the differences in their molecular properties, e.g., viscosity, thermal conductivity, etc.

In mathematical aspects, the continuum hypothesis gives us the ability to use techniques of continuous functions, differential and integral calculus.

The continuum hypothesis is the main assumption (but not the only) in models of gas dynamics. There are also some general assumptions that space is Euclidean space, the speeds encountered are far smaller than the speed of light, and so on. For particular models there are some special assumptions about the properties of continuum media (presence or absence of viscosity, heat conduction, gravity, etc.).

Characteristics of Continuum Media. Elements of Thermodynamics

For a quantitative description of the processes in gas dynamics, we need to introduce some characteristics.

The position of fluid particles is characterized by its radius-vector $\vec{r}$ in a coordinate system. The change in position of a particle with time is described by its velocity

$$\vec{W} = \frac{d\vec{r}}{dt}, \tag{1.1}$$

where t is time.

The state of the gas is also characterized by other parameters that are called thermodynamic parameters. The *gas density* ρ is the mass of media that is contained in the unit volume. The force which acts on the unit surface that is perpendicular to it is called *pressure*. There are other thermodynamic parameters, such as temperature T, specific volume $\eta = 1/\rho$, and entropy S. The accurate definition of this notion is the subject of

dV

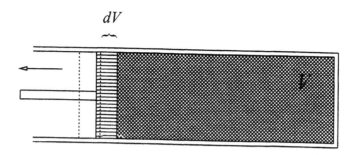

Figure 1.1:

thermodynamics (see, for example [33]). Here we will restrict our discussion to only the facts that we need to accomplish our purpose.

In thermodynamics, the notion of external parameters is introduced. These parameters characterize the position of the external, with respect to gas bodies, with which gas interacts during its motion. This notion can be demonstrated in the following example. Let us consider a homogeneous gas enclosed in a cylindrical vessel, which is closed on one end by a stationary wall and by a piston on the other end. In this case, the role that the external body plays is on the walls of the vessel and on the surface of the piston, which is in contact with the gas. They determine the boundary of volume V, which is occupied by the gas. This volume is the external parameter. In a general case, some force fields created by the internal forces must be included in the external parameters, for example, the electromagnetic field.

When gas interacts with external bodies, the external parameters are changed and the gas delivers work.

The example in Figure 1.1 illustrates that when the piston makes an infinitesimal movement, there is an infinitesimal change to the external parameter volume dV, and the gas performs *work*:

$$dA = -p\,dV, \tag{1.2}$$

or, with respect to the unit of mass of the gas,

$$da = -p\,d\eta, \tag{1.3}$$

where $\eta = 1/\rho$ is a specific volume, that is the volume of the unit of mass of the gas. Let us note that work is different from zero only when the external bodies are moving. For example, when the gas is elapsing from the vessel to a vacuum, there is no movement of the external bodies and work computed by the previous formula is equal to zero. This is because the pressure on the interface between the gas and vacuum state is equal to zero. When the gas is performing work, its energy is changing. The change in gas energy can also be due to the heat exchange with the external bodies.

The relationship between gas energy, work which is performed by the gas, and heat from external sources is determined by the first thermodynamics principle, which is the general law of energy conservation. For a mass unit, it looks as follows:

$$d\varepsilon = dQ - p\,d\eta, \qquad (1.4)$$

where ε is the specific internal energy, Q is the amount of heat (with respect to the unit mass) and is supplied by external bodies. The previous relationship is valid not only for the considered example, but also for any *equilibrium process* in the gas. Recall that the process is called thermodynamic equilibrium if it is going infinitesimally slow, and each intermediate stage is equilibrium. In its turn, the equilibrium thermodynamical state of the system is where, for a given external parameter and absence of the exchange of heat with external bodies, the state of the system does not change.

The real processes run with finite speeds, and strictly speaking, are not equilibrium. But idealization, related to the concept of equilibrium, is good enough for the description of a wide range of practical applications.

From the first principle of thermodynamics, we can conclude that internal energy ε is a one-valued function for the state of the gas. The quantities Q and a are not only functions of the state of the system, but are also dependent on the history of the system. That is, on the process which moves the system in its current state. In another words, $d\varepsilon$ is the full differential, in contrast to dQ and da, which are not, in general, full differentials of some expressions.

The existence of another full differential, and consequently, a one-valued function for the state of the gas (called entropy S), is postulated by the second principle of thermodynamics:

$$dS = \frac{dQ}{T} \qquad (1.5)$$

or

$$T\,dS = d\varepsilon + p\,d\eta. \qquad (1.6)$$

Here T is the temperature. The temperature is an internal parameter, which is characterized by the state of the gas. From a statistical point of view, the temperature is determined by mean kinetic energy of a chaotic movement of the molecules of the gas.

It is evident that entropy is determined by an arbitrary constant. The value of this constant follows from the third principle of thermodynamics: when absolute temperature is equal to zero then entropy is also equal to zero:

$$S \to 0, \quad \text{when } T \to 0 . \qquad (1.7)$$

Let us note that in the general case for processes that are not equilibrium, but allow the introduction of functions like temperature, entropy, and so on, the second principle of thermodynamics looks as follows:

$$dS \geq \frac{dQ}{T},$$

(1.8)

and the sign of equality is valid for equilibrium processes. In particular, for thermally insulated systems when $dQ = 0$ (adiabatic systems), the second principle of thermodynamics is written in the following form:

$$dS \geq 0.$$

(1.9)

This inequality means that the processes in the system run in such a way that entropy of the system is not vanishing.

For equilibrium processes, there are only two independent functions among thermodynamical functions p, ρ, ε, η, T, S, etc. If we choose temperature T and density ρ as the independent functions, then the other functions can be expressed as follows:

$$p = p(\rho, T), \quad \varepsilon = \varepsilon(\rho, T),$$

(1.10)

and so on. Such equations are called *equations of state*.

In the simplest case of the *ideal gas*, these equations have the following form:

$$p = \rho R T, \quad \varepsilon = \varepsilon(T),$$

(1.11)

where R is called the *gas constant*. If internal energy depends on the temperature as the linear function, then the second equation can be written as follows:

$$\varepsilon = \frac{R T}{\gamma - 1},$$

(1.12)

where γ is a dimensionless quantity, which is equal to the ratio of heat capacities for a given pressure and a given volume.

Fluid Particles

In considering the motion of fluids, it is helpful to keep an infinitesimal volume of fluid as a geometrical point in a mathematical continuum of numbers and call it a *fluid particle*. Each fluid particle, at a given moment of time, can be associated with an ordered triple of numbers, e.g., (a, b, c), and its motion followed in time. The state properties at the position (a, b, c), are called the state properties of the fluid particle itself at $t = t_0$, thus giving a unique identity to this particle. As this particle moves about, its state properties will be understood to be the same as the local state properties of the continuum.

Lagrangian and Eulerian Approach for Description of Movement of Continuum Media

The finite-difference method for calculation of time dependent flows has mostly been based on either the Eulerian or the Lagrangian form of the gas dynamics equations. In the Eulerian approach, the independent space variables refer to a coordinate system fixed in space through which the fluid is thought of as moving. The flow, being characterized by the time-dependent velocity field, is found by solving the initial value problem. In the Lagrangian approach, the independent space variables refer to the co-ordinate system fixed in fluid undergoing all the motion and distortion of the fluid in order that the particles of the fluid are permanently identi-fied by their *Lagrangian variables*, while their actual positions in space are among the dependent quantities to be solved. These two approaches are equivalent, except that the Lagrangian form gives more information: it tells where each particle of fluid originally came from. This considerably facilitates the calculations of flows when two or more fluids, with different equations of state, are present. In this book, we will consider only the numerical methods that are based on the Lagrangian descriptions.

The role of the *Lagrangian variables* can play coordinates of the particles in the initial moment. In this case we can write

$$x_i = g_i(x_1^0, x_2^0, x_3^0, t), \quad i = 1, 2, 3, \tag{1.13}$$

where x_i is the current coordinate of the particle, and x_i^0 is the coordinate of the particle at the initial moment of time $t = 0$, that is, the Lagrangian variables.

If we fixed x_1^0, x_2^0, x_3^0 in the previous equation, we would obtain the law for motion of a particular particle, or its trajectory. If we fix t and consider g_i as a function of x_1^0, x_2^0, x_3^0, we obtain the distribution of all particles in space at this moment of time.

Formula 1.13 gives us a one-to-one correspondence between x_i and x_i^0. Let us note that if, with time, some vacuum zones appeared, then there would be no images in space for the initial positions x_i^0 for the points of physical space x_i which correspond to these vacuum zones. We will not consider these cases. From a computational point of view, such problems need special procedures to treat these situations. Therefore we can express x_i^0 as a function of x_i and t:

$$x_i^0 = h_i(x_1, x_2, x_3, t), \quad i = 1, 2, 3$$

and the Jacobian of this transformation,

$$\Delta = \frac{D(x_1, x_2, x_3)}{D(x_1^0, x_2^0, x_3^0)}$$

is assumed to be not equal to zero for any time moment.

1.1 The Integral Form of Gas Dynamics Equations

Let us consider the volume which is formed by some fixed set of fluid particles. This volume is moving with a medium and changes its configuration, called *fluid volume*. Let us denote this volume as $V(t)$. For time interval $\Delta t = t_2 - t_1$, fluid particles are moved to new positions and volume $V(t)$ is deformed. The number of fluid particles in this volume will remain the same and the mass of gas will not be changed:

$$\int_{V(t_2)} \rho(\vec{r}, t_2)\, dV = \int_{V(t_1)} \rho(\vec{r}, t_1)\, dV, \qquad (1.14)$$

where $\vec{r}$ is the radius-vector of the fluid particle. The equation 1.14 is the *law of mass conservation* in the Lagrange form.

Let us now consider the *conservation law for momentum*. Because during their movement particles do not come to this fluid volume and do not go out, then there is no flux of momentum through the surface of the volume. We will denote this surface as $\Sigma(t)$. Variation of momentum occurs only because of external body and surface forces:

$$\int_{V(t_2)} \rho(\vec{r}, t_2)\, \vec{W}(\vec{r}, t_2)\, dV - \int_{V(t_2)} \rho(\vec{r}, t_2)\, \vec{W}(\vec{r}, t_2)\, dV \quad (1.15)$$

$$= -\int_{t_1}^{t_2} \int_{\Sigma(t)} p\, \vec{n}\, d\Sigma\, dt + + \int_{t_1}^{t_2} \int_{V(t)} \vec{F}\, dV,$$

where $\vec{F}$ is the density of body force, such as gravity force, and so on. To explain the origin of the first term in the right-hand side of the previous equation, we need to understand that there is also some interaction between the fixed volume and the rest of the gas. This interaction determines the surface force. In general, the direction of the vector of this force $\vec{P}$ does not coincide with the direction of normal to the surface. The tangent component of this force is determined by the presence of viscosity in the media, which is friction force. If we consider the inviscid medium, then the surface force is reduced to its normal component, which can be presented as follows:

$$(\vec{P}, \vec{n}) = -p\vec{n}, \qquad (1.16)$$

where p is the usual gas dynamics pressure. The total momentum of this force is

$$-\int_{t_1}^{t_2} \int_{\Sigma(t)} p\, \vec{n}\, d\Sigma\, dt. \qquad (1.17)$$

Let us now consider the *law of conservation of full energy*. The total energy of a gas (internal plus kinetic) in volume $V(t)$ is

$$\int_{V(t)} \rho \left(\varepsilon + \frac{|\vec{W}|^2}{2} \right) dV. \qquad (1.18)$$

Its variation occurs because of the work of the external bodies and surface forces and also because of external sources. If we denote Q for the density of body sources of energy (for example, the intensity of Joule's heating by electrical current), and $\vec{H}$ for the vector of density of heat flux,

$$\vec{H} = -\kappa \operatorname{grad} T, \tag{1.19}$$

where $\kappa = \kappa(\rho, T)$ is heat conductivity, then the conservation law for full energy can be written in the following form:

$$
\begin{aligned}
&\int_{V(t_2)} \rho(\vec{r}, t_2) \left[\epsilon(\vec{r}, t_2) + \frac{|\vec{W}(\vec{r}, t_2)|^2}{2} \right] dV \\
&- \int_{V(t_2)} \rho(\vec{r}, t_2) \left[\epsilon(\vec{r}, t_1) + \frac{|\vec{W}(\vec{r}, t_1)|^2}{2} \right] dV = \\
&- \int_{t_1}^{t_2} \int_{\Sigma(t)} p\,(\vec{W}, \vec{n})\, d\Sigma\, dt + \int_{t_1}^{t_2} \int_{V(t)} (\vec{F}, \vec{W})\, dV + \\
&+ \int_{t_1}^{t_2} \int_{V(t)} Q\, dV + - \int_{t_1}^{t_2} \int_{\Sigma(t)} (\vec{H}, \vec{n})\, d\Sigma\, dt.
\end{aligned}
\tag{1.20}
$$

Three equations for gas dynamics contain the following five unknown functions: ρ, p, ϵ, T, and $\vec{W}$. To complete this system of equations, we need to introduce the thermodynamical equations of state:

$$p = p(\rho, T), \quad \epsilon = \epsilon(\rho, T).$$

The equations for gas dynamics express the conservation laws and are general and the same for all gases; on other hand, equations of state give us information about the concrete properties of media. Let us recall that we have an additional equation

$$\frac{d\vec{r}}{dt} = \vec{W}, \tag{1.21}$$

which is the definition of velocity.

1.2   Integral Equations for the One-Dimensional Case

For future considerations, it is useful to consider one-dimensional flows, where all parameters depend only on one spatial variable and time. We will consider only the case for Cartesian coordinates and will assume that all variables depend only on $x_1 = x$. The general equations for the laws of conservation are also valid for the one-dimensional case. Let us choose the parallelepiped shown in Figure 1.2 as volume V in a general formula for conservation laws. The bases for this parallelepiped are unit squares,

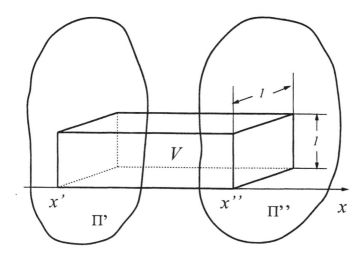

Figure 1.2: *Unit parallelepiped.*

which lie on planes Π' and Π'', drawn through points $x\prime$ and x'', and are perpendicular to the x axis (see Figure 1.2).

In the one-dimensional case, it is sufficient to obtain equations for such a parallelepiped. The elementary volume is

$$dV = 1 \cdot 1 \cdot dx,$$

where the units have dimensions of its length and its product is the area of the cross-section of the parallelepiped. If we assume that the velocity vector has only one component WX, which, for simplicity, we will denote as W, then we can rewrite the continuity equation for the one-dimensional case as follows:

$$\int_{x'}^{x''} [\rho(x, t_2) - \rho(x, t_1)] \, dx +$$
$$\int_{t_1}^{t_2} [\rho(x'', t) \, W(x'', t) - \rho(x', t) \, W(x', t)] \, dt. \qquad (1.22)$$

If we consider phase plane (x, t) and select contour C in this plane as the rectangle with sides parallel to the coordinate lines (see Figure 1.3), then it easy to see that the following equation,

$$\oint_C (\rho \, dx - \rho W \, dt) = 0, \qquad (1.23)$$

and equation 1.22 are equivalent. The same result remains valid for any contour C that is formed by any number of segments which are parallel

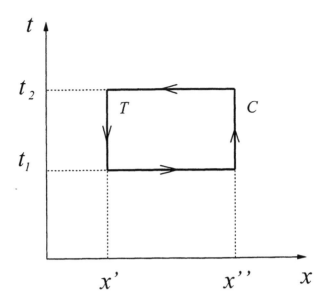

Figure 1.3: *Simple contour in* (x, t) *plane.*

to the coordinate lines (see Figure 1.4). In fact, we can divide domain T into some number of rectangles and for each of them write equation 1.23. By summation of the contour integrals, we obtain that the terms that are related to the internal boundaries will cancel because each segment of the internal boundaries is passed twice in opposite directions. Finally, we again use formula 1.23, but for a more complicated contour C, as shown in Figure 1.4. For some arbitrary contour C, we can cover plane (x, t) with lines that are parallel to the coordinate lines (see Figure 1.5). Then we can choose a minimal domain T^* with contour C^* which contains the original domain T with contour C. For contour C^*, formula 1.23 is valid. Refining the rectangular grid and passing to the limit, we obtain that $C^* \to C$, and formula 1.23 is valid for any closed piecewise smooth contour C on plane (x, t). It is also assumed that integrand functions are bounded, piecewise continuous, and on contour C, can be discontinuous only in isolated points.

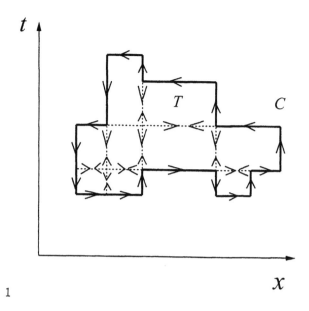

Figure 1.4: *Piecewise linear contour in* (x, t) *plane.*

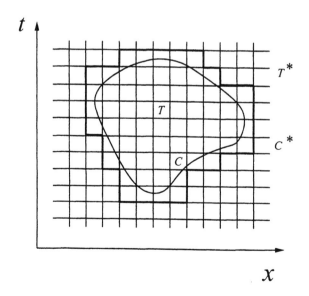

Figure 1.5: *Arbitrary contour in* (x, t) *plane.*

Similar to 1.23, we can present laws of conservation of momentum and energy:

$$\oint_C \left[\rho\, W\, dx - (p + \rho W^2)\, dt \right] = \int_T F\, dx dt \qquad (1.24)$$

$$\oint_C \left\{ \rho \left(\varepsilon + \frac{W^2}{2} \right) dx - \left[\rho W \left(\varepsilon + \frac{W^2}{2} + \frac{p}{\rho} \right) + H \right] dt \right\} =$$

$$\int_T F\, W\, dx\, dt + \int_T Q\, dx\, dt \qquad (1.25)$$

1.3 Differential Equations of Gas Dynamics in Lagrangian Form

The differential equations of gas dynamics can be obtained from integral equations 1.14, 1.15, and 1.20.

If we divide equation 1.14 by Δt and then pass to the limit for $\Delta t \to 0$, we get

$$\frac{d}{dt} \int_{V(t)} \rho\, dV = 0 . \qquad (1.26)$$

Let us note that the differentiation made here is along the trajectories of gas particles. Now in the previous formula $dV = dx_1\, dx_2\, dx_3$, if we make a change in the variables from (x_1, x_2, x_3) to (x_1^0, x_2^0, x_3^0), we get

$$dV = dx_1\, dx_2\, dx_3 = \Delta\, dx_1^0\, dx_2^0\, dx_3^0 = \Delta\, dV^0 ,$$

where Δ is the Jacobian of the transformation, which is assumed to be positive. The variable volume $V(t)$ is transformed to some volume V^0 in the space of the initial configuration, and this volume is not dependent on time. Therefore,

$$\frac{d}{dt} \int_{V(t)} \rho\, dV = \frac{d}{dt} \int_{V^0} \rho\, \Delta\, dV^0 = \int_{V^0} \frac{d}{dt} (\rho\, \Delta)\, dV^0 = 0 . \qquad (1.27)$$

Because V^0 is an arbitrary volume, we get

$$\frac{d}{dt} (\rho\, \Delta) = 0 . \qquad (1.28)$$

This is the differential continuity equation in the Lagrangian form. Let us note that the differential continuity equation can also be written in the following form:

$$\frac{d\rho}{dt} + \rho\, \mathrm{div}\, \vec{W} = 0 . \qquad (1.29)$$

The momentum equation 1.15 can be transformed as follows:

$$\frac{d}{dt} \int_{V(t)} \rho \vec{W} dV = \int_{V(t)} \left(-\mathrm{grad}\, p + \vec{F} \right) dV . \qquad (1.30)$$

Changing the variables as previously mentioned, we get

$$\frac{d}{dt} \int_{V(t)} \rho \vec{W} \, dV = \frac{d}{dt} \int_{V^0} \rho \vec{W} \, \Delta \, dV^0 = \tag{1.31}$$

$$\int_{V^0} \frac{d}{dt} \left(\rho \vec{W} \, \Delta \right) dV^0 = \int_{V^0} \rho \Delta \frac{d\vec{W}}{dt} \, dV^0 = \int_{V(t)} \rho \frac{d\vec{W}}{dt} \, dV.$$

Here, we took into account that $\rho \Delta$ is a constant and can be a factor in the differentiation. If we substitute 1.31 for 1.30, we get

$$\rho \frac{d\vec{W}}{dt} = -\operatorname{grad} p + \vec{F}. \tag{1.32}$$

Similarly, we can obtain the differential form of the energy equation.

Finally, the system for differential equations of gas dynamics in the Lagrangian form is

$$\frac{d\rho}{dt} + \rho \operatorname{div} \vec{W} = 0,$$

$$\rho \frac{d\vec{W}}{dt} = -\operatorname{grad} p + \vec{F}, \tag{1.33}$$

$$\rho \frac{d}{dt} \left(\varepsilon + \frac{|\vec{W}|^2}{2} \right) = -\operatorname{div}(p\vec{W}) + Q + \vec{F} \cdot \vec{W} - \operatorname{div} \vec{H},$$

$$p = p(\rho, T), \quad \varepsilon = \varepsilon(\rho, T).$$

1.4 The Differential Equations in 1-D Lagrangian Mass Coordinates

Let us consider a motion of gas in 1-D where all variables depend only on $x_1 = x$. In this section, for simplicity, we will assume that external force $\vec{F}$ and source Q are absent.

Differential equations for this case follow from the general form considered in the previous section and has the following form:

$$\frac{d\rho}{dt} + \rho \frac{\partial W}{\partial x} = 0,$$

$$\rho \frac{dW}{dt} = -\frac{\partial p}{\partial x}, \tag{1.34}$$

$$\rho \frac{d}{dt} \left(\varepsilon + \frac{|W|^2}{2} \right) = -\frac{\partial (pW)}{\partial x} - \frac{\partial H}{\partial x}$$

$$p = p(\rho, T), \quad \varepsilon = \varepsilon(\rho, T),$$

where W and H are projections of the corresponding vectors on the x axis.

Let us introduce the *Lagrangian mass coordinates*. We will explain this notion using the concrete example of the problem of a one-dimensional,

plane gas elapsing to the vacuum. The statement of the problem can be
formulated as follows. Let us consider two parallel planes and assume that
the space between them is filled with gas. The left plane is a stationary
wall and its Eulerian coordinate x_0 is constant and does not change during
the process. The right plane, which separates the gas from the vacuum, is
removed momentarily in the initial time moment. After this, the process
of the gas elapsing to vacuum commences. We assume that this process
will be one-dimensional. In particular, this means that interface between
the gas and the vacuum will remain a plane for all time moments (its
coordinate $x_v(t)$), and the trajectories of all particles are straight lines
which are parallel to the x axis.

Let us consider the gas behavior in "unit parallelepiped" (see Figure
1.2) $x_0 \leq x \leq x_v(t)$. Let us note that although the volume of this parallelepiped is changing with time, the fluid particles are not flowing out and
are not coming to this volume. Therefore, the mass in this volume remains
constant. Let us denote this mass as M.

Let us make a cut of the parallelepiped by some plane which is parallel
to the bases of the parallelepiped. We will denote this cut as A. It is clear
that all the particles that were in this cut at the initial time moment will
move identically and form at any time moment $t > 0$ plane cut (called
fluid cut). The Eulerian coordinate of this cut $x_A(t)$ is changing with time,
corresponding to the motion of the gas particles. The particles which were
on the left from cut A can not be on the right from cut A during the
movement because of the assumption about the one-dimensional nature of
the process. Therefore the mass of medium which is left from cut A will be
constant during the process and equal to

$$s_A = \int_{x_0}^{x_A(t)} \rho(y,t) \cdot 1 \cdot 1 \, dy = \int_{x_0}^{x_A(0)} \rho(y,0) \cdot 1 \cdot 1 \, dy. \qquad (1.35)$$

The product of the units here correspond to the unit area of the cross-
section of the parallelepiped. Later, we will omit this product. To avoid
any misunderstanding related to this dimension, we will consider that s
is a mass which corresponds to the unit of the cross-section and has the
dimension *gram/centimeter²*.

Let us consider the general formula

$$s = \int_{x_0}^{x} \rho(y,t) \, dy, \qquad (1.36)$$

which gives us a correspondence between x and s. When x goes from x_0
to $x_v(t)$, then s changes from 0 to M. If we fix x, then function $s(t)$ given
by 1.36 will present a mass which occupies the left with respect to the cut
$x = const$ part of the parallelepiped. If we fix s ($0 \leq s \leq M$), then the
upper limit in 1.36 will be the function of time $x(t)$, and this function will

correspond to the trajectory of the particles from the left where mass s is contained.

Therefore, for each fluid particle, s is constant and is not changed during movement. This means that s can be chosen as a Lagrangian coordinate. And because s has the dimension of mass, it is called the *Lagrangian mass coordinate*.

Using coordinate s we can write equations for gas dynamics in the following form:

$$\frac{d}{dt}\left(\frac{1}{\rho}\right) = \frac{\partial W}{\partial s}, \tag{1.37}$$

$$\frac{dW}{dt} = -\frac{\partial p}{\partial s}, \tag{1.38}$$

$$\frac{d}{dt}\left(\varepsilon + \frac{W^2}{2}\right) = -\frac{\partial}{\partial s}(pW) - \frac{\partial H}{\partial s}, \tag{1.39}$$

$$\frac{dx}{dt} = W, \tag{1.40}$$

$$p = p(\rho, T), \quad \varepsilon = \varepsilon(\rho, T). \tag{1.41}$$

1.5 Statements for Gas Dynamics Problems in Lagrange Variables

The advantages of using the Lagrangian approach for solutions to gas dynamics equations are related to the statement for some boundary conditions, and also for treating problems where many different gases must be considered.

Let us consider an example of the statement for the boundary condition for the problem of gas elapsing to the vacuum. In this case, the domain of variable s is known in advance as $(0 \leq s \leq M)$ and does not change with time. Therefore, the boundary conditions look as follows:

$$W(0, t) = 0, \quad p(M, t) = 0. \tag{1.42}$$

That is, the left wall does not move and pressure on the interface of the gas and vacuum is equal to zero. In the Eulerian coordinates, we do not know the position of the interface between the gas and vacuum, and these unknown coordinates participate in the statement of the conditions for pressure and lead to additional difficulties. In general, it is convenient to solve one-dimensional problems, regarding the movement of gas, which occupy some finite domain and where the mass remains constant by using the Lagrangian variable.

It is also convenient to use the Lagrangian variables for problems where many different gases need to be considered. In these cases, the interfaces between the different gases have the given values of mass variables. In the

Eulerian coordinates, these interfaces move in space and special procedures must be used for tracking them.

A general discussion about the advantages and disadvantages of using the Lagrangian and Eulerian approaches for solving equations for continuum media and examples of concrete problems can be found in [114].

1.6　Different Forms of Energy Equations

For future consideration, it is important to introduce some different forms of energy equations. If we multiply the equation of motion by W, we get the following equation:

$$\frac{d}{dt}\left(\frac{W^2}{2}\right) = -W\frac{\partial p}{\partial s}, \tag{1.43}$$

which determines the variation in time for kinetic energy. If we subtract this equation from the equation for energy 1.39 (where $H = 0$), we get

$$\frac{d\varepsilon}{dt} = -p\frac{\partial W}{\partial s}. \tag{1.44}$$

If we use the continuity equation, we get

$$\frac{d\varepsilon}{dt} = -p\frac{d}{dt}\left(\frac{1}{\rho}\right). \tag{1.45}$$

Equations 1.44 and 1.45 have a direct physical sense where they show that the variation of internal energy occurs due to the work of the pressure force.

Using the second principle of thermodynamics and 1.45, we get

$$T\frac{dS}{dt} = \frac{d\varepsilon}{dt} + p\frac{d}{dt}\left(\frac{1}{\rho}\right) = 0, \tag{1.46}$$

where $S(\rho, T)$ is a specific entropy. Therefore, for smooth adiabatic flows, we get another formulation of the energy equation:

$$\frac{dS}{dt} = 0. \tag{1.47}$$

All previous considerations are valid for the general equations of state.

Let us now consider the ideal gas. If we express internal energy in terms of pressure and density, we get

$$\varepsilon = \frac{1}{\gamma - 1}\frac{p}{\rho}. \tag{1.48}$$

From this formula and equation 1.45, we can obtain the following equation:

$$\frac{d\varepsilon}{dt} + p\frac{d}{dt}\left(\frac{1}{\rho}\right) = \frac{\rho^{\gamma-1}}{\gamma - 1}\frac{d}{dt}\left(\frac{p}{\rho^\gamma}\right) = 0. \tag{1.49}$$

Therefore, for adiabatic equilibrium flows of the ideal gas for each fluid particle, the ratio p/ρ^γ does not change with time:

$$\frac{p}{\rho^\gamma} = const.$$ (1.50)

From previous equations, we can obtain an explicit formula for the entropy of the ideal gas

$$S = \frac{R}{\gamma - 1} \ln \frac{p}{\rho^\gamma}.$$ (1.51)

If at some moment, all fluid particles have the same entropy, then corresponding to 1.47, they will have the same entropy for all moments of time. Such flows where the entropy of all particles is the same for all time are called *isentropic*.

The different formulations for the energy equation were obtained by assuming that there are no dissipative processes. If there is heat conductivity, then entropy of the particles will vary. This variation can be described by the equation

$$\frac{dS}{dt} = -\frac{1}{T}\frac{\partial H}{\partial s}, \quad H = -\kappa(\rho, T)\rho\frac{\partial T}{\partial s}.$$ (1.52)

Real gases always have some heat conductivity. Therefore adiabatic flows can be considered only as a limited case when the coefficient κ is small.

There is an opposing situation when $\kappa \to 0$. Because the value of heat flux H remains finite, the temperature must be homogeneous in space: $\partial T/\partial s \to 0$. From a physical point of view this is evident, and because of high conductivity, the inhomogeneity in temperature actively smoothes out. In practice, this situation occurs in a phenomenon which takes place with a very high temperature because the coefficient of heat conductivity, $\kappa(\rho, T)$, increases very rapidly with temperature.

This type of flow is called *isothermal* and in this case equation

$$T = const.$$ (1.53)

plays the role of energy equation. The equation of state for the ideal gas in an isothermal case is

$$\frac{p}{\rho} = A,$$ (1.54)

where A is constant.

If we compare the last equation with 1.50, we can conclude that some results for isothermal flows for the ideal gas can be obtained formally from corresponding formulas for the adiabatic case if we use $\gamma = 1$.

Thus, we obtained some different forms for the energy equation for gas dynamics flows. For smooth flows all of these forms are equivalent in the sense that they can be obtained from each other by some transformation.

This means that for the statement and solution for real problems, one can use any of these forms. From a physical point of view, the meaning of each form can be different in that each form corresponds to some physical law. Also, some forms may be more convenient from a computational point of view.

1.7   Acoustic Equations

Equations for gas dynamics are non-linear (quasi-linear) equations and this fact is the reason for the primary difficulties of their investigation and numerical solutions for particular problems. Until now, there was no proof for the existence and uniqueness of solutions to gas dynamics equations for general cases. This is the reason why the properties of these equations are investigated for more simple, special cases. The investigations of these examples give information about qualitative phenomenon for gas dynamics flows. One widespread special case is the *acoustic approximation*.

The acoustics are dealt with using "small" motions in the gas. That is, motions where all the gas parameters have small deviations from some initial value. This assumption gives us the ability to make linearization of the original non-linear equation.

Let us consider the unbounded region which is filled with a homogeneous gas at rest and consider a case where there is no viscosity or heat conductivity.

We will denote the initial values of the gas parameters as p_0, ρ_0, and $\vec{W}_0 = 0$. Then a small perturbation is introduced. The problem is to investigate the behavior of this small perturbation. It is natural to expect that the perturbation, which is small at the initial moment, will be small all the time. This means that the solution to the problem can be found in the form of a small deviation from the initial values:

$$p = p_0 + \tilde{p}, \quad \rho = \rho_0 + \tilde{\rho}. \tag{1.55}$$

This condition, where the deviation of parameters are small, means that

$$\left| \frac{\tilde{p}}{p} \right| \ll 1, \quad \left| \frac{\tilde{\rho}}{\rho} \right| \ll 1. \tag{1.56}$$

Similar inequalities are assumed to be valid not only for the functions but also for its derivatives.

If we substitute expression 1.55 into the continuity equation and take into account that the terms of the second order of smallness may be neglected, we get

$$\frac{d\tilde{\rho}}{dt} + \rho_0^2 \frac{\partial \tilde{W}}{\partial s} = 0. \tag{1.57}$$

Now if we consider pressure p as a function of the two parameters, density and temperature, we can write

$$\frac{dp}{dt} = \left(\frac{dp}{d\rho}\right)_S \frac{d\rho}{dt} + \left(\frac{\partial p}{\partial S}\right) \frac{dS}{dt} = c^2 \frac{d\rho}{dt}, \tag{1.58}$$

where

$$c^2 = \left(\frac{dp}{d\rho}\right)_S .$$

(As is usual in thermodynamics, subscript S or ρ by the derivative means that this derivative is taken when this parameter is fixed.) The second term in the right-hand side is omitted because the movement of gas in acoustics is adiabatic and the energy equation can be written in the form

$$\frac{dS}{dt} = 0.$$

Using 1.58, we can rewrite equation 1.57 as follows:

$$\frac{d\tilde{p}}{dt} = -c_0^2 \rho_0^2 \frac{\partial \tilde{W}}{\partial s}, \tag{1.59}$$

where subscript 0 means that the value of this parameter corresponds to the initial data.

The linearization of the momentum equation gives us the following result:

$$\frac{d\tilde{W}}{dt} = -\frac{\partial \tilde{p}}{\partial s}. \tag{1.60}$$

The system of equations 1.59 and 1.60 are the acoustics equations.

If we eliminate one of the unknown functions, for example, $\tilde{W}$, we obtain the simplest equation of the hyperbolic type for the description of the behavior for the deviation of pressure:

$$\frac{d\tilde{p}}{dt} = (c_0 \rho_0)^2 \frac{\partial^2 \tilde{p}}{\partial s^2} . \tag{1.61}$$

The other variables satisfy similar equations.

The solution to this equation can be found by decomposing the solution into traveling waves:

$$\tilde{p}(s, t) = f_1(s - c_0 \rho_0 t) + f_2(x + c_0 \rho_0 t), \tag{1.62}$$

where f_1 and f_2 are determined by the form of the initial perturbations. The physical meaning of the solution given by formula 1.62 is as follows: the small perturbation travels from left to right with speed $c_0 \rho_0$ without any distortion. In gas dynamics, by definition, the speed of propagation of the small perturbations is called the *speed of sound*. In our case, when we use the Lagrangian mass variable, the speed $a = c_0\,\rho_0$ is the mass speed of sound. If we will use the Eulerian coordinates, then the speed of sound role will be played by c_0. For adiabatic flows for an ideal gas c is

$$c = \sqrt{\gamma\,R\,T}\,. \tag{1.63}$$

All the results are valid for the isothermal case if we formally use $\gamma = 1$. Then the *isothermal speed of sound* is

$$c_T = \sqrt{R\,T}\,. \tag{1.64}$$

As it follows from formula 1.62, perturbation does not increase with time. It is a justified assumption that the solution at all times is small if the initial perturbation is small.

The straight lines

$$x = \pm\,c_0\,\rho_0\,t + const \tag{1.65}$$

on plane x, t, on which the perturbations are travelling, are called the *characteristics* (see the general definition of characteristics described later in this book).

When we consider a plane travel wave, the gas particles stay at rest until a disturbance causes them to move. The particles stop at the moment the disturbance goes away. The direction of particle displacement can be found using equations 1.60 and 1.62. For example, for a disturbance that is moving to the right, we get

$$\tilde{W} = \frac{c_0}{\rho_0}\,\tilde{\rho}\,. \tag{1.66}$$

In other words, in a compression wave, particles are displaced in the direction that coincides with the direction of wave propagation, and in a rarefaction wave, particles are displaced in a direction that is opposite to the direction of wave propagation. Using equation 1.66, we can estimate the speeds at which acoustic approximations are valid:

$$|\tilde{W}/c_0| = |\tilde{\rho}/\rho_0| \ll 1\,. \tag{1.67}$$

That is to say, acoustic approximation can be used if the disturbance in the velocity of the particles is small with respect to the speed of sound.

1.8   Reference Information

In this section we recall some definitions for the system of quasi-linear equations of the first order:

$$\frac{\partial \vec{u}}{\partial t} + A\,\frac{\partial \vec{u}}{\partial x} = \vec{b}\,, \tag{1.68}$$

where
$$\vec{u}(x,t) = \{u_1(x,t), \ldots, u_n(x,t)\}$$
is an unknown vector-function that is formed by n scalar functions,
$$A = \{a_{i,j}(x,t,\vec{u}), \ i,j = 1,2,\ldots n\},$$
is a matrix of order n, and
$$\vec{b} = \{b_1(x,t,\vec{u}), \ldots, b_n(x,t,\vec{u})\}$$
is a vector of the right-hand side.

System 1.68 is quasi-linear because matrix A and vector $\vec{b}$, in general, depend on solution $\vec{u}$.

The number $\lambda(x,t,\vec{u})$, is called the *eigenvalue*, and vector $\vec{l} = \{l_1, \ldots, l_n\}$ is called the *left eigenvector* for matrix A if

$$\vec{l}A = \lambda\vec{l} \tag{1.69}$$

and it is assumed that at least one component of vector $\vec{l}$ is not zero. As it follows from 1.69, to find all eigenvalues λ, we need to solve the equation of the n-th degree:

$$\det(A - \lambda E), \tag{1.70}$$

where E is the identity matrix.

System 1.68 is called *strictly hyperbolic* in the region of space $(x,t,\vec{u})$, if everywhere in this region, eigenvalues $\lambda_1, \ldots, \lambda_n$ of matrix A are real and different. Because in this book we will consider only case, of strictly hyperbolic system we, for simplicity, will call it hyperbolic.

Suppose that system 1.68 is hyperbolic. Let us multiply it from the left side by eigenvector $\vec{l}^{(k)}$, which corresponds to eigenvalue λ_k. Taking into account equation 1.69, we get:

$$\vec{l}^{(k)}\frac{\partial\vec{u}}{\partial t} + \vec{l}^{(k)}A\frac{\vec{u}}{\partial x} = \vec{l}^{(k)}\left(\frac{\partial\vec{u}}{\partial t} + \lambda_k\frac{\partial\vec{u}}{\partial x}\right) = \vec{l}^{(k)}\vec{b} = f_k. \tag{1.71}$$

The expression in braces in 1.71 is the derivative along the line which is determined by equation

$$\frac{dx}{dt} = \lambda(x,t,\vec{u}), \quad k = 1,2,\ldots,n. \tag{1.72}$$

Then, we will use the following notation:

$$\partial/\partial t + \lambda_k\,\partial/\partial x = (d/dt)_k. \tag{1.73}$$

For example, if
$$dx/dt = W,$$

then
$$d/dt = \partial/\partial t + W\,\partial/\partial x$$
and differentiation is made on the trajectory of the particle.

Lines, which are determined by the equations in 1.72, are called *characteristics*. The equations in 1.71, which are obtained by transformation of the original system of equations using 1.72, can be rewritten in the following form:

$$\vec{l}^{(k)}\left(\frac{d\vec{u}}{dt}\right)_k = f_k, \quad k = 1, 2, \ldots, n\,. \tag{1.74}$$

These equations are called the *characteristic form* of the system 1.68. The main feature of this system is that in each equation in 1.74, the differentiation is made along only one characteristic.

1.9   The Characteristic Form of Gas Dynamics Equations

Let us now investigate the system of gas dynamics equations for the one-dimensional case in the Lagrangian mass variables. We can express the derivative of pressure with respect to s as follows:

$$\frac{\partial p}{\partial s} = \left(\frac{\partial p}{\partial \rho}\right)_S \cdot \frac{\partial \rho}{\partial s} + \left(\frac{\partial p}{\partial S}\right)_\rho \cdot \frac{\partial S}{\partial s} = -c^2\rho^2\frac{\partial}{\partial s}\left(\frac{1}{\rho}\right) + \alpha\frac{\partial S}{\partial s}\,, \tag{1.75}$$

where $c = \sqrt{(\partial p/\partial \rho)_S}$ - speed of sound, $\alpha = (\partial p/\partial S)_\rho$.

Let us now rewrite the main equation for the gas dynamics equation using 1.75. If $H = 0$, we get

$$\frac{\partial}{\partial t}\left(\frac{1}{\rho}\right) - \frac{\partial W}{\partial s} = 0\,, \tag{1.76}$$

$$\frac{\partial W}{\partial t} - c^2\rho^2\frac{\partial}{\partial s}\left(\frac{1}{\rho}\right) + \alpha\frac{\partial S}{\partial s} = 0\,, \tag{1.77}$$

$$\frac{\partial S}{\partial s} = 0\,. \tag{1.78}$$

If we introduce notations

$$\vec{u} = \begin{pmatrix} 1/\rho \\ W \\ S \end{pmatrix}, \quad A = \begin{pmatrix} 0 & -1 & 0 \\ -c^2\rho^2 & 0 & \alpha \\ 0 & 0 & 0 \end{pmatrix}, \tag{1.79}$$

then systems 1.76, 1.77, and 1.78 can be rewritten in the form of 1.68 where the role of x is played by s, and vector $\vec{b}$ is equal to zero. To find the eigenvalues, we must solve the equation

$$\det\left(A - \lambda\,E\right) = -\lambda\left(\lambda^2 - c^2\rho^2\right) = 0\,, \tag{1.80}$$

which gives us the following eigenvalues:

$$\lambda_1 = 0, \quad \lambda_2 = c\rho, \quad \lambda_3 = -c\rho. \tag{1.81}$$

These values are real and different from each other if condition $(\partial p/\partial \rho)_S > 0$ is satisfied, which is true for most practical and interesting materials. Therefore, the system of one-dimensional equations for gas dynamics is a system of hyperbolic type.

The equations

$$\frac{ds}{dt} = 0, \quad \frac{ds}{dt} = c\rho, \quad \frac{ds}{dt} = -c\rho \tag{1.82}$$

define the characteristics of gas dynamics equations.

Let us note that for the Eulerian variables equations, the characteristics are as follows:

$$\frac{dx}{dt} = 0, \quad \frac{dx}{dt} = W + c, \quad \frac{dx}{dt} = W - c \tag{1.83}$$

To find all the eigenvectors, we need to solve equation 1.69 for all three values of λ. Finally, we get

$$\vec{l}^{(1)} = \left\{ \begin{array}{c} 0 \\ 0 \\ 1 \end{array} \right\}, \quad \vec{l}^{(2)} = \left\{ \begin{array}{c} -c\rho \\ 1 \\ \alpha/(c\rho) \end{array} \right\}, \quad \vec{l}^{(3)} = \left\{ \begin{array}{c} c\rho \\ 1 \\ -\alpha/(c\rho) \end{array} \right\}. \tag{1.84}$$

Using these vectors, we construct the *characteristic form of gas dynamics equations*:

$$\frac{\partial S}{\partial t} = 0, \tag{1.85}$$

$$\left(\frac{\partial W}{\partial t} + c\rho \frac{\partial W}{\partial s} \right) - c\rho \left[\frac{\partial}{\partial t} \left(\frac{1}{\rho} \right) + c\rho \frac{\partial}{\partial s} \left(\frac{1}{\rho} \right) \right] +$$

$$\frac{\alpha}{c\rho} \left(\frac{\partial S}{\partial t} + c\rho \frac{\partial S}{\partial s} \right) = 0, \tag{1.86}$$

$$\left(\frac{\partial W}{\partial t} - c\rho \frac{\partial W}{\partial s} \right) + c\rho \left[\frac{\partial}{\partial t} \left(\frac{1}{\rho} \right) - c\rho \frac{\partial}{\partial s} \left(\frac{1}{\rho} \right) \right] -$$

$$\frac{\alpha}{c\rho} \left(\frac{\partial S}{\partial t} - c\rho \frac{\partial S}{\partial s} \right) = 0. \tag{1.87}$$

1.10   Riemann's Invariants

Let us consider isentropic flows where $S = const$, then the characteristic form of equations is

$$\left(\frac{\partial W}{\partial t} + c\rho \frac{\partial W}{\partial s} \right) - c\rho \left[\frac{\partial}{\partial t} \left(\frac{1}{\rho} \right) + c\rho \frac{\partial}{\partial s} \left(\frac{1}{\rho} \right) \right] = 0, \tag{1.88}$$

$$\left(\frac{\partial W}{\partial t} - c\rho \frac{\partial W}{\partial s} \right) + c\rho \left[\frac{\partial}{\partial t} \left(\frac{1}{\rho} \right) - c\rho \frac{\partial}{\partial s} \left(\frac{1}{\rho} \right) \right] = 0, \tag{1.89}$$

which follows from the general case, and equation 1.85 is satisfied automatically.

If we introduce the function

$$\varphi(\rho) = \int_{\rho_0}^{\rho} c\rho \, d\left(\frac{1}{\rho}\right) = \int_{\rho_0}^{\rho} \frac{c}{\rho} \, d\rho, \tag{1.90}$$

then it is only a function of ρ because we consider the isentropic case $S = S_0 = const$ and $c(\rho, S) = c(\rho, S_0) = c(\rho)$. Using this function, we can make changes to unknown functions and instead of using W and $1/\rho$, we introduce two new functions:

$$r^+(W, \rho) = W - \varphi(\rho) = W + \int_{\rho_0}^{\rho} \frac{c}{\rho} \, d\rho, \tag{1.91}$$

$$r^-(W, \rho) = W + \varphi(\rho) = W - \int_{\rho_0}^{\rho} \frac{c}{\rho} \, d\rho. \tag{1.92}$$

Functions r^+ and r^- are called *Riemann's invariants*. Using this notion, we can rewrite the characteristic form of gas dynamics equations in a more compact form:

$$\frac{\partial r^+}{\partial t} + c\rho \frac{\partial r^+}{\partial s} = 0, \tag{1.93}$$

$$\frac{\partial r^-}{\partial t} - c\rho \frac{\partial r^-}{\partial s} = 0. \tag{1.94}$$

This system of gas dynamics equations is Riemann's invariants.

The second and third characteristics are usually called C^+ *and* C^- *characteristics of gas dynamics equations.* We can write these as

$$r^+ = const \text{ along } C^+ : \frac{ds}{dt} = +c\rho, \tag{1.95}$$

$$r^- = const \text{ along } C^- : \frac{ds}{dt} = -c\rho. \tag{1.96}$$

The invariants $r^{\pm}$ are determined up to the constants. Therefore, the lower limit ρ_0 in the definition of function $\varphi(\rho)$ can be chosen as the value of density in an arbitrary fixed particle.

For the ideal gas, we can write an explicit form for the Riemann invariants:

$$r^+ = W + \frac{2}{\gamma - 1} c, \quad r^- = W - \frac{2}{\gamma - 1} c. \tag{1.97}$$

Riemann invariants have graphic interpretation. Let us consider the gas dynamics flow, for which the initial state is given. This means that for $t = 0$ (see Figure 1.6), we know the distribution of all the physical parameters in space, and in particular, we know the gas velocity $W(s, 0)$

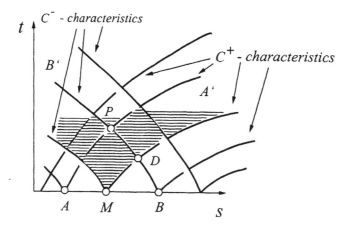

Figure 1.6: *Domain of dependence and range of influence.*

and speed of sound $c(s, 0)$. Therefore, using the formulas in 1.97, we can also compute the initial values for invariants r^+ and r^-. There is a grid formed by C^+ and C^- characteristics on the plane (s, t) because only two characteristics from these families intersect each point on the plane (x, t) (see Figure 1.6). On each characteristic, the value of each corresponding invariant is transferred without changing: on the C^+ characteristic AA', corresponding invariant r^+ is constant, and on the C^- characteristic BB', invariant r^- is constant. Hence, we can write the following equations:

$$r_P^+ = W_P + \frac{2}{\gamma - 1} c_P = r_A^+ = W_A + \frac{2}{\gamma - 1} c_A, \qquad (1.98)$$

$$r_P^- = W_P - \frac{2}{\gamma - 1} c_P = r_B^- = W_B - \frac{2}{\gamma - 1} c_B, \qquad (1.99)$$

where the subscript shows at what point this function is taken (see Figure 1.6). Using previous equations, we can find the values of W_P and c_P in any point P:

$$W_P = \frac{1}{2} (W_A + W_B) + \frac{c_A - c_B}{\gamma - 1},$$

$$\qquad (1.100)$$

$$c_P = \frac{\gamma - 1}{4} (W_A - W_B) + \frac{1}{2} (c_A + c_B).$$

If we know the gas velocity W_P and the speed of sound c_P, we can find values of all the other parameters in point P.

However, from the equations in 1.100, we cannot conclude that the parameters in point P are completely determined by the initial values in only two points A, B. In fact, point P is the intersection of two characteristics, which are dependent on the solution and are not known beforehand. The behavior of the characteristics, that is the slope in each point (for example, the slope of characteristic BB' in point D), is determined by the parameter of the gas in this point. In its turn, the values of the gas parameters in this point depend on Riemann's invariants r^+ and r^-, which come to this point from points M and B. Finally, the position of node P on the plane (s, t) and the values of the gas parameters in this point depend on the initial values on segment AB and do not depend on the values of the gas parameters outside of segment AB. In this sense, segment AB is called the *domain of dependence of point* P. The shaded "curvilinear angle" in figure 1.6 is called the *range of influence of point* M. That is the zone where the state of the gas depends on the parameters in point M (but is not completely determined by them).

1.11 Discontinuous Solutions

As we mentioned in the introduction to this chapter, gas dynamics flows are frequently characterized by internal discontinuities, such as shocks. In this section we present derivation of special boundary conditions on such discontinuities.

Let us consider the discontinuous solutions for gas dynamics equations. To do this, we need to consider the integral form of equations which allows discontinuous solutions.

Suppose that the flow under consideration is one-dimensional and we consider the case of Cartesian coordinates where all the parameters depend only on one coordinate x. The surface where the flow parameters have discontinuity is a plane which moves in space with velocity $\mathcal{D}(t)$.

For the description of such a flow, we can use integral equations 1.23, 1.24, and 1.25. If we omit terms related to the external forces, sources of energy, and heat fluxes, we get:

$$\oint_C (\rho \, dx - \rho W \, dt) = 0, \tag{1.101}$$

$$\oint_C \left[\rho W \, dx - (p + \rho W^2) \, dt \right] = 0, \tag{1.102}$$

$$\oint_C \left\{ \rho \left(\varepsilon + \frac{W^2}{2} \right) dx - \left[\rho W \left(\varepsilon + \frac{W^2}{2} + \frac{p}{\rho} \right) \right] dt \right\} = 0, \tag{1.103}$$

where C is a contour in the (x, t) plane. Note that the direction of gas velocity W is perpendicular to the plane of discontinuity. Let us choose contour C as $ABB'A'$ in Figure 1.7.

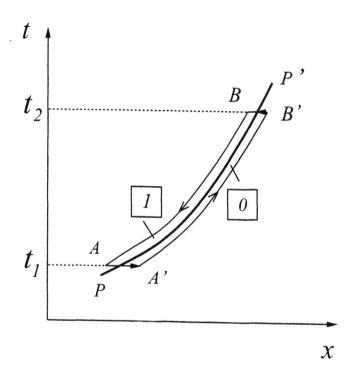

Figure 1.7: *Derivation of Hugoniot conditions.*

Line PP' is the trajectory of discontinuity $x(t)$. Therefore, on this line we have

$$\frac{dx}{dt} = \mathcal{D}(t).$$ (1.104)

The values of the parameters, which correspond to the right and left from discontinuity, will be denoted by subscripts 0 and 1 respectively. Our goal is to obtain relations which connect these values. To do this, we write equations 1.101, 1.102, and 1.103 for contour $ABB'A'$ and bring together sides $A'B'$ and AB, but keeping the contour on different sides from the discontinuity line $x(t)$.

Integrals over the top and bottom bases are going to zero such that, the first equation in the limit comes

$$\int_{P'}^{P} (\rho_1 \, dx - \rho_1 \, W_1 \, dt) + \int_{P}^{P'} (\rho_0 \, dx - \rho_0 \, W_0 \, dt) = 0.$$ (1.105)

Taking into account that on the discontinuity line PP'

$$dx = \mathcal{D} \, dt,$$

we can rewrite 1.105 as follows:

$$-\int_{t_1}^{t_2} (\rho_1 \mathcal{D} - \rho_1 \, W_1) \, dt + \int_{t_1}^{t_2} (\rho_0 \mathcal{D} - \rho_0 \, W_0) \, dt = 0.$$ (1.106)

Now because of arbitrary t_1, t_2, we get

$$\rho_1 (W_1 - \mathcal{D}) = \rho_0 (W_0 - \mathcal{D}).$$ (1.107)

Similarly, we get

$$\rho_1 (W_1 - \mathcal{D})^2 + p_1 = \rho_0 (W_0 - \mathcal{D})^2 + p_0,$$ (1.108)

$$\rho_1 (W_1 - \mathcal{D}) \left(\varepsilon_1 + \frac{p_1}{\rho_1} + \frac{(W_1 - \mathcal{D})^2}{2} \right) =$$

$$\rho_0 (W_0 - \mathcal{D}) \left(\varepsilon_0 + \frac{p_0}{\rho_0} + \frac{(W_0 - \mathcal{D})^2}{2} \right).$$ (1.109)

Equations 1.107, 1.108, and 1.109 are the general form of relations which connect the values of parameters on both sides of the discontinuity and speed of propagation of the discontinuity front.

These relations are called *Hugoniot relations*, which express the laws of conservation for fluxes of mass, momentum, and energy through the discontinuity surface.

Sometimes it is useful to have these relations in a coordinate system which is moving with discontinuity with speed $\mathcal{D}$. In this case, all velocities are transformed as

$$u = W - \mathcal{D}$$

and equations 1.107, 1.108, and 1.109 become

$$\rho_1 u_1 = \rho_0 u_0 , \tag{1.110}$$

$$\rho_1 u_1^2 + p_1 = \rho_0 u_0^2 + p_0 , \tag{1.111}$$

$$\rho_1 u_1 \left(\varepsilon_1 + \frac{p_1}{\rho_1} + \frac{u_1^2}{2} \right) =$$

$$\rho_0 u_0 \left(\varepsilon_0 + \frac{p_0}{\rho_0} + \frac{u_0^2}{2} \right) . \tag{1.112}$$

1.11.1 Contact Discontinuity

Expression $m = \rho_0 u_0 = \rho_1 u_1$ is the flux of mass through discontinuity. Depending on the value of this flux, discontinuities in gas dynamics can be divided into two groups.

If

$$m = 0 , \tag{1.113}$$

then discontinuity is called *contact discontinuity*. The condition $m = 0$ means that there is no flux of mass through discontinuity. Because of the physical meaning, densities ρ_1 and ρ_0 are not equal to zero, and condition 1.113 can be satisfied only if $u_1 = u_0 = 0$. This means that

$$W_1 = W_0 = \mathcal{D} . \tag{1.114}$$

That is, the discontinuity line on the phase plane coincides with the trajectory of the particle. Therefore, contact discontinuity does not move through the mass, is related to the same particles and moves with them. This is the reason why it is convenient to consider contact discontinuity in Lagrangian variables.

Because of 1.113, the first and last of the Hugoniot relations are satisfied automatically and the second relation gives

$$p_1 = p_0 . \tag{1.115}$$

Thus, on the surface of the contact discontinuity, the pressure and normal component of velocity must be continuous. Other functions have a jump.

Let us note that if there are different materials which have different equations of state, then the interfaces between them are contact discontinuities.

1.11.2 Shock Waves. The Hugoniot Adiabat

If $m \neq 0$ then the discontinuity is called *shock waves*.

Let us rewrite Hugoniot conditions 1.110, 1.111, and 1.112 in the following form:

$$\frac{\eta_0}{\eta_1} = \frac{u_0}{u_1}, \tag{1.116}$$

$$\frac{u_0^2}{\eta_0} - \frac{u_1^2}{\eta_1} = p_1 - p_0, \tag{1.117}$$

$$\varepsilon_1 - \varepsilon_0 = \frac{u_0^2 - u_1^2}{2} + p_0 \eta_0 - p_1 \eta_1, \tag{1.118}$$

where $\eta = 1/\rho$.

If we express u_1 from equation 1.116 and substitute it in 1.117, we get

$$u_0^2 = \eta_0^2 \frac{p_1 - p_0}{\eta_0 - \eta_1}. \tag{1.119}$$

Similarly, we get

$$u_1^2 = \eta_1^2 \frac{p_1 - p_0}{\eta_0 - \eta_1}. \tag{1.120}$$

Using these formulas, we can find the jump for specific kinetic energy:

$$\frac{1}{2}\left(u_0^2 - u_1^2\right) = \frac{(p_1 - p_0)\left(\eta_0^2 - \eta_1^2\right)}{2\left(\eta_0 - \eta_1\right)} = \frac{1}{2}\left(p_1 - p_0\right)\left(\eta_1 + \eta_0\right). \tag{1.121}$$

Now from this formula and equation 1.118, we get the following relation:

$$\varepsilon_1(p_1, \eta_1) - \varepsilon_0(p_0, \eta_0) = \frac{1}{2}\left(p_1 + p_0\right)\left(\eta_0 - \eta_1\right), \tag{1.122}$$

which is called the *Hugoniot adiabat*. This equation connects the values of the thermodynamical parameters of a gas (pressure and specific volume) from both sides of the discontinuity.

If a shock wave propagates through the gas with a given state (parameters p_0 and η_0 are given), then to determine the parameters of the gas beyond the front of discontinuity, we need to specify one of the parameters p_1 or η_1.

1.11.3 Shock Waves in an Ideal Gas

Let us consider a special case of ideal gas where

$$\varepsilon = \frac{p\,\eta}{\gamma - 1}.$$

Using this relation, we can transform the Hugoniot adiabat to one of the following forms:

$$\frac{p_1}{p_0} = \frac{(\gamma + 1)\,\eta_0 - (\gamma - 1)\,\eta_1}{(\gamma + 1)\,\eta_1 - (\gamma - 1)\,\eta_0} \tag{1.123}$$

or

$$\frac{\eta_1}{\eta_0} = \frac{(\gamma + 1)\,p_0 - (\gamma - 1)\,p_1}{(\gamma + 1)\,p_1 - (\gamma - 1)\,p_0}. \tag{1.124}$$

For future consideration, we will derive the formulas for comparison of velocity u and speed of sound c. For an ideal gas $c^2 = \gamma p \eta$, we can use 1.119 and find the ratio of the squares of these velocities:

$$\frac{u_0^2}{c_0^2} = \frac{\eta_0^2\,(p_1 - p_0)}{\gamma\,p_0\,\eta_0\,(\eta_0 - \eta_1)} = \frac{p_1 - p_0}{\gamma\,p_0\,(1 - \eta_1/\eta_0)}. \tag{1.125}$$

If we eliminate the ratio of specific volumes using 1.124, we get

$$\frac{u_0^2}{c_0^2} = \frac{1}{2\gamma}\left[(\gamma - 1) + (\gamma + 1)\frac{p_1}{p_0}\right]. \tag{1.126}$$

And in a similar way, we can obtain

$$\frac{u_1^2}{c_1^2} = \frac{1}{2\gamma}\left[(\gamma - 1) + (\gamma + 1)\frac{p_0}{p_1}\right]. \tag{1.127}$$

Now using all these formulas, we can describe the properties of shock waves in an ideal gas. The intensity of the shocks is defined by its amplitude. That is, by the magnitude of jump for the gas parameters. Let us assume that the initial state of the gas is given by the parameters p_0 and η_0. Then, the graphical representation of the Hugoniot adiabat on the plane of thermodynamical state (p, η) (line $\mathcal{H}$ in Figure 1.8), is the geometric locus where the gas can be transformed from its initial state p_0, η_0 by the shock wave. If we specify the specific volume η_1 beyond the front of the shock (by this we determine the intensity of the shock $\eta_0 - \eta_1$) by using the Hugoniot adiabat, we can determine the pressure p_1 and then all the other parameters.

The adiabat $\mathcal{H}$ on plane (p, η) has horizontal and vertical asymptotes. In fact, as it follows from 1.123, when $\eta_1/\eta_0 \to \infty$, the ratio of the pressures is $p_1/p_0 \to -(\gamma - 1)/(\gamma + 1)$. Similarly, from 1.124, it follows that if $p_1/p_0 \to \infty$, then the ratio of the specific volumes is going to the finite value

$$\frac{\eta_1}{\eta_0} = \frac{\gamma - 1}{\gamma + 1}. \tag{1.128}$$

From a physical point of view, this means that we cannot compress gas more than $(\gamma + 1)/(\gamma - 1)$ times by using shock waves.

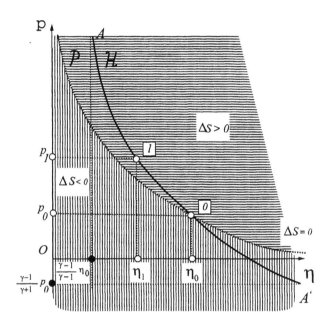

Figure 1.8: *The Hugoniot and Poisson adiabat.*

Now let us consider another limiting case where $p_1/p_0 \to 1$. We have $\eta_1/\eta_0 \to 1$ and consequently, the state of the gas, beyond of the shock front, differs slightly from the initial state p_0, η_0. In particular, if $c_0 \to c_1$ we have

$$\frac{u_0^2}{c_0^2} \to 1, \quad \frac{u_1^2}{c_1^2} \to 1. \tag{1.129}$$

Therefore, the shock wave of very small intensity $((\eta_0 - \eta_1) \to 0)$ propagates with the speed of sound.

1.11.4 Zemplen Theorem

The Hugoniot adiabat consists of two branches: OA for $\eta < \eta_0$ and OA' for $\eta > \eta_0$. The signs of the increments $\Delta p = p_1 - p_0$ and $\Delta \eta = \eta_1 - \eta_0$ on the adiabat are opposite each other. Thus, branch OA corresponds to shock waves where pressure ($\Delta p > 0$) and compression ($\Delta \eta < 0, \Delta \rho > 0$) increase, but branch OA' corresponds to shock waves where pressure ($\Delta p < 0$) and rarefaction ($\Delta \eta > 0, \Delta \rho < 0$) decrease. For compression shocks ($p_1/p_0 > 1$), from 1.126 and 1.127, we get

$$\frac{u_0^2}{c_0^2} > 1, \quad \frac{u_1^2}{c_1^2} < 1. \tag{1.130}$$

For rarefaction shocks we have the opposite relations

$$\frac{u_0^2}{c_0^2} < 1, \quad \frac{u_1^2}{c_1^2} > 1. \tag{1.131}$$

Let us now consider the variation of entropy in shock waves. Let us draw a line on plane (p, η) through point (p_0, η_0) where entropy remains constant and equal to S_0, that is, the value in point (p_0, η_0). Such a line is called the *Poisson adiabat* and its equation follows from 1.50:

$$p_1 \eta_1^\gamma = p_0 \eta_0^\gamma. \tag{1.132}$$

This dependence is shown graphically in Figure 1.8 (line $\mathcal{P}$).

The Poisson adiabat is the geometrical locus on plane (p, η) where we can transform the gas from its initial state p_0, η_0 adiabatically without changing the entropy. That is, on line $\mathcal{P} : \Delta S = S_1 - S_0 = 0$, above $\mathcal{P}$ (the domain shaded by horizontal lines) $\Delta S > 0$, and below the adiabat (the domain shaded by vertical lines) $\Delta S < 0$.

The relative disposition of the Hugoniot and Poisson adiabats is such that when $\eta_1 < \eta_0$, the Poisson adiabat goes lower than the Hugoniot one. It is easy to see that from 1.132 for the Poisson adiabat for $p \to \infty$, we have $\eta_1 \to 0$, but the entire Hugoniot adiabat is placed on the left side from the asymptote

$$\eta_1 = \frac{\gamma - 1}{\gamma + 1} \eta_0 > 0.$$

Thus, for branch OA of the Hugoniot adiabat, $\Delta S > 0$. That is, in compressed shocks, the entropy of the gas increases in comparison with the values of the initial state.

Branch OA' of the Hugoniot adiabat for $\eta_1 > \eta_0$ is placed below the Poisson adiabat, and consequently, for rarefaction shocks, entropy decreases. This contradicts the second principle of thermodynamics: $dS \geq 0$. Therefore, rarefaction shocks cannot exist. We came to this conclusion for an ideal gas, but it remains true, in the general case, for very weak restrictions on the form of equations of state and is called the *Zemplen Theorem*.

1.11.5 Approximation of the "Strong Wave".

We already mentioned the case of the wave for compression with limit amplitude

$$\frac{p_1}{p_0} = \infty, \quad \eta_1 = \frac{\gamma - 1}{\gamma + 1} \eta_0 . \tag{1.133}$$

This case is called *approximation of the "strong wave"* and is realized when we neglect pressure on the initial state in comparison with the pressure beyond the shock. For example, we can use this approximation when the shock is propagating on a "cold" background where $p \approx 0$. In this case, we can obtain explicit formulas for all the parameters beyond the front of the shock in terms of parameters on the background and speed of the shock front. If we take into account that $u = W - \mathcal{D}$, and assume that the gas on the background is at rest ($W_0 = 0$), we get the following equation from 1.133 and 1.116:

$$W_1 = \frac{2}{\gamma + 1} \mathcal{D} . \tag{1.134}$$

Pressure p_1 can be obtained from 1.117:

$$p_1 = \frac{2}{\gamma + 1} \frac{\mathcal{D}^2}{\eta_0} , \tag{1.135}$$

and temperature from the equation of state:

$$T_1 = \frac{2(\gamma - 1)}{(\gamma + 1)^2} \frac{\mathcal{D}^2}{R} . \tag{1.136}$$

Now we can summarize the results. If the shock is propagating on an ideal gas which is at rest ($W_0 = 0$) and "cold" ($p_0 = 0$, $T_0 = 0$), with density ρ_0, then, for the parameters beyond the shock front, we have the following formulas:

$$\eta_1 = \frac{\gamma - 1}{\gamma + 1} \eta_0 , \quad \rho_1 = \frac{\gamma + 1}{\gamma - 1} \rho_0 , \quad p_1 = \frac{2}{\gamma + 1} \frac{\mathcal{D}^2}{\eta_0} , \tag{1.137}$$

$$W_1 = \frac{2}{\gamma + 1} \mathcal{D} , \quad T_1 = \frac{2(\gamma - 1)}{(\gamma + 1)^2} \frac{\mathcal{D}^2}{R} . \tag{1.138}$$

In a more general case, when $p_0 \neq 0$, the formulas are as follows:

$$\eta_1 = \frac{\gamma - 1}{\gamma + 1} \eta_0 + \frac{2\gamma}{\gamma + 1} \frac{p_0 \eta_0^2}{\mathcal{D}^2}, \quad \rho_1 = \frac{1}{\eta_1}, \tag{1.139}$$

$$p_1 = \frac{2}{\gamma + 1} \frac{\mathcal{D}^2}{\eta_0} - \frac{\gamma - 1}{\gamma + 1} p_0, \tag{1.140}$$

$$W_1 = \frac{2}{\gamma + 1} \mathcal{D} - \frac{2\gamma}{\gamma + 1} \frac{p_0 \eta_0}{\mathcal{D}}, \quad T_1 = \frac{p_1}{\rho_1 R}. \tag{1.141}$$

1.11.6 Structure of the Shock Front

In previous sections we were considering discontinuous solutions without taking into account the presence of dissipative processes. However, some dissipative processes (viscosity, heat conductivity, etc.) are present in all real processes. The main influence of such dissipative processes is observed in such zones of flows, where gas parameters vary rapidly along spatial coordinates. Namely, such conditions are realized on the shock front. It is well known that dissipative processes lead from the non-adiabatic of flows to thermodynamically irreversible transformation of energy to its lowest form – heat. *Dissipation is the reason for the increase of entropy on the shock front.*

Therefore, the discontinuous solutions which we have been considering are a sort of idealized situation for gases with very small coefficients of heat conductivity and viscosity. Heat conductivity was taken into account when we were deriving equations for gas dynamics.

However, we were not taking viscosity into account. This is because we construct our finite-difference scheme for modeling a real process where we are able to neglect *real viscosity*. To understand the influence of *scheme viscosity*, which arises from the finite-difference scheme itself (because the finite-difference scheme is a discrete model of gas dynamics equations), and *artificial viscosity*, which we will introduce in the finite-difference scheme, we need to understand the main consequences of the presence of viscosity. From a mathematical point of view in a one-dimensional case, we can model the presence of viscosity by a viscous "addition" ω ("viscous pressure") to the gas dynamics pressure p, in the following form:

$$\omega = -\nu \frac{\partial W}{\partial x}, \tag{1.142}$$

where ν is the coefficient of viscosity.

The gas dynamics equations become:

$$\frac{d}{dt} \left(\frac{1}{\rho} \right) = \frac{\partial W}{\partial s}, \tag{1.143}$$

$$\frac{dW}{dt} = -\frac{\partial \tilde{p}}{\partial s}, \tag{1.144}$$

$$\frac{d}{dt}\left(\varepsilon + \frac{W^2}{2}\right) == -\frac{\partial}{\partial s}\left(\tilde{p}\,W\right) - \frac{\partial H}{\partial s}, \qquad (1.145)$$

where $\tilde{p}$ is the sum of gas dynamics and viscous pressure: $\tilde{p} = p + \omega$. In the Lagrangian mass variable, viscous pressure and heat flux are

$$\omega = -\nu\,\rho\,\frac{\partial W}{\partial s}, \quad H = \kappa\,\rho\,\frac{\partial T}{\partial s}. \qquad (1.146)$$

We will consider the simplest case where coefficients ν and κ are constants and the gas is ideal.

Let the shock front move with constant speed $\mathcal{D}$ on the background gas with known parameters p_0, ρ_0, W_0, where the gas stretches for s until $+\infty$. It is natural to assume that $W_0 = 0$. We can also assume that gas beyond the shock front stretches until $-\infty$. Let us assume that on both "infinities" the parameters are in a steady state, that is

$$\left(\frac{\partial f}{\partial s}\right)_{s\to\pm\infty} = 0.$$

We chose such an idealization of the problem statement because we do not know in advance how wide the front will be. Also, we do not want to consider the influence of concrete boundary conditions. We will assume that the profiles of all the gas parameters are stationary, and that the profiles propagate through the mass with constant speed D. The mass speed D and Eulerian speed $\mathcal{D}$ are satisfied by the following relation:

$$D = (\mathcal{D} - W)\,\rho.$$

Now it is possible to show that all the parameters can be found in the following form:

$$f(s,t) = f(\xi), \quad \xi = s - D\,t. \qquad (1.147)$$

Function $f(\xi)$ gives the profile of parameter f on variable s in a given moment of time ($t = const$), or the law of variation of this parameter with time for a given fluid particle ($s = const$).

The detailed description of the solution to this problem is given in many courses of computational fluid dynamics (see, for example, [102]). We present here only the final results. We note that the solution, for example, $\eta = \eta(\xi)$, has the form which is presented in Figure 1.9 and $\eta_1 \leq \eta \leq \eta_0$, where value η_1 coincides with the value which follows from the Hugoniot condition and is equal to

$$\eta_1 = \frac{\gamma - 1}{\gamma + 1}\eta_0 + \frac{2\gamma}{\gamma + 1}\frac{p_0}{D^2}.$$

It also important to note that values η_0 and η_1 do not depend on the concrete form of dissipation. A particular type of viscosity or heat conductivity determines only the behavior of smoothing of parameters, width of the front, etc., but does not determine how intensive the shock is.

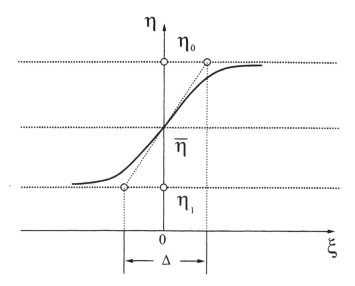

Figure 1.9: *The profile for $\eta(\xi)$ for linear viscosity.*

Structure of the Shock Front in Viscous Media.
Linear Viscosity. Quadratic Viscosity

To distinguish from other possible approaches, we will call viscosity ω in the form of 1.146 as *linear viscosity*. If we consider only the presence of viscosity $(H = 0)$, then

$$\eta(\xi) = \frac{\eta_1 + \eta_0\, e^{a\,\xi}}{1 + e^{a\,\xi}}, \quad a = \frac{(\gamma + 1)\, D\,(\eta_0 - \eta_1)}{2\,\nu} > 0. \tag{1.148}$$

This curve has a point of inflection in $\xi = 0$, $\bar{\eta} = (\eta_0 + \eta_1)/2$ (see Figure 1.9). From a formal point of view, the width of the shock front is equal to infinity. However, the main variation of the function occurs in some finite domain. Let us determine the *effective width of the front*. This can be done in different ways, but all of them give the same qualitative results. We will use the following definition. Let us draw a tangent to curve $\eta(\xi)$ in point $\xi = 0$. The points of intersection of this tangent with lines $\eta = \eta_0$ and $\eta = \eta_1$ give us the width of the front which we will denote as Δ (see Figure 1.9). The formula for Δ is

$$\Delta = \frac{8\,\nu}{(\gamma + 1)\, D\,(\eta_0 - \eta_1)}. \tag{1.149}$$

From this formula, we can conclude that this width is determined by the *intensity of the shock and the coefficient of viscosity*. If we diminish the

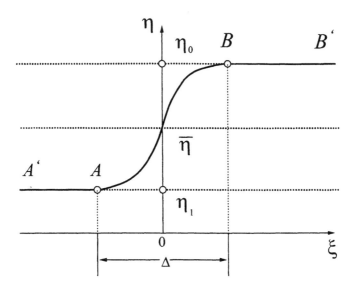

Figure 1.10: *The profile for $\eta(\xi)$ for quadratic viscosity.*

coefficient of viscosity ν then the width of the front also diminishes, and in the limit, when $\nu = 0$, we obtain the discontinuous solution.

Another way to introduce viscosity is called *quadratic viscosity*:

$$\omega = \mu \rho \left(\frac{\partial W}{\partial s}\right)^2 ,   \tag{1.150}$$

which is proportional to the square of the derivative of the velocity.

For this type of viscosity, we get the following solution:

$$\eta(\xi) = \frac{\eta_0 + \eta_1}{2} + \frac{\eta_0 - \eta_1}{2} \, \cos\left(\sqrt{\frac{\gamma + 1}{2\mu}}\,\xi + \frac{\pi}{2}\right) .   \tag{1.151}$$

We are only interested in the part of the curve between points A and B (see Figure 1.10). Namely, this part of the curve $\eta(\xi)$ and parts of the straight lines $\eta = \eta_0$ and $\eta = \eta_1$ (line $B'BAA'$) give us the solution to the problem for the structure of the shock front. The important feature of this solution in comparison to the case of linear viscosity is that the width of the front is finite and

$$\overline{\Delta} = \sqrt{2\mu/(\gamma + 1)}\,\pi .   \tag{1.152}$$

It is important to note that the width of the front *does not depend on the intensity of the shock*. When the viscous coefficient μ is going to zero, the smooth solution also goes to the discontinuous solution.

We will use the qualitative results obtained in this section when we introduce artificial viscosity into the finite-difference schemes.

2   Conservation Laws and Properties of Operators

Let us consider how the method of support-operators can be used for the construction of finite-difference schemes for equations of gas dynamics in the Lagrangian form. As we mentioned in the introduction, there are many different possible forms of energy equations. We will use the non-conservative form of the energy equation, which corresponds to the form 1.44 in 1-D. We will also assume heat flux $\vec{H}$, density of body forces $\vec{F}$, and density of body sources of energy Q are equal to zero. We do this because the concrete form of these terms and consequently, the method of its approximation, depends on a particular problem, and special consideration has to be given for each problem. Thus, we will construct our finite-difference scheme for the following differential equations:

$$\frac{d\rho}{dt} + \rho \operatorname{div} \vec{W} = 0, \qquad (2.1)$$

$$\rho \frac{d\vec{W}}{dt} = -\operatorname{grad} p, \qquad (2.2)$$

$$\rho \frac{d\varepsilon}{dt} = -p \operatorname{div} \vec{W}. \qquad (2.3)$$

In the initial time moment $t = 0$, the distribution of velocity, pressure, density, and internal energy are given as

$$\vec{W}(x, y, 0) = \vec{W}^0(x, y), \qquad (2.4)$$

$$\rho(x, y, 0) = \rho^0(x, y), \qquad (2.5)$$

$$\varepsilon(x, y, 0) = \varepsilon^0(x, y), \qquad (2.6)$$

and pressure can be found from the equation of state

$$p(x, y, 0) = P(\rho(x, y, 0), \varepsilon(x, y, 0)). \qquad (2.7)$$

This system of equations is formulated in terms of the same invariant differential operators of the first order as the elliptic and heat equations, that is, operators div and grad. Therefore, in order to construct a finite-difference scheme for these equations, we can use the same discrete operators as for the elliptic and heat equations.

Here we have some new features which are related to the nonlinear nature of gas dynamic equations: the presence of multiplication of the scalar and vector functions (the second equation), and multiplication of two scalar functions (the third equation). Therefore, it is necessary to introduce the discrete analogs of these operations. In the case where the discrete functions belong to different discrete spaces, the definition of these operations is not an easy problem.

Let us now consider in detail what conservation laws we need to preserve in the discrete case and how these conservation laws depend on the properties of the differential operators. First, we will consider the law of variation of volume:

$$\frac{dV}{dt} = \oint_S (\vec{W}, \vec{n})\, dS\,, \tag{2.8}$$

where V is a fluid volume with fluid boundary S, and $\vec{n}$ is the outward normal to the boundary. Equation 2.8 shows that the variation of fluid volume depends only on the component of the velocity vector which is normal to the boundary of V.

Let us consider how the equation in 2.8 can be obtained from equations for gas dynamics. Let us consider a fluid particle with mass δm and volume δV. The mass of the fluid particle does not change with time, that is, $\delta m = const$ for each particle, and density in this particle is

$$\rho = \delta m / \delta V\,.$$

If we substitute this expression for density in the continuity equation, taking into account that $d(\delta m)/dt = 0$, then we get

$$\frac{d}{dt}(\delta V) = \delta V \operatorname{div} \vec{W}\,. \tag{2.9}$$

Now, if we do a summation of all the fluid particles in volume V and pass to the limit with $\delta V \to 0$, we get

$$\frac{dV}{dt} = \int_V \operatorname{div} \vec{W}\, dV\,. \tag{2.10}$$

For operator div, we have

$$\int_V \operatorname{div} \vec{W}\, dV = \oint_S (\vec{W}, \vec{n})\, dS\,. \tag{2.11}$$

Then from 2.10 and 2.11, we obtain the law of variation for volume as shown in 2.8.

Therefore, to preserve the law of variation of volume in a discrete case, we need to have the analog of property 2.11 for the discrete operator DIV. Let us note that the same requirements we already obtained when we considered the elliptic equations and discrete operators DIV, which we constructed in previous chapters, satisfy this property.

Let us now consider the conservation law for momentum. It follows from equation 2.2 by its integration over domain V:

$$\frac{d}{dt}\left(\int_V \rho \vec{W}\, dV\right) = -\int_V \operatorname{grad} p\, dV = -\oint_S p\vec{n}\, dS\,, \qquad (2.12)$$

where the last equality is the divergence property of operator grad. That is,

$$-\int_V \operatorname{grad} p\, dV = -\oint_S p\vec{n}\, dS\,. \qquad (2.13)$$

Therefore, to obtain conservative FDS we have to preserve the analog of this property in the discrete case. In previous chapters we did not consider this property of operator grad.

Let us show how this property is related to properties of operator div and the integral identity for divergence and gradient:

$$\int_V p\operatorname{div}\vec{A}\, dV + \int_V (\vec{A}, \operatorname{grad} p)\, dV = \oint_S p\,(\vec{A}, \vec{n})\, dS\,. \qquad (2.14)$$

If vector $\vec{A}$ is one of basis vectors $(1,0)$ or $(0,1)$, then div of such vectors is equal to zero and the previous equation becomes 2.13.

Therefore, if the discrete operators DIV and GRAD satisfy some discrete analog of 2.14 and DIV is equal to zero on constant vectors, then we will have a discrete analog of property 2.13 for GRAD.

Let us now consider the conservation law for total energy. Conservation of total energy can be derived from the momentum equation and the equation for specific internal energy. Namely, if we form the scalar product of the momentum equation with vector $\vec{W}$, then sum the resulting equation with the equation for specific internal energy, and then integrate the resulting equation over V, we obtain

$$\frac{d}{dt}\left[\int_V \left(\frac{|\vec{W}|^2}{2} + \varepsilon\right)\rho\, dV\right] = \qquad (2.15)$$

$$-\left[\int_V (\operatorname{grad} p,\, \vec{W})\, dV + \int_V p\operatorname{div}\vec{W}\, dV\right] = -\oint_{\partial V} p\,(\vec{W}, \vec{n})\, dS.$$

The last equality is a consequence of the integral identity related to operators div and grad. Equation 2.15 implies that, when there is no exterior influence, the change in the total energy is equal to the work done by surface forces. This is a very important conservation law and we want to preserve it in the discrete case. It is clear that if the difference analogs of operators div and grad are constructed independently, then the analog of the integral identity, in general, fails to hold, and the corresponding FDS will not be conservative. Therefore, if we want to construct a conservative FDS, then the finite-difference analogs of div and grad must satisfy a difference analog of integral identity.

We also need to consider the conservation of mass. For the Lagrangian methods, media is presented by a set of fluid particles, and the mass of each particle does not change with time. This is the law of mass conservation and is satisfied automatically. However, if we want to follow the ideology of the support-operator method, we have to have the ability to write a discrete equation of mass conservation in the form which is the analog of the continuity equation 2.1.

In fact, this requirement gives us the formula for prime operator DIV. That is, in the case of the equation for gas dynamics, we do not have freedom to choose the prime operator. For the Lagrangian description of media, we have

$$\delta m = \rho \, \delta V. \tag{2.16}$$

As a result of the summation of the previous equation over all particles passed to the limit with $\delta V \to 0$, we get

$$m = \sum \delta m = \int_V \rho \, dV.$$

Therefore, equation 2.16 for the Lagrangian description of the media plays the role of the continuity equation. Let us transform equation 2.16 to a form that is analogous to the continuity equation 2.1. After differentiating equation 2.16 with respect to time and taking into consideration that $d(\delta m)/dt = 0$, we get

$$\frac{d\rho}{dt} + \rho \left[\frac{1}{\delta V} \frac{d}{dt} (\delta V) \right] = 0. \tag{2.17}$$

Compare this equation and equation 2.1 and we can conclude that

$$\operatorname{div} \vec{W} = \lim_{\delta V \to 0} \frac{\frac{d}{dt} (\delta V)}{\delta V} . \tag{2.18}$$

Here we need to understand that δV is in fact a function of velocity.

Hence, in the discrete case, when particles have a finite size, we need to use equation 2.18 for construction of the prime operator DIV.

Let us show that such a definition for the discrete operator DIV does not contradict the requirements which we have previously formulated. Recall that for operator DIV, we have two conditions: the first is that it has to satisfy the divergence property and second, DIV must be equal to zero on the constant vectors. From 2.18, we get

$$\int_V \operatorname{div} \vec{W} \, dV = \lim_{\delta V \to 0} \left(\sum \delta V \operatorname{div} \vec{W} \right) = \lim_{\delta V \to 0} \sum \frac{d}{dt} (\delta V) =$$

$$\frac{d}{dt} \lim_{\delta V \to 0} \sum \delta V = \frac{d}{dt} (V) = \oint_S \left(\vec{W}, \vec{n} \right) dS.$$

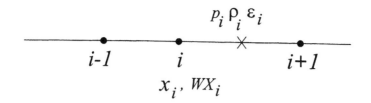

Figure 3.11: *Discretisation for gas dynamics equations in 1-D.*

Since we will determine the discrete operator DIV on the basis of 2.18, then similar considerations will also be valid in the discrete case. Therefore, for such prime operator the discrete analog of the law of variation of volume will be valid. Now we need to check that for constant vectors, DIV is equal to zero. If all particles in media move with constant velocity, then any fluid volume does not change and δV in formula 2.18 is equal to zero. That is, divergence of the constant vector is equal to zero. Thus, we have checked that the definition of 2.18 does not contradict previously formulated requirements.

Now we need to show how the definition of prime operator DIV, on the basis of formula 2.18, is related to the operator DIV which we considered in previous chapters. In the next section, we will show that for the case where the discrete vector $\vec{W}$ belongs to $\mathcal{H}N$, that is, nodal discretisation for a vector field, formula 2.18 gives us the same prime operator as we had obtained for discretisation of elliptic equations. Consequently, we will obtain the same operator GRAD.

Now because discrete operators DIV and GRAD constructed for elliptic equations satisfy all properties mentioned in this section, we can use them for approximations of gas dynamics equations.

3   Finite-Difference Algorithms in 1-D

3.1   Discretisation in 1-D

The following discretisation is generally accepted for the Lagrangian numerical methods [114], [100], [102], and [101]. For coordinates and Cartesian components of velocity vectors, nodal discretisation is used, that is, $\vec{W} \in \mathcal{H}N$. For pressure, specific internal energy, and density, cell-valued discretisation is used, that is p, ε, $\rho \in HC$. This discretisation is presented in Figure 3.11.

Because we use the Lagrangian description of media, we need to associate some mass m_i^C with each fluid volume V_i which will not be changed in time. The following is the usual way to determine this mass. We have the initial condition for density. In other words, we have density at the initial time moment as a function of the space variable $\rho(x, 0)$, and we have the positions of the grid nodes at the initial time moment. Hence, we can compute all volumes V_i and determine mass m_i^C as follows:

$$m_i^C = \rho(x_i^*, 0) \, V_i \,, \tag{3.1}$$

where x_i^* is the coordinate of the center of the cell.

3.2 Discrete Operators in 1-D

3.2.1 The Prime Operator

As we already mentioned, the expression for prime operator DIV has to be constructed on the basis of the definition in 2.18. Now we have to decide what volume will play the role of δV. For our type of discretisation, it is natural to use volume $V_i = x_{i+1} - x_i$, which is the length of segment $[x_i, x_{i+1}]$. Because V_i does not depend explicitly on time, we get

$$(\text{DIV}\vec{W})_i = \frac{1}{V_i} \frac{dV_i}{dt} = \tag{3.2}$$

$$\frac{1}{V_i} \sum_{k=0}^{1} \frac{\partial V_i}{\partial x_{i+k}} \frac{\partial x_{i+k}}{dt} \,.$$

Derivatives of the coordinates with respect to time are the velocities of related nodes, therefore

$$\frac{dx_{i+k}}{dt} = WX_{i+k} \,. \tag{3.3}$$

Finally, we obtain the following expression for DIV:

$$(\text{DIV}\vec{W})_i = \frac{1}{V_i} \sum_{k=0}^{1} \frac{\partial V_i}{\partial x_{i+k}} WX_{i+k} \,. \tag{3.4}$$

Since $V_i = x_{i+1} - x_i$, we have the following formulas for derivatives of volume with respect to coordinates:

$$\frac{\partial V_i}{\partial x_i} = -1, \quad \frac{\partial V_i}{\partial x_{i+1}} = 1 \,. \tag{3.5}$$

Using these expressions, we can transform formula 3.4 into

$$(\text{DIV}\vec{W})_i = \frac{WX_{i+1} - WX_i}{x_{i+1} - x_i} \,. \tag{3.6}$$

This is exactly the same operator as we had for the 1-D case for elliptic equations.

Therefore, the prime operator is the same, and consequently, operator GRAD will be the same and all considerations from the chapter related to elliptic equations are still valid. In particular, operator DIV has divergence property.

As we know for the case of the cell-valued discretisation for scalar functions, the prime operator B is operator DIV in internal cells and in the approximation of the normal component of the vector on the boundary. Therefore, to complete the definition for the prime operator, we need to determine the approximation for $(\vec{n}, \vec{W})$ on the boundary. In the 1-D case, we get an approximation for $(\vec{n}, \vec{W})$ that is equal to $-WX_1$ or WX_M on the correspondent boundary.

3.2.2 The Derived Operator

The derived operator GRAD for all nodes can be written in the following form:

$$GX_i = \frac{u_i - u_{i-1}}{VN_i}, \tag{3.7}$$

where $VN_i = 0.5\,(x_{i+1} - x_{i-1})$, and fictitious nodes $i = 0$ and $i = M + 1$ have following coordinates:

$$x_0 = x_1, \ x_{M+1} = x_M. \tag{3.8}$$

It would be useful to use another form of formulas for GRAD that is similar to formula 3.4 for operator DIV such as

$$GX_i = -\frac{1}{VN_i} \sum_{k=0}^{1} \frac{\partial V_{i-k}}{\partial x_i} u_{i-k}. \tag{3.9}$$

It is obvious that operator GRAD satisfies the divergence property.

3.2.3 Boundary Conditions and Discretisations

We will consider two types of boundary conditions. The first type of boundary condition is the *free boundary*, where we have pressure p as a given function of the coordinate and time on the boundary. In 1-D, this means that pressure is given at $x = x_1$ and $x = x_M$. The second type of boundary is the condition where we have a normal component of velocity as a given function on the boundary. This means that in 1-D, values WX_1 and WX_M are given.

The case for free boundary conditions is similar to the case for Dirichlet boundary conditions for elliptic equations. We introduce the values of the pressure on the boundary, and these values will be associated with fictitious cells at $i = 0$ and $i = M$ (see Figure 3.12). For this type of boundary condition, pressure on the boundary is given, and these given values participate in formulas for GRAD in the boundary nodes.

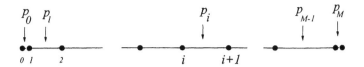

Figure 3.12: *Pressure discretisation in 1-D.*

In the case where velocity is given on the boundary, values WX_1, WX_M are given and it is not necessary to compute these values from the equation of motion.

3.3   Semi-Discrete Finite-Difference Schemes in 1-D

First, let us consider the *semi-discrete* finite-difference scheme, where time is continuous and only discretisation in space is made.

For the continuity equation, we have the following approximation:

$$\rho_i(t) = \frac{m_i^C}{V_i(t)}, \tag{3.10}$$

and because of our definition of DIV, this equation can be rewritten as follows:

$$\frac{d\rho_i}{dt} + \rho_i \left(\text{DIV } \vec{W}\right)_i = 0. \tag{3.11}$$

Let us note that we do not have a problem with multiplication in the second term in the previous equation because ρ and DIV $\vec{W}$ belong to the same discrete space HC.

The discrete equation for specific internal energy is as follows:

$$\rho_i \frac{d\varepsilon_i}{dt} = -p_i \left(\text{DIV } \vec{W}\right)_i. \tag{3.12}$$

There is also no problem here with multiplication because all quantities belong to space HC.

When approximating the momentum equation, we have the problem of how to approximate the multiplication of ρ and $d\vec{W}/dt$ because these quantities belong to different spaces. Since the right-hand side in the momentum equation is GRAD p, then we need to have the projection operator from space HC in space HN. Let us denote this operator as M:

$$M : HC \rightarrow HN.$$

Then the discrete momentum equation looks as follows:

$$(M\,\rho)_i \frac{d\vec{W}_i}{dt} = -(\text{GRAD}\,p)_i. \tag{3.13}$$

Let us consider what requirements we have for operator M. First, the previous discrete equation must express conservation of momentum. Recall that conservation of momentum in the differential case follows from the equation

$$\rho \frac{d\vec{W}}{dt} = -\operatorname{grad} p$$

by multiplication of the elementary volume and then integration over the domain. In addition, it is taken into account that the mass of fluid particle $\rho\, dV$ does not depend on time and it is possible to factor the differentiation with respect to time outside the integral sign. In the discrete case, conservation of momentum must be obtained in a similar way. Therefore, expression

$$(M\,\rho)_i\, VN_i\,, \tag{3.14}$$

which is the analog of $\rho\, dV$, does not have to depend on time. Expression 3.14 has meaning of mass and is related to the node. We have only one quantity which has meaning of mass and does not depend on time: the mass of cell $m_i^C = \rho_i\, V_i$. Therefore, it is natural to use the following relation to determine operator M:

$$(M\,\rho)_i\, VN_i = \frac{m_i^C + m_{i-1}^C}{2} = \frac{1}{2}\sum_{k=0}^{1} m_{i-k}^C\,.$$

In the right-hand side of the last equation, we have the sum of masses of cells which have node i as a vertex. From the previous equation, we obtained the following definition for operator M:

$$(M\,\rho)_i = \frac{1}{2\,VN_i}\sum_{k=0}^{1} \rho_{i-k}\, V_{i-k}\,. \tag{3.15}$$

For this definition of operator M, the discrete analog for the conservation of momentum has the following form:

$$\frac{d}{dt}\left((M\rho)\,\vec{W},\vec{I}\right)_{\mathcal{H}N} = -(\operatorname{GRAD} p,\vec{I})_{\mathcal{H}N} \approx -p\big|_{x_1}^{x_M}\,. \tag{3.16}$$

Let us now consider the discrete analog for the conservation of total energy. Using the same procedure as for the differential case, we get

$$\left(M\rho, \frac{d}{dt}\left(\frac{\vec{W}^2}{2}\right)\right)_{\mathcal{H}N} + \left(\rho, \frac{d\varepsilon}{dt}\right)_{HC} = \tag{3.17}$$
$$-(\operatorname{GRAD} p,\vec{W})_{\mathcal{H}N} - (p, \operatorname{DIV} \vec{W})_{HC}\,.$$

The notation $(\bullet,\bullet)_{HC}$ means that we use the summation only over the interior cells and it is not a real inner product when p is not equal to zero

on the boundary. From the definition of operator GRAD, we can conclude
that the expression on the right-hand side is divergent. Now using the
definition of the adjoint operator for the transformation of the first term
on the left-hand side, we get

$$\left(\rho, \frac{d}{dt} \left[M^* \left(\frac{\vec{W}^2}{2} \right) + \varepsilon \right] \right)_{\overset{H}{C}} \approx - WX \cdot p|_{x_1}^{x_M} . \qquad (3.18)$$

When deriving this equation, we took into account that

$$(M^* \varphi)_i = \frac{1}{2} \sum_k^1 \varphi_{i+k} . \qquad (3.19)$$

In other words, the coefficients of this operator do not depend on time and
we can put operator M^* under time differentiation.

3.4 Fully Discrete, Explicit, Computational Algorithms

Let us at first consider the general sequence of computations.

Initial Conditions

First, at the initial moment of time, we generate a grid in our domain.
In 1-D it must be some set of nodes with the following coordinates:

$$x_1 \leq x_2 \leq \ldots \leq x_{M-1} \leq x_M .$$

There are many publications devoted to different aspects of generating 1-D
grids. We do not have enough space here to discuss this problem so we refer
readers to the book [70], where you can find the necessary references. We
just have to note that we use the Lagrangian description of the media, and
if we are solving a multi-material problem, then the positions of the nodes
must coincide with the positions of the interfaces between the different
materials.

The grid divides the domain into segments with endpoints in the grid
nodes. These segments (meshes) identify as Lagrangian (fluid) particles,
and we observe the behavior of all the parameters related to these particles.

Since we use the Lagrangian description of media, we need to associate
some mass m_i^C with each fluid volume V_i which will not be changed in time.
The usual way in which to determine this mass is the following. We have
the initial condition for density, that is, we have density at the initial time
moment as a function of x $\rho(x, 0)$, and we have the positions of grid nodes
at the initial time moment. Thus, we can compute all volumes V_i which
are the lengths of the segments, and then determine mass m_i^C as follows:

$$m_i^C = \rho(x_i^*, 0) \, V_i .$$

From the statement of the original differential problem, we have the initial conditions for velocity $\vec{W}$, ρ, and ε. For pressure, we have the equation of state $p = P(\rho, \varepsilon)$ or, more generally speaking, $F(p, \rho, \varepsilon) = 0$. Then, we can assume that for any point, we also have the given values for pressure p. The computational domain can contain different materials with different equations of state; therefore, functions P or F can depend on the fluid particles.

Since we now have the initial conditions for the velocity vector, we can compute the initial values for velocity in the nodes

$$WX_i^0 = wx^0\left(x_{i,}\right). \tag{3.20}$$

Values of density and specific internal energy at the initial time moment are computed from the initial condition:

$$\rho_i^0 = \rho^0(x_i^*, 0), \tag{3.21}$$
$$\varepsilon_i^0 = \varepsilon^0(x_i^*, 0). \tag{3.22}$$

Values of pressure at the initial time moment can be computed from the equation of state:

$$p_i^0 = P(\rho_i^0, \varepsilon_i^0). \tag{3.23}$$

New Time Step

The sequence of computation for the explicit finite-difference scheme is as follows. Suppose we know all the quantities on the n-th time level and we need to compute all the quantities on the next $n + 1$ time level.

First, we compute the new velocity field from the equation

$$(M\,\rho)_i^n\,\frac{\vec{W}_i^{n+1} - \vec{W}_i^n}{\Delta t} = -\left(\text{GRAD}\,p^n\right)_i. \tag{3.24}$$

It is important to note that the coefficients of operator GRAD depend on the coordinates of the grid point. In the previous formula, these coordinates are taken from the n-th time level, and a particular volume VN_i which is the denominator in the formula for GRAD, is taken from the n-th time level and is VN_i^n.

Second, we compute new values for the specific internal energy:

$$\rho_i^n\,\frac{\varepsilon_i^{n+1} - \varepsilon_i^n}{\Delta t} = -p_i^n\left(\text{DIV}\,\vec{W}^n\right)_i. \tag{3.25}$$

Now we can compute new positions for the grid points:

$$\frac{x_i^{n+1} - x_i^n}{\Delta t} = WX_i^n. \tag{3.26}$$

Using these new coordinates, we can compute the volumes of the cells on the new time level:

$$V_i^{n+1} = F_V\left(x_{i+k}^{n+1}; k = 0, 1\right) = x_{i+1}^{n+1} - x_i^{n+1}. \tag{3.27}$$

If we know the volume of the cells, we can compute density on the new time level:

$$\rho_i^{n+1} = \frac{m_i^C}{V_i^{n+1}}. \tag{3.28}$$

Let us be reminded that the mass of the cell is constant and does not depend on time, and therefore does not have a time index.

Finally, we can compute pressure on the new time level from the equation of state:

$$p_i^{n+1} = P(\rho_i^{n+1}, \varepsilon_i^{n+1}). \tag{3.29}$$

In the case where the equation of state is given in implicit form,

$$F_i(p_i^{n+1}, \rho_i^{n+1}, \varepsilon_i^{n+1}) = 0, \tag{3.30}$$

we need to solve one non-linear equation of one unknown in each cell to find the pressure. Function F_i has index i, because for each Lagrangian particle, the equation of state, in principle, can be different.

This is a purely explicit scheme, which has some restrictions on the ratio of time and space steps that ensure stability of the finite-difference scheme.

Let us consider what will happen with the conservation laws. First, we consider the law for variation of volume. We need to note that volume of the cell is a bilinear function of the coordinates of its vertices and the coordinates of these vertices are linear functions of the corresponding component of the velocities (formula 3.26). Then, the following formula is valid:

$$(\mathrm{DIV}\vec{W}^n)_i = \frac{1}{V_i^n} \frac{V_i^{n+1} - V_i^n}{\Delta t},$$

and, in general,

$$(\mathrm{DIV}\vec{W}^\sigma)_i = \frac{1}{V_i^n} \frac{V_i^{n+1} - V_i^n}{\Delta t} \tag{3.31}$$

if we use velocity $\vec{W}^\sigma = \sigma \vec{W}^{n+1} + \vec{W}^n$ instead of velocity $\vec{W}^n$ in the formulas in 3.26. If we rewrite the previous equation as

$$\frac{V_i^{n+1} - V_i^n}{\Delta t} = V_i^n \cdot (\mathrm{DIV}\vec{W}^\sigma)_i, \tag{3.32}$$

then it looks like the analog of the differential equation 2.9.

Now let us show that by using 3.31, we can rewrite the continuity equation in a form which is analogous to the differential case. The discrete analog of the continuity equation is

$$\frac{\rho_i^{n+1} - \rho_i^n}{\Delta t} = \frac{1}{\Delta t}\left(\frac{m_i^C}{V_i^{n+1}} - \frac{m_i^C}{V_i^n}\right) = \tag{3.33}$$

$$-\frac{m_i^C}{V_i^{n+1}} \frac{1}{\Delta t}\left[\frac{V_i^{n+1} - V_i^n}{V_i^n}\right] =$$

$$-\rho_i^{n+1}\,\mathrm{DIV}\,\vec{W}^\sigma.$$

Now let us consider the law of conservation for momentum. Let us be reminded that the integral form for the conservation of momentum is

$$\int_{V(t_2)} \rho(x,t_2)\,\vec{W}(x,t_2)\,dx - \int_{V(t_1)} \rho(x,t_1)\,\vec{W}(x,t_1)\,dx =$$
$$-\int_{t_1}^{t_2} \left(p(x_M,t) - p(x_1,t) \right) dt \approx -\Delta t\,\left(p(x_M,t) - p(x_1,t) \right).$$

In the discrete case, conservation of momentum follows from equation 3.24 if we multiply it by volume $V N_i^n$ and take the summation of that over all nodes:

$$\frac{\sum_i (M\rho)_i^n\,\vec{W}_i^{n+1}\,V N_i^n - \sum_i (M\rho)_i^n\,\vec{W}_i^n\,V N_i^n}{\Delta t} = \qquad (3.34)$$
$$-\sum_{i,j} (\mathrm{GRAD}\,p^n)_i\,V N_i \approx \left(p(x_M,t) - p(x_1,t) \right).$$

Since $(M\,\rho)_i\,V N_i$ does not depend on time, then

$$(M\,\rho)_i^n\,V N_i^n = (M\,\rho)_i^{n+1}\,V N_i^{n+1}.$$

Hence, we can replace the time index n at $(M\,\rho)_i$ and $V N_i$ in the first term in equation 3.34 by $n+1$. After this, the equation 3.34 looks similar to the differential case:

$$\sum_i (M\rho)_i^{n+1}\,\vec{W}_i^{n+1}\,V N_i^{n+1} - \sum_i (M\rho)_i^n\,\vec{W}_i^n\,V N_i^n = \qquad (3.35)$$
$$-\Delta t\,\sum_i (\mathrm{GRAD}\,p^n)_i\,V N_i.$$

It is easy to see that for the constructed finite-difference scheme, the discrete analog for conservation of full energy is not satisfied. Let us recall that at first we need to obtain the balance equation for kinetic energy. This can be done by multiplication of the left-hand side of equation 3.24 by

$$\vec{W}^{n+0.5} = \frac{\vec{W}_i^{n+1} + \vec{W}_i^n}{2}$$

and then making the summation over the nodes. That is,

$$\sum_i (M\,\rho)_i^n\,\frac{\frac{(\vec{w}_i^{n+1})^2 - (\vec{w}_i^n)^2}{2}}{\Delta t}\,V N_i^n = \qquad (3.36)$$
$$-\left(\mathrm{GRAD}\,p^n,\,\vec{W}^{n+0.5} \right)_{\mathcal{H}N}.$$

If we use the same consideration for $(M \rho)_i \, V N_i$ as for the momentum equation, we can rewrite the previous equation in the following form:

$$\frac{\sum_i (M \rho)_i^{n+1} \frac{(\vec{W}_i^{n+1})^2 \, V N_i^{n+1}}{2} - \sum_i (M \rho)_i^n \frac{(\vec{W}_i^n)^2 \, V N_i^n}{2}}{\Delta t} = \quad (3.37)$$

$$- \left(\mathrm{GRAD} \, p^n, \vec{W}^{n+0.5} \right)_{\mathcal{HN}} .$$

Similar to the differential case, we can write the following balance equation for specific internal energy as

$$\sum_i \rho_i^n \frac{\varepsilon_i^{n+1} - \varepsilon_i^n}{\Delta t} \, V_i^n = \quad (3.38)$$

$$- \sum_i p_i^n \, (\mathrm{DIV} \vec{W}^n)_i \, V_i^n = - \left(p^n, \mathrm{DIV} \vec{W}^n \right)_{HC} ,$$

where the summation is over the cells and again $\rho_i^n \, V_i^n$ is a mass of cells and does not depend on time. Therefore, we can replace the time index n by $n+1$, or in other words, we can put this quantity under time differentiation. The conservation law for full energy will be satisfied if the sum of the right-hand sides of the equations 3.37 and 3.38 can be reduced to the analog of the boundary integral. Using the discrete identity for operators DIV and GRAD, we can show that it is possible only when the discrete functions p and $\vec{W}$ in both equations are the same. The conservation law for full energy is not satisfied when only function p is the same in both equations.

Thus, to preserve the conservation law of full energy for schemes that are discrete in space and time, we need to make additional considerations. One possible way to construct a finite-difference scheme which will satisfy the conservation law of full energy is to take the velocity in the equation for internal energy as half the sum of the values from the n and $n + 1$ time steps. That is, instead of equation 3.25, we consider the following equation:

$$\rho_i^n \frac{\varepsilon_i^{n+1} - \varepsilon_i^n}{\Delta t} = -p_i^n \left(\mathrm{DIV} \, \vec{W}^{n+0.5} \right)_i . \quad (3.39)$$

It is important to note that the finite-difference scheme still remains explicit in the sense that when we compute the specific internal energy from equation 3.39, we already know velocity $\vec{W}^{n+1}$.

The natural explicit conservative finite-difference scheme is

$$(M \rho)_i^n \frac{\vec{W}_i^{n+1} - \vec{W}_i^n}{\Delta t} = - (\mathrm{GRAD} \, p^n)_i , \quad (3.40)$$

$$\rho_i^n \frac{\varepsilon_i^{n+1} - \varepsilon_i^n}{\Delta t} = -p_i^n \left(\mathrm{DIV} \, \vec{W}^{n+0.5} \right)_i , \quad (3.41)$$

$$\frac{x_i^{n+1} - x_i^n}{\Delta t} = WX_i^{n+0.5} , \qquad (3.42)$$

$$V_i^{n+1} = F_V\left(x_{i+k}^{n+1}; k = 0, 1\right) , \qquad (3.43)$$

$$\rho_i^{n+1} = \frac{m_i^c}{V_i^{n+1}} , \qquad (3.44)$$

$$p_i^{n+1} = P(\rho_i^{n+1}, \varepsilon_i^{n+1}). \qquad (3.45)$$

The explicit form of these equations is

$$m_i^N \frac{WX_i^{n+1} - WX_i^n}{\Delta t} = -\left(p_i^n - p_{i-1}^n\right) , \qquad (3.46)$$

$$m_i^C \frac{\varepsilon_i^{n+1} - \varepsilon_i^n}{\Delta t} = -p_i^n \left(WX_{i+1}^{n+0.5} - WX_i^{n+0.5}\right) , \qquad (3.47)$$

$$\frac{x_i^{n+1} - x_i^n}{\Delta t} = WX_i^{n+0.5} , \qquad (3.48)$$

$$V_i^{n+1} = x_{i+1}^{n+1} - x_i^{n+1} , \qquad (3.49)$$

$$\rho_i^{n+1} = \frac{m_i^c}{V_i^{n+1}} , \qquad (3.50)$$

$$p_i^{n+1} = P(\rho_i^{n+1}, \varepsilon_i^{n+1}), \qquad (3.51)$$

where $m_i^N = 0.5\left(m_{i+1}^C + m_i^C\right)$ is *mass of node*.

These equations are presented in the sequence that the actual computations are made.

For this FDS all computations can be made explicitly without the solutions to any system of linear or non-linear equations.

Realization of Boundary Conditions

At first we consider the case of "free boundary", where the pressure on the boundary is given as a function of time and position on the boundary. This is the simplest case because it can be considered in the same way as the Dirichlet boundary conditions for elliptic equations. Namely, we take the pressure on the boundary in fictitious cells at $i = 0$ and $i = M$ that are equal to a given pressure. Boundary values are used in equation 3.46 to compute velocities of boundary nodes.

Now let us consider the condition of impermeability. In 1-D, it is a very simple case because the velocities WX_1 and WX_M are given and do not need to be computed. We will use these velocities in the equation for internal energy 3.47 and for moving boundary nodes.

3.5   Fully Discrete, Implicit, Computational Algorithm

We will consider the following general form of the implicit finite-difference scheme:

$$m_i^N \frac{WX_i^{n+1} - WX_i^n}{\Delta t} = -\left(p_i^{(\sigma_1)} - p_{i-1}^{(\sigma_1)}\right), \tag{3.52}$$

$$m_i^C \frac{\varepsilon_i^{n+1} - \varepsilon_i^n}{\Delta t} = -p_i^{(\sigma_1)}\left(WX_{i+1}^{(\sigma_3)} - WX_i^{(\sigma_3)}\right), \tag{3.53}$$

$$\frac{x_i^{n+1} - x_i^n}{\Delta t} = WX_i^{(\sigma_2)}, \tag{3.54}$$

$$V_i^{n+1} = x_{i+1}^{n+1} - x_i^{n+1}, \tag{3.55}$$

$$\rho_i^{n+1} = \frac{m_i^C}{V_i^{n+1}}, \tag{3.56}$$

$$p_i^{n+1} = P(\rho_i^{n+1}, \varepsilon_i^{n+1}). \tag{3.57}$$

The full set of conservation laws will be valid for these equations when $\sigma_3 = 0.5$, σ_1, and σ_2 can be chosen arbitrarily. The finite-difference scheme will be implicit if $\sigma_1 > 0$.

Let us analyze the structure of the system for non-linear equations 3.52, 3.53, 3.54, 3.55, and 3.56. Originally, it is the system where the role of unknowns is played by all the quantities: pressure, velocity, specific internal energy, coordinates of nodes, and density. However, different variables play different roles in this system. To understand the structure of this system, let us suppose that we have found pressure p_i^{n+1} on a new time level, then using equation 3.52, we can determine the velocity vector WX_i^{n+1} on the new time level. Now using this velocity, we can compute the values of specific internal energy on the new time level from equation 3.53. We then compute new coordinates from equation 3.54, and using these coordinates, we compute the new volume and new density from equation 3.56. This means that all variables can be considered as a composite function of pressure. Consequently, it is possible to eliminate all variables except for pressure from the original system of equations and obtain a new system of non-linear equations which will contain only pressure.

To show the structure of this new system of equations, let us introduce some stencils. The first stencil, which we will call $St_p(i)$, is a stencil for values of pressure which participate in the approximation for the equation of motion in node i. It contains two cells, i and $i-1$, that is, $St_p(i) = \{i, i-1\}$. The second stencil, which we will denote as St_ε, is a stencil for values of velocities which participate in the approximation for the equation for specific internal energy in cell i. It contains two nodes, i and $i+1$, or $St_\varepsilon = \{i, i+1\}$. Coordinates from the the same stencil participate in the expression for volumes of the cell i.

Using these stencils, the structure of dependence in each equation can be written in the following form. For the equation of motion we have:

$$WX_i^{n+1}(p_s^{n+1} : s \in St_p(i)) = 0. \tag{3.58}$$

For the energy equation, we have

$$\varepsilon_i^{n+1}(p_i^{n+1}, ((WX_k^{n+1}(p_s^{n+1} : s \in St_p(k))) : k \in St_\varepsilon(i))) = 0. \tag{3.59}$$

The equation for the new coordinates is as follows:

$$x_i^{n+1}(WX_i^{n+1}) = 0. \tag{3.60}$$

The dependence of volume on the coordinates is:

$$V_i^{n+1}\left(x_k^{n+1} : k \in St_\varepsilon(i)\right) = 0. \tag{3.61}$$

The equation for density is:

$$\rho_i^{n+1}(V_i^{n+1}) = 0. \tag{3.62}$$

And finally, for the equation of state, we have:

$$F_i(p_i, \rho_i, \varepsilon_i) = 0. \tag{3.63}$$

Using these dependencies, the system of non-linear equations, we can express all variables in terms of pressure and substitute them into the equation of state 4.32. Finally, the system of non-linear equations, which contains only pressure, can be written in the schematic form as follows:

$$\Phi_i\,(p_k : k \in St(i)) = \tag{3.64}$$
$$F_i\,(p_i, \rho_i(V_i(x_k(WX_k(p_s : s \in St_p(k))))) :$$
$$k \in St_\varepsilon(i)),$$
$$\varepsilon_i(p_i, ((WX_k(p_s : s \in St_p(k))) : k \in St_\varepsilon(i)))) = 0.$$

We omit here the index $n+1$ for pressure, because all values are taken from the new time step. The resulting stencil $St(i)$ for pressure for this system of nonlinear equations contains three cells for each i, $St(i) = \{i-1, i, i+1\}$.

We can now use Newton's method and some of its modifications (for a general explanation see the Introduction) to solve the system of non-linear equations in 3.64.

Newton's Method

If some approximation $p_i^{(s)}$ for p_i^{n+1} is given (here, s is the iteration number), then the increment $\delta p_{(s+1)_i} = p_i^{(s+1)} - p_i^{(s)}$ must be determined from the linear system of equations:

$$\sum_{k \in St(i)} \left(\frac{\partial \Phi_i}{\partial p_k}\right)^{(s)} \delta p_k^{(s+1)} = -\Phi_i^{(s)}. \tag{3.65}$$

Let us note that index (s) at derivatives $\frac{\partial \Phi_i}{\partial p_k}$ means that it must be computed by using values of its arguments on the previous (s) iteration.

Now we need to find the formulas for the derivatives

$$\frac{\partial \Phi_i}{\partial p_k}$$

and investigate the properties of the system of linear equations.

To obtain formulas for the derivatives, we will use the structure of dependence for the function Φ_i on corresponding pressures:

$$\frac{\partial \Phi_i}{\partial p_i} = \tag{3.66}$$

$$\frac{\partial F_i}{\partial p_i} +$$

$$\frac{\partial F_i}{\partial \rho_i} \frac{\partial \rho_i}{\partial V_i} \sum_{\alpha=0}^{1} \left(\frac{\partial V_i}{\partial x_{i+\alpha}} \frac{\partial x_{i+\alpha}}{\partial W X_{i+\alpha}} \frac{\partial W X_{i+\alpha}}{\partial p_i} \right) +$$

$$\frac{\partial F_i}{\partial \varepsilon_i} \left[\frac{\partial \varepsilon_i}{\partial p_i} + \sum_{\alpha=0}^{1} \left(\frac{\partial \varepsilon_i}{\partial W X_{i+\alpha}} \frac{\partial W X_{i+\alpha}}{\partial p_i} \right) \right],$$

$$\frac{\partial \Phi_i}{\partial p_{i+1}} = \tag{3.67}$$

$$\frac{\partial F_i}{\partial \rho_i} \frac{\partial \rho_i}{\partial V_i} \frac{\partial V_i}{\partial x_{i+1}} \frac{\partial x_{i+1}}{\partial W X_{i+1}} \frac{\partial W X_{i+1}}{\partial p_{i+1}} +$$

$$\frac{\partial F_i}{\partial \varepsilon_i} \frac{\partial \varepsilon_{i,j}}{\partial W X_{i+1}} \frac{\partial W X_{i+1}}{\partial p_{i+1}},$$

and use a similar formula for the derivative with respect to p_{i-1}.

Let us note here that all the derivatives are taken from the (s) iteration.

Formulas for derivatives which participate in equations 3.66 and 3.67 follow from the equations of finite-difference schemes 3.52, 3.53, 3.54, 3.55, and 3.56 and are as follows:

$$\frac{\partial \rho_i}{\partial V_i} = -\frac{m_i^C}{V_i^2} = -\frac{m_i^C}{(x_{i+1} - x_i)^2}, \tag{3.68}$$

$$\frac{\partial x_{i+\alpha}}{\partial W X_{i+\alpha}} = \Delta t\, \sigma_2, \tag{3.69}$$

$$\frac{\partial W X_{i+\alpha}}{\partial p_i} = \frac{\sigma_1 \Delta t}{m_{i+\alpha}^N} \left(\frac{\partial V_i}{\partial x_{i+\alpha}} \right)^n = (-1)^{\alpha+1} \frac{\sigma_1 \Delta t}{m_{i+\alpha}^N}, \tag{3.70}$$

$$\frac{\partial \varepsilon_i}{\partial p_i} = -\frac{\sigma_1 \Delta t}{m_i^C} \sum_{\alpha=0}^{1} \left(\left(\frac{\partial V_i}{\partial x_{i+\alpha}} \right)^n W X_{i+\alpha}^{(\sigma_3)} \right), \tag{3.71}$$

$$\frac{\partial \varepsilon_i}{\partial W X_{i+\alpha}} = -\frac{\sigma_3\, p_i^{(\sigma_1)} \Delta t}{m_i^C} \left(\frac{\partial V_i}{\partial x_{i+\alpha}} \right)^n = (-1)^{\alpha} \frac{\sigma_3\, p_i^{(\sigma_1)} \Delta t}{m_i^C}. \tag{3.72}$$

Using these expressions, we obtain the following formulas for derivatives $d\Phi_i/dp_k$:

$$\frac{\partial \Phi_i}{\partial p_i} = \tag{3.73}$$

$$\frac{\partial F_i}{\partial p_i} -$$

$$\frac{m_i^C}{V_i^2} \frac{\partial F_i}{\partial \rho_i} \sigma_1 \sigma_2 (\Delta t)^2 \sum_{\alpha=0}^{1} \left[\left(\frac{\partial V_i}{\partial x_{i+\alpha}} \right)^{(s)} \left(\frac{\partial V_i}{\partial x_{i+\alpha}} \right)^n \Big/ m_{i+\alpha}^N \right] -$$

$$\frac{\partial F_i}{\partial \varepsilon_i} \left[\frac{\sigma_1 \Delta t}{m_i^C} \left(\sum_{\alpha=0}^{1} \left(\frac{\partial V_i}{\partial x_{i+\alpha}} \right)^n W X_{i+\alpha}^{(\sigma_3)} \right) + \right.$$

$$\left. \sigma_3 \Delta t \, p_i^{(\sigma_1)} \sum_{\alpha=0}^{1} \left(\left(\frac{\partial V_i}{\partial x_{i+\alpha}} \right)^n \right)^2 \Big/ m_{i+\alpha}^N \right],$$

$$\frac{\partial \Phi_i}{\partial p_{i+1}} = \tag{3.74}$$

$$-\frac{m_i^C}{V_i^2} \frac{\partial F_i}{\partial \rho_i} \sigma_1 \sigma_2 (\Delta t)^2 \left(\frac{\partial V_i}{\partial x_{i+1}} \right)^{(s)} \left(\frac{\partial V_{i+1}}{\partial x_{i+1}} \right)^n \Big/ m_{i+1}^N -$$

$$\frac{\partial F_i}{\partial \varepsilon_i} \frac{\sigma_1 \sigma_3}{m_i^C} (\Delta t)^2 p_i^{(\sigma_1)} \left(\frac{\partial V_i}{\partial x_{i+1}} \right)^{(s)} \left(\frac{\partial V_{i+1}}{\partial x_{i+1}} \right)^n \Big/ m_{i+1}^N.$$

Because the derivatives of volume with respect to coordinates are equal to $+1, -1$, we get

$$\frac{\partial \Phi_i}{\partial p_i} = \tag{3.75}$$

$$\frac{\partial F_i}{\partial p_i} -$$

$$\frac{m_i^C}{V_i^2} \frac{\partial F_i}{\partial \rho_i} \sigma_1 \sigma_2 (\Delta t)^2 \left[\frac{1}{m_i^N} + \frac{1}{m_{i+1}^N} \right] -$$

$$\frac{\partial F_i}{\partial \varepsilon_i} \frac{\sigma_1 \Delta t}{m_i^C} \left[\left(W X_{i+1}^{(\sigma_3)} - W X_i^{(\sigma_3)} \right) + \sigma_3 \Delta t \, p_i^{(\sigma_1)} \left(\frac{1}{m_i^N} + \frac{1}{m_{i+1}^N} \right) \right],$$

$$\frac{\partial \Phi_i}{\partial p_{i+1}} = \tag{3.76}$$

$$\frac{m_i^C}{V_i^2} \frac{\partial F_i}{\partial \rho_i} \sigma_1 \sigma_2 (\Delta t)^2 \frac{1}{m_{i+1}^N} +$$

$$\frac{\partial F_i}{\partial \varepsilon_i} \frac{\sigma_1 \sigma_3}{m_i^C} (\Delta t)^2 p_i^{(\sigma_1)} \frac{1}{m_{i+1}^N}.$$

To find $\delta p_i^{(s+1)}$, it is necessary to solve, at each iteration of Newton's method, the three-point linear system of equations whose coefficients are calculated by the previous formulas from the values of the quantities at the preceding s-iteration. There are many methods for solving the three-diagonal system of linear equations (see for instance [111]) and there is no problem in solving this system. Actually, in 1-D, we have only one problem with the classical Newton's method, namely, at each iteration, it is necessary to recalculate the coefficients of the linear system. This difficulty can be avoided by using the method of Newton-Kantorovich, or by the method of parallel chord [94], where instead of using matrix $||\partial \Phi_i^{(s)}/\partial p_k||$, another matrix G which is close to matrix $||\partial \Phi_i^{(s)}/\partial p_k||$, is used. We chose matrix G to be independent of the iteration number, self-adjoint, and positive definite.

The Method of Parallel Chords

The general idea for the method of parallel chords was discussed in the introduction. In our case, we will do the following. First, we take all the quantities in formulas 3.75 and 3.76 from the previous time step. That is, instead of using the values from the (s) iteration, we will use the values from time step n. Second, we will divide equation 3.65 by

$$-\left(\frac{m_i^C}{V_i^2} \frac{\partial F_i}{\partial \rho_i} \sigma_1 \sigma_2 (\Delta t)^2 + \frac{\partial F_i}{\partial \varepsilon_i} \frac{p_i}{m_i^C} \sigma_1 \sigma_3 (\Delta t)^2 \right) .$$

Finally, we obtain the following equations in the internal nodes:

$$A_{i-1}^i \, \delta p_{i-1}^{(s+1)} + A_i^i \, \delta p_i^{(s+1)} + A_{i+1}^i \, \delta p_{i+1}^{(s+1)} = -\tilde{\Phi}_i , \qquad (3.77)$$

where

$$A_i^i = \qquad\qquad\qquad\qquad\qquad\qquad\qquad\qquad\qquad (3.78)$$

$$\left[-\frac{\partial F_i}{\partial p_i} + \frac{\partial F_i}{\partial \varepsilon_i} \frac{\sigma_1 \Delta t}{m_i^C} \left(W X_{i+1}^{(\sigma_3)} - W X_i^{(\sigma_3)} \right) \right] \Bigg/$$

$$\left(\frac{m_i^C}{V_i^2} \frac{\partial F_i}{\partial \rho_i} \sigma_1 \sigma_2 (\Delta t)^2 + \frac{\partial F_i}{\partial \varepsilon_i} \frac{p_i}{m_i^C} \sigma_1 \sigma_3 (\Delta t)^2 \right) +$$

$$\left(\frac{1}{m_i^N} + \frac{1}{m_{i+1}^N} \right) ,$$

$$A_{i-1}^i = -\frac{1}{m_{i-1}^N} , \qquad\qquad\qquad\qquad (3.79)$$

$$A_{i+1}^i = -\frac{1}{m_{i+1}^N} , \qquad\qquad\qquad\qquad (3.80)$$

$$\tilde{\Phi}_i = -\frac{\Phi_i}{-\left(\frac{m_i^C}{V_i^2} \frac{\partial F_i}{\partial \rho_i} \sigma_1 \sigma_2 (\Delta t)^2 + \frac{\partial F_i}{\partial \varepsilon_i} \frac{p_i}{m_i^C} \sigma_1 \sigma_3 (\Delta t)^2 \right)} . \qquad (3.81)$$

Operator A can be presented in the following form:

$$A = \mathcal{D}\,\overline{M}\,D + R,\qquad (3.82)$$

where

$$(D\,p)_i = p_i - p_{i-1},\ (\mathcal{D}\,u)_i = -(u_{i+1} - u_i),\qquad (3.83)$$

$$(\overline{M}\,u)_i = \frac{1}{m_i^N}\,u_i,\qquad (3.84)$$

$$(R\,p)_i = \frac{-\dfrac{\partial F_i}{\partial p_i} + \dfrac{\partial F_i}{\partial \varepsilon_i}\dfrac{\sigma_1\,\Delta t}{m_i^C}\left(W X_{i+1}^{(\sigma_3)}\right)}{\dfrac{m_i^C}{V_i^2}\dfrac{\partial F_i}{\partial \rho_i}\sigma_1\,\sigma_2(\Delta t)^2 + \dfrac{\partial F_i}{\partial \varepsilon_i}\dfrac{p_i}{m_i^C}\sigma_1\,\sigma_3\,(\Delta t)^2}\,p_i\,.\qquad (3.85)$$

It is easy to see that

$$\mathcal{D} = D^*$$

and because $m_i^N > 0$, the diagonal operator $\overline{M}$ is symmetric and positive definite $\overline{M} = \overline{M}^* > 0$. Therefore, the first term in the expression for operator A is symmetric and positive definite. Operator R is a diagonal operator, therefore it is symmetric. If the natural requirements,

$$\frac{\partial F_i}{\partial p_i} > 0,\ \frac{\partial F_i}{\partial \rho_i} < 0,\ \frac{\partial F_i}{\partial \varepsilon_i} < 0,\qquad (3.86)$$

are satisfied, then operator R is positive definite for reasonable constraints on the time step.

Boundary Conditions

In the case of a free boundary, all formulas remain the same because the equation of motion for boundary nodes can be written in the same form as for the internal nodes using fictitious cells.

For the case of impermeability boundary conditions, we have to take into account that velocities $W X_1$ and $W X_M$ are given, and consequently, all derivatives such as

$$\frac{\partial W X_1}{\partial p_k},\ \frac{\partial W X_M}{\partial p_k}$$

are equal to zero. In particular, this means that

$$\frac{\partial \Phi_1}{\partial p_0} = 0$$

and

$$\frac{\partial \Phi_{M-1}}{\partial p_M} = 0\,.$$

That is, the stencil in the boundary cells does not contain fictitious pressures.

3.6   Stability Conditions

3.6.1   General Remarks

In the case of finite-difference schemes for equations of Lagrangian gas dynamics, all general considerations about the stability of finite-difference schemes which were discussed in the introduction are valid. Let us only note that in the case of Lagrangian gas dynamics, the particles move with the media, and this means that we cannot control the spatial distribution of nodes. In another words, at each moment of time, the grid is given. The stability condition is usually formulated as a condition on the time step and some effective length related to spatial discretisation. These conditions also include some characteristics of the physical process, for example, maximum gas velocity, which is also given and cannot be controlled. This means that from a practical point of view, for a given FDS, there is only one parameter which we can change to satisfy the stability conditions: the time step.

Therefore, the main question for a given finite-difference scheme is: Is it possible to choose a time step which will give us a stable solution?

Let us recall that the finite-difference scheme which is stable for any time and spatial step is called *unconditionally stable*, the finite-difference scheme which is stable if some relation between the time and spatial step is satisfied is called *conditionally stable*, and if it is not possible to obtain a stable solution by changing the time step, the finite-difference scheme is called *unconditionally unstable*.

Therefore, from a practical point of view, it is enough to know: Is the finite-difference scheme unconditionally stable, conditionally stable, or unconditionally unstable? The exact formulation of the stability condition is not so important, because for real problems, when we do not know the structure of the solution, any estimation may be useless, and there is only one way to obtain a stable solution, and that is to reduce the time step.

Moreover, because of the difficulties related to non-linearity, the nature of the equation of gas dynamics is that almost all stability conditions of the finite-difference scheme are obtained for linear approximation of gas dynamics equations. That is, acoustic equations, and usually the boundary conditions, are not considered. This approach is generally accepted and gives us the ability to obtain the main qualitative results about stability of finite-difference schemes for equations of gas dynamics.

Therefore, the practical way to understand main stability properties of finite-difference schemes for equations of gas dynamics is as follows. First, we need to understand the main characteristic properties of finite-difference schemes, such as implicit or explicit schemes, the type of approximation of spatial derivatives, and so on. Then, we need to construct a similar type of scheme for simple equations, for example, acoustic equations, and investigate the stability properties for this simple case. Usually, this investigation allows us to make qualitative conclusions about the stability properties of finite-difference schemes for the original equations. If you are planning to solve a lot of problems which have some common features, then it makes sense to spend some time in trying to obtain more accurate stability conditions for this special class of problems.

Stability conditions for FDS for Lagrangian gas dynamics were the subject of special consideration in the following papers [49], [104], [5], and [4]. It is shown that for equations of Lagrangian gas dynamics FDS will be conditionally stable if discrete operators DIV and GRAD satisfy conditions which we formulated in the previous chapter. That is, both operators satisfy the divergence property and satisfy the discrete analog of integral identity.

Readers who are interested in questions related to investigations of stability of finite-difference schemes for gas dynamics equations can find useful information in the following papers: [58], [110], [46], and [100].

3.6.2 One-Dimensional Transport Equations

To understand the essence of the problem of investigation of stability for finite-difference schemes, it is also very useful to consider the simplest case: the one-dimensional linear transport equation (see, for example, [110], [46], and [100]),

$$\frac{\partial u}{\partial t} + a\frac{\partial u}{\partial x} = 0, \tag{3.87}$$

where a is constant. We will consider the Cauchy problem for this equation in the region $-\infty < x < \infty$, $t > 0$, with the initial condition at $t = 0$:

$$u(x, 0) = v_0(x). \tag{3.88}$$

The solution to this problem is the traveling wave

$$u(x, t) = v_0(x - a t), \tag{3.89}$$

which propagates with speed a. The profile of this wave is given by the initial data. Lines $x - at = const$, on which the constant values of the solution are transported, are characteristics. To solve this problem, we do not need numerical methods, but the investigation of different finite-difference schemes for this problem can give us useful information about the properties of finite-difference schemes for the general equations of gas dynamics.

3.6.3 Finite-Difference Schemes for the One-Dimensional Transport Equation

Let us consider some finite-difference schemes for problems 3.87 and 3.88. We will consider a case where the spatial grid is uniform when the mesh size is equal to h. The coordinates of the nodes are $x_k = k\,h$, $k = \pm 1, \pm 2, \ldots$.

Let us consider two explicit finite-difference schemes. The first scheme is

$$\frac{u_k^{n+1} - u_k^n}{\Delta t} + a \frac{u_{k+1}^n - u_k^n}{h} = 0, \; k = \pm 1, \pm 2, \ldots \quad (3.90)$$

$$u_k^0 = v_0(x_k).$$

The second scheme is

$$\frac{u_k^{n+1} - u_k^n}{\Delta t} + a \frac{u_k^n - u_{k-1}^n}{h} = 0, \; k = \pm 1, \pm 2, \ldots \quad (3.91)$$

$$u_k^0 = v_0(x_k).$$

Formally, these two schemes are distinguished only by the approximation of spatial derivatives, so called "right" and "left" derivatives. But in the sense of stability, these schemes are absolutely different: when $a > 0$, the first scheme is unconditionally unstable and the second scheme is conditionally stable.

To show this, we will use the method of harmonics which was described in the previous chapter on the example of investigation of stability of finite-difference schemes for heat equations. Let us consider a particular solution,

$$u_k^n = q^n e^{i k \varphi}, \quad (3.92)$$

where i is an imaginary unit, φ is an arbitrary real number, and $q = q(\varphi)$ is a complex number whose value must be found.

If we substitute 3.92 into the finite-difference scheme 3.90 with the "right" derivative, we get

$$\frac{q^{n+1} e^{i\varphi k} - q^n e^{i\varphi k}}{\Delta t} + \frac{q^n e^{i\varphi(k+1)} - q^n e^{i\varphi k}}{h} = 0. \quad (3.93)$$

Cancel factors $q^n e^{i\varphi k}$, and we get

$$q = 1 + \gamma(1 - e^{i\varphi k}) = (1 + \gamma - \gamma \cos \varphi) - i\gamma \sin \varphi, \quad (3.94)$$

where

$$\gamma = \frac{a \, \Delta t}{h} \quad (3.95)$$

is the parameter which characterizes the finite-difference scheme. Therefore, function 3.92 will be the particular solution to the finite-difference scheme if, for a given φ, the value of q is determined by formula 3.94. The square of the module of q is

$$|q|^2 = (1 + \gamma - \gamma \cos \varphi)^2 + \gamma^2 \sin^2 \varphi = 1 + 4\gamma(\gamma + 1) \sin^2(\varphi/2). \quad (3.96)$$

Since $a > 0$, then the parameter γ is also positive. This means that independent of the ratio of h and Δt for all φ, and that $\sin \varphi/2 \neq 0$, we have

$|q| > 1$. That is, the necessary condition of stability is not valid, and the finite-difference scheme 3.90 is unconditionally unstable.

Let us now consider the scheme 3.91. Similar considerations give us the following expression for $|q|^2$:

$$|q|^2 = 1 - 4\gamma(1 - \gamma)\sin^2 \varphi/2. \tag{3.97}$$

Therefore,

$$|q(\varphi)| \leq 1 \quad \text{when } \gamma \leq 1,$$
$$\tag{3.98}$$
$$|q(\varphi)| > 1 \quad \text{when } \gamma > 1.$$

The last inequality means that for $\gamma > 1$, the necessary condition for stability is not satisfied. Therefore, in this case, the finite-difference scheme 3.91 is not stable. From the first inequality, we can only conclude that for $\gamma \leq 1$, the finite-difference scheme can be stable, but because the method of harmonics gives us only the necessary conditions, then obtaining sufficient conditions, we need to use another method. The sufficient condition can be found using the following considerations. Let us rewrite the finite-difference scheme in the following form:

$$u_k^{n+1} = (1 - \gamma)u_k^n + \gamma u_{k-1}^n. \tag{3.99}$$

Since $\gamma \leq 1$, the coefficients in the right-hand side of this equation are positive and we can write the following inequality:

$$\left|u_k^{n+1}\right| \leq (1 - \gamma)\left|u_k^n\right| + \gamma\left|u_{k-1}^n\right|. \tag{3.100}$$

If we take the maximum of both sides with respect to k on the corresponding time level, we get

$$||u^{n+1}||_{max} \leq ||u^n||_{max}$$

or

$$||u^{n+1}||_{max} \leq ||u^n||_{max} \leq \cdots \leq ||u^0||_{max}. \tag{3.101}$$

This inequality, for our case, means stability of the finite-difference scheme.

Therefore, the necessary and sufficient condition for stability of finite-difference scheme 3.91 is

$$\Delta t \leq \frac{h}{a}, \ a > 0, \tag{3.102}$$

and finite-difference scheme 3.91 is conditionally stable. The condition in 3.102 is called the Courant condition and is a typical condition for stability of a finite-difference scheme for hyperbolic equations. Let us note that for the case of a non-uniform grid and non-constant coefficient $a = a(x, t)$, the Courant condition must be satisfied for all grid nodes and for all time moments:

$$\Delta t \leq \frac{x_{k+1} - x_k}{a(x_k, t^n)}. \tag{3.103}$$

It can be explained in a similar way for the case of the heat equation by using the principle of frozen coefficients.

The Finite-Difference Scheme with "Central Difference"

Along with "one-sided" finite-difference schemes that we have considered, it is logical to analyze the so-called "central difference" scheme:

$$\frac{u_k^{n+1} - u_k^n}{\Delta t} + a \frac{u_{k+1}^n - u_{k-1}^n}{h} = 0, \; k = \pm 1, \pm 2, \ldots \tag{3.104}$$

$$u_k^0 = v_0(x_k).$$

This scheme is in contrast to the previous ones that have a second-order truncation error in the spatial variable. Let us show that this scheme is unconditionally unstable. Using the method of harmonics, we obtain the following expressions for number q:

$$q = 1 - \gamma \frac{e^{i\varphi} - e^{-i\varphi}}{2} = 1 - i\gamma \sin \varphi$$

and

$$q = 1 + \gamma^2 \sin^2 \varphi.$$

This means that for all γ and φ, $\sin \varphi \neq 0$, and we get $|q(\varphi)| > 1$. Therefore, the finite-difference scheme 3.104 is unstable for any relation for steps h and Δt.

Implicit Finite-Difference Scheme

All schemes that we have investigated were explicit schemes. From a practical point of view, it is important to understand what happens if we use implicit schemes. Let us consider an implicit scheme with a "left" difference. Hence, we have

$$\frac{u_k^{n+1} - u_k^n}{\Delta t} + a \frac{u_k^{n+1} - u_{k-1}^{n+1}}{h} = 0, \; k = \pm 1, \pm 2, \ldots \tag{3.105}$$

$$u_k^0 = v_0(x_k).$$

For the approximation of the spatial derivative, we use values of the solution from a new $(n + 1)$ time level. Here, we will not consider the question of how to arrange the computation procedure for this type of approximation. We will only concentrate on stability properties.

The method of harmonics gives us the following expression for q:

$$q = \frac{1}{1 + \gamma(1 - e^{-i\varphi})} = \frac{1}{[1 + \gamma(1 - \cos\varphi)] + i\gamma\sin\varphi},$$

and because $\gamma > 0$, we have

$$|q| < 1$$

for all values of γ and φ, or in other words, the implicit scheme 3.105 is unconditionally stable.

Therefore, it makes sense to use the implicit scheme in practical computations. Although, when we use implicit schemes, we need to take into account that, in general, for equations of gas dynamics, the implicit finite-difference schemes give us the system of non-linear equations and to solve this system of equations, we need to use an iteration method. The convergence conditions for the method of solving the system of non-linear equations, usually can also be formulated in terms of the relation between time and spatial steps. In some cases, these conditions can be more restrictive for the time step than conditions for the explicit finite-difference scheme. This means that use of the implicit scheme does not solve all problems with stability. We also need to choose the appropriate iteration method for solving the system of non-linear equations.

3.6.4   Stability Conditions for 1-D Acoustic Equations

Let us consider one-dimensional acoustic equations

$$\rho_0 \frac{\partial W X}{\partial t} = -\frac{\partial p}{\partial x} \tag{3.106}$$

and

$$\frac{\partial p}{\partial t} = -c_0^2 \rho_0 \frac{\partial W X}{\partial x}, \tag{3.107}$$

where we omit the sign ˜ for the sake of brevity.

The analog of our explicit finite-difference scheme for gas dynamics equations for this type of equation, is as follows:

$$\rho_0 \frac{W X_k^{n+1} - W X_k^n}{\Delta t} = -\frac{p_k^n - p_{k-1}^n}{h}, \tag{3.108}$$

$$\frac{p_k^{n+1} - p_k^n}{\Delta t} = -\rho_0 c_0^2 \frac{W X_{k+1}^{n+1} - W X_k^{n+1}}{h}. \tag{3.109}$$

We will find the necessary stability conditions for this finite-difference scheme by using the method of harmonics. In the case where we have two variables and two equations, we will try to find the solution in the following form:

$$p_k^n = P q^n e^{ik\varphi}, \tag{3.110}$$

$$W X_k^n = W q^n e^{ik\varphi}, \tag{3.111}$$

where quantities q and φ have the same meaning as in previous considerations: quantities P and W are the amplitudes of harmonics. After substitution of this particular solution in the finite-difference scheme, we get

$$\rho_0 \, W \, e^{ik\varphi} \frac{q^{n+1} - q^n}{\Delta t} = -P \, q^n \frac{e^{ik\varphi} - e^{i(k-1)\varphi}}{h}, \qquad (3.112)$$

$$P \, e^{ik\varphi} \frac{q^{n+1} - q^n}{\Delta t} = -\rho_0 \, c_0^2 \, W \, q^{n+1} \frac{e^{i(k+1)\varphi} - e^{ik\varphi}}{h}. \qquad (3.113)$$

Or after cancellation by common factors, we get

$$\rho_0 \, W \, (q - 1) + P \, \frac{\Delta t}{h} \, (1 - e^{-i\varphi}) = 0, \qquad (3.114)$$

$$P \, (q - 1) + W \, \rho_0 \, c_0^2 \, \frac{\Delta t}{h} \, (e^{i\varphi} - 1) = 0. \qquad (3.115)$$

This system of equations, with respect to W and P, will have a nontrivial solution if the determinant will be equal to zero. This condition gives us the following equation for q:

$$q^2 - 2 \left(1 - 2\gamma^2 \sin^2(\varphi/2)\right) q + 1 = 0, \qquad (3.116)$$

where

$$\gamma = \frac{c_0 \, \Delta t}{h}.$$

The roots of this equation are

$$q_\pm = 1 - 2\gamma \sin(\varphi/2) \left[\gamma \sin(\varphi/2) \pm \sqrt{\gamma^2 \sin^2(\varphi/2) - 1} \right]. \qquad (3.117)$$

Let us consider a case where the Courant condition is not satisfied, that is, $\gamma > 1$. Then there are values of φ where the expression under the square root is positive. Hence, roots q_+ and q_- are real, and the absolute value of q_+ is bigger than 1. Therefore, we can conclude that our finite-difference scheme is not stable if the Courant condition is not satisfied.

It is possible to show (see, for example, [110]) that the Courant condition $\gamma \le 1$ or

$$\Delta t \le \frac{h}{c_0} \qquad (3.118)$$

is a necessary condition for stability of the finite-difference scheme being considered. We do not have enough space to present proof of this fact.

3.7   Homogeneous Finite-Difference Schemes. Artificial Viscosity

If we consider the flows of compressible gases in the absence of viscosity and heat conductivity, then it follows, as shown in the introduction, that they are described by discontinuous quantities. Therefore, there is no system of differential equations applicable to the description of a discontinuous flow for the simple reason that the parameters of discontinuous flows are not differentiable. The singularities arising in the flow parameters are the following: a weak discontinuity (discontinuity of the derivatives), a contact discontinuity (a boundary separating gases with different thermodynamic parameters), and finally, a strong discontinuity (a shock wave). The set of all flows with singularities indicated can be described in a unified way – the conservation laws of mass, momentum, and energy must be satisfied for any distinguished portion of the gas or space.

In general, these conservation laws are not sufficient for a complete description of the flow: they must be augmented by requiring a nondecrease of the entropy of any fixed mass of gas. This requirement excludes the appearance of non-physical solutions. The integral conservation laws with the condition of nondecrease of the entropy, form the *conservative system of equations of gas dynamics* that is applicable to any flows including those with discontinuous parameters.

The system of integral conservation laws is extremely inconvenient to approximate by finite-differences on a fixed mesh in the case where there is the presence of discontinuities of the solution, since it requires explicitly distinguishing the lines of discontinuity, and satisfying the integral conservation laws on the lines of discontinuity (the Hugoniot conditions).

Ignoring the discontinuities that arise in the flow and approximating the integral conservation laws directly in the same way as for smooth flows lead, as a rule, to the solution where the shock wave is replaced by oscillations of large amplitude (see Figure 3.13).

It is important to note that such behavior of the approximate solution is not a result of instability of the finite-difference scheme. The refinement of the grid in space and time gives the same solution. The mechanism of such oscillations can be roughly explained as follows.

The discrete model of media, which corresponds to the finite-difference scheme, can be interpreted as a system that contains concentrated masses (thin parallel plates with mass that is equal to the mass of grid meshes), which interact between each other without friction by means of some elastic media that fills the space between the plates. In the mechanical sense, this system is equivalent to a set of balls with different masses, which can only move in a horizontal direction, and are connected by strings of different stiffness (see Figure 3.14). When a strong disturbance (analog of a shock wave) propagates through this system, the balls begin to move. This movement has oscillatory character and is not dumped because of the absence of friction in the system.

Direct approximation of the integral conservation laws can lead to the right description of the flow parameters only in the case where the finite-difference scheme used contains dissipation or "approximation viscosity".

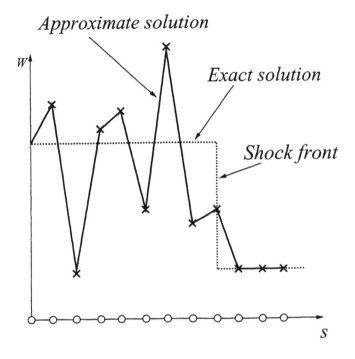

Figure 3.13: *Structure of the solution*

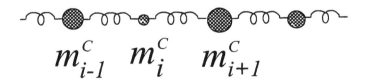

Figure 3.14: *Mechanical interpretation*

In approximating the non-dissipative terms of the equations, the discrete approximation contributes small increments (and large increments for the discontinuous solution) which can stabilize the numerical solution. The finite-difference scheme that we consider in this book does not contain enough "approximation viscosity" so it is necessary to introduce *artificial viscosity* explicitly in the finite-difference scheme.

In the method for artificial viscosity, we avoid a detailed consideration of shock waves and consider the flow of gases that possess some viscosity which sometimes is quite unrelated to physical viscosity (the reason why it is called artificial viscosity). By introducing this viscosity into the equations of gas dynamics, we approximately describe shock waves as smooth shock transition, which was considered in the first section of this chapter.

Introducing artificial viscosity gives us the ability to compute all types of flows using the same discrete equations. The finite-difference scheme of this type is called the *homogeneous finite-difference scheme*.

3.8   Artificial Viscosity in 1-D

For one-dimensional flows, artificial viscosity is usually introduced as an addition to pressure, that is, by replacing pressure p with quantity $\tilde{p}$:

$$\tilde{p} = p + \omega . \tag{3.119}$$

There are two main types of artificial viscosity: *linear viscosity*, or *Landshoff viscosity* [76], and *quadratic viscosity*, or *von Neumann and Richtmyer viscosity* [93].

The introduction of these types of viscosity is based on an analysis which was made in the introduction for differential equations (see section 1.11.6).

Since artificial viscosity is introduced as an addition to pressure, then it also related to cells, or $\omega \in HC$.

In the case of linear viscosity, it is necessary to introduce the analog of the expression $\nu \, \partial W / \partial x$ in cells. Coefficient ν can be chosen using a dimensional consideration:

$$\nu = \nu_* \, L \, \rho \, U , \tag{3.120}$$

where ν_* is a dimensionless parameter, L, the characteristic for linear size, and U, the characteristic for speed. It is common for U to be taken as equal to the local adiabatic speed of sound, C_A, the role of L plays the size of the cell, and ρ is equal to density in the cell, and instead of product $L \rho$, we can use the mass of the cell. Finally, the discrete linear artificial viscosity is as follows:

$$\omega_i = \nu_* \, m_i^C \, (C_A)_i \, \frac{W_{i+1} - W_i}{x_{i+1} - x_i} . \tag{3.121}$$

In the case of quadratic viscosity, it is necessary to introduce the analog of $\omega = \mu \, (\partial W / \partial x)^2$. Coefficient μ can be chosen using dimensional considerations:

$$\mu = \mu_* \, L^2 \, \rho \,, \qquad (3.122)$$

where μ_* is the dimensionless constant. Now, because $L \, \rho$ has the dimension of the mass, we can write the discrete analog for quadratic viscosity as follows:

$$\omega_i = \mu_* \, (x_{i+1} - x_i) \, m_i^C \, (\frac{W_{i+1} - W_i}{x_{i+1} - x_i})^2 \,. \qquad (3.123)$$

A drawback of linear viscosity is that it acts throughout the flow, so that strong smoothing of the shock wave corresponding to a large ν, is always connected with the reduction of the accuracy of the computation in the region of the smooth flow. Again, this is a consequence of the fact that the characteristic width of the front, in this case, depends on the intensity of the shock.

For quadratic viscosity, the width of the shock transition is order $O(h)$ and does not depend on the strength of the shock wave. The viscous term ω in a smooth part of the flow has order $O(h^2)$ and hence, does not have a strong affect on the accuracy of the computations. These are the advantages of quadratic viscosity.

From the introduction, we know another special feature of quadratic artificial viscosity. In contrast to linear viscosity, in which the approach of the limiting profile $\eta(\xi)$ to the asymptotic values η_0 and η_1 occurs analytically, in the case of quadratic viscosity, this approach does not occur analytically: at conjugate points A and B, the second derivative has discontinuity. Since the gradients are large in the zone of the shock wave, this discontinuity of the derivatives leads to a constant source of perturbations which cause strong oscillations of the hydrodynamical quantities in the neighborhood of the front. The amplitude of the oscillations decreases as the coefficient of viscosity decreases. The strong dependence of the profile of the shock wave on coefficient μ_* is a characteristic feature of quadratic viscosity, which in some cases, hampers interpretation of the numerical results.

Since viscosity is introduced to smooth existing shock waves and those arising from compression waves while in rarefaction waves, the gradients decrease even in the absence of viscosity. It follows that in the numerical algorithm to increase the accuracy, it is expedient to eliminate the effect of the viscosity in the region of rarefaction waves. That is, take the viscosity coefficient equal to zero in such regions. In the planar case in compression waves and shock waves, the inequality $\Delta W / \Delta x < 0$ is satisfied, while for the rarefaction waves, $\Delta W / \Delta x \geq 0$. Therefore, we will use the following:

$$\omega_i = 0 \,, \quad \text{if } \Delta W \geq 0 \,. \qquad (3.124)$$

In the framework of our considerations, it is important to note that our main goal in this book is to construct good approximations for spatial differential operators which participate in the original differential equations.

From this point of view, the spatial derivative $\partial W/\partial x$ which participates in the expression for artificial viscosity, is the 1-D analog of div W, and linear viscosity is just the analog of the scalar part of the real viscosity term in the equations of gas dynamics for viscous compressible flows. Hence, to be consistent with the differential case, we need to use the discrete analog of operator div for the approximation of viscosity, which we used for the approximation of the energy equation. In 1-D, the approximation of the differential part of the viscosity term looks trivial, but for 2-D, it is a real problem.

Usually, artificial viscosity is computed using parameters from a previous time step, that is, *explicitly*, for both explicit and implicit finite-difference schemes. For the implicit finite-difference scheme, the work of viscous forces is taken into account before starting the non-linear iteration for pressure.

Readers who are interested in the details of introducing of artificial viscosity can find them in the following papers: [93], [76], [77], [117], [100], [101], [151], [102], [106], [110], [122], [88].

3.9 The Numerical Example

As an example of a numerical computation, we will consider a solution to a problem regarding the formation of a strong shock wave by a piston. The statement of the problem is as follows. The working fluid is assumed to be an ideal gas with $\gamma = 5/3$, compressed by a piston (left end of the segment) moving to the right with velocity of 1. The initial conditions involve a stationary gas which occupies segment $[0, 1]$, with a density of 1 and an internal energy of 10^{-4}. The right end of the segment is a stationary wall. The expected post shock conditions are given by a pressure of 1.333, a density of 4, an internal energy of 0.5, and a shock speed of 1.333.

In our computation, we used a linear artificial viscosity with coefficient $\nu_* = 1.5$. For an explicit finite-difference scheme, we will use $dt = 0.00025$, and for an implicit one, we will use $dt = 0.001$. For the solution to the system of linear equations for dp in the framework of an implicit scheme, we have used the Gauss-Seidel iteration method, the tolerance for the linear and non-linear iteration was 10^{-4}. The numerical solution for the explicit and implicit schemes are practically the same, and we present here only the numerical results for the explicit scheme. The number of nodes is 101.

In Figures 3.15, 3.16, 3.17, and 3.18, we present the exact solution (solid line) and the approximate solution (broken line) for velocity, pressure, density, and energy, respectively.

The behavior of density and internal energy near the piston is called *entropy trace* and is explained in detail in [102]. Briefly, it can be explained as follows. The main reason is that the exact formulas that describe the discontinuous solution as shock waves are valid only when we consider media

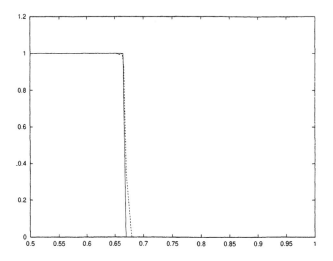

Figure 3.15: *Velocity, explicit algorithm, $t = 0.5$.*

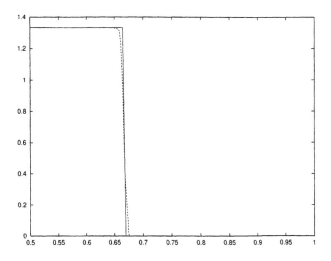

Figure 3.16: *Pressure, explicit algorithm, $t = 0.5$.*

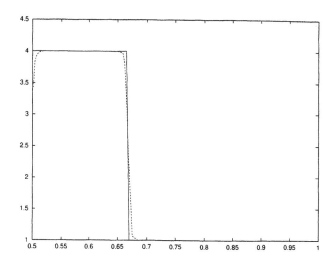

Figure 3.17: *Density, explicit algorithm, $t = 0.5$.*

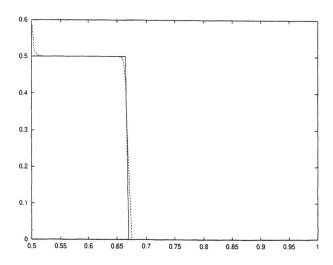

Figure 3.18: *Internal energy, explicit algorithm, $t = 0.5$.*

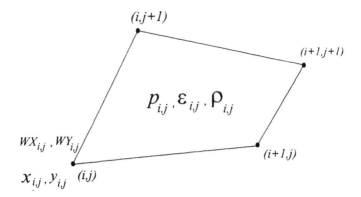

Figure 4.19: *Discretisation for gas dynamics equations*

without dissipation. In the presence of dissipation, the solution becomes continuous, and there is some structure to the wave front. For the discrete model of media, which is described by the finite-difference scheme, the jump-like initial profiles do not correspond to the profiles in the discrete shock. Therefore, some reconstruction of the profiles begins, and as a result, the solution asymptotically reaches the regime of shock with a "viscous" structure. The "entropy trace" at the point of initial discontinuity is the consequence of this reconstruction. The pressure and velocity in this region are smooth, therefore, the trace cannot be smoothed out by gas dynamics mechanisms. If initial data corresponds to the "viscous" structure of the discrete shock, then entropy trace will not appear.

4   The Finite-Difference Algorithm in 2-D

4.1   Discretisation in 2-D

We will use discretisation that is generally accepted for the Lagrangian numerical methods. For coordinates and Cartesian components of velocity vectors, nodal discretisation is used. That is, $\vec{W} \in \mathcal{H}N$. For pressure, specific internal energy, and density, cell-valued discretisation is used. That is, p, ε, $\rho \in HC$. The discretisation is presented in Figure 4.19.

Since we use the Lagrangian description of media, we need to associate some mass $m_{i,j}^C$ with each fluid volume $V_{i,j}$ that will not be changed in time. The usual way to determine this mass is as follows. We have the initial condition for density, that is to say, we have density at the initial time moment as a function of space variables $\rho(x, y, 0)$, and we have the positions of the grid nodes at the initial time moment. Therefore, we can compute all volumes $V_{i,j}$ and then determine the mass $m_{i,j}^C$ as follows:

$$m_{i,j}^C = \rho(x_{i,j}^*, y_{i,j}^*, 0)\, V_{i,j}\,, \tag{4.1}$$

where $(x_{i,j}^*, y_{i,j}^*)$ are the coordinates for the center of the cell. Hence, for each cell, we have mass $m_{i,j}^C$.

4.2 Discrete Operators in 2-D

4.2.1 The Prime Operator

As we have already mentioned, the expression for prime operator DIV has to be constructed on the basis of the definition in 2.18. Now we must decide what volume will play the role of δV. For this type of discretisation, it is natural to use volume $V_{i,j} = \Omega_{i,j}$, which is the area of quadrangle $(i, j), (i + 1, j), (i + 1, j + 1), (i, j + 1)$, as shown in Figure 4.19. Since $V_{i,j}$ does not depend explicitly on time, we get

$$(\mathrm{DIV}\vec{W})_{i,j} = \frac{1}{V_{i,j}} \frac{dV_{i,j}}{dt} = \tag{4.2}$$

$$\frac{1}{V_{i,j}} \sum_{k,l=0}^{1} \left(\frac{\partial V_{i,j}}{\partial x_{i+k,j+l}} \frac{\partial x_{i+k,j+l}}{dt} + \frac{\partial V_{i,j}}{\partial y_{i+k,j+l}} \frac{\partial y_{i+k,j+l}}{dt} \right).$$

Now, because the derivatives of the coordinates with respect to time are the velocities of the related vertices, we get

$$\frac{dx_{i+k,j+l}}{dt} = WX_{i+k,j+l}, \quad \frac{dy_{i+k,j+l}}{dt} = WY_{i+k,j+l}. \tag{4.3}$$

Finally, we get the following expression for DIV:

$$(\mathrm{DIV}\vec{W})_{i,j} = \frac{1}{V_{i,j}} \sum_{k,l=0}^{1} \left(\frac{\partial V_{i,j}}{\partial x_{i+k,j+l}} WX_{i+k,j+l} + \frac{\partial V_{i,j}}{\partial y_{i+k,j+l}} WY_{i+k,j+l} \right).$$

$$\tag{4.4}$$

For $V_{i,j}$, we have the following formula:

$$V_{i,j} = 0.5 \left[(x_{i,j} - x_{i+1,j+1})(y_{i+1,j} - y_{i,j+1}) - \right.$$

$$\tag{4.5}$$

$$\left. (x_{i+1,j} - x_{i,j+1})(y_{i,j} - y_{i+1,j+1}) \right].$$

Then, for the derivatives of volume with respect to the coordinates, we get

$$\frac{\partial V_{i,j}}{\partial x_{i,j}} = 0.5\,(y_{i+1,j} - y_{i,j+1}), \quad \frac{\partial V_{i,j}}{\partial x_{i+1,j+1}} = -0.5\,(y_{i+1,j} - y_{i,j+1}),$$

$$\frac{\partial V_{i,j}}{\partial x_{i+1,j}} = -0.5\,(y_{i,j} - y_{i+1,j+1}), \quad \frac{\partial V_{i,j}}{\partial x_{i,j+1}} = 0.5\,(y_{i,j} - y_{i+1,j+1}),$$

$$\frac{\partial V_{i,j}}{\partial y_{i,j}} = -0.5\,(x_{i+1,j} - x_{i,j+1}), \quad \frac{\partial V_{i,j}}{\partial y_{i+1,j+1}} = 0.5\,(x_{i+1,j} - x_{i,j+1}),$$

$$\frac{\partial V_{i,j}}{\partial y_{i+1,j}} = 0.5\,(x_{i,j} - x_{i+1,j+1}), \quad \frac{\partial V_{i,j}}{\partial y_{i,j+1}} = -0.5\,(x_{i,j} - x_{i+1,j+1}).$$

Using these expressions, we can transform formula 4.4 into

$$(\text{DIV}\vec{W})_{i,j} =$$
$$\frac{1}{V_{i,j}}\,0.5\,[((y_{i+1,j} - y_{i,j+1})\,(WX_{i,j} - WX_{i+1,j+1}) - \qquad (4.6)$$
$$(y_{i,j} - y_{i+1,j+1})\,(WX_{i+1,j} - WX_{i,j+1})) -$$
$$((x_{i+1,j} - x_{i,j+1})\,(WY_{i,j} - WY_{i+1,j+1}) -$$
$$(x_{i,j} - x_{i+1,j+1})\,(WY_{i+1,j} - WY_{i,j+1}))]\,.$$

If we compare the obtained formula with formula 3.14 for operator DIV from the chapter related to discretisation of elliptic equations, we can see that it is the same formula. Therefore, the prime operator is the same, and consequently, the same will be operator GRAD and all considerations from the chapter related to elliptic equations are still valid.

In particular, operator DIV has divergence property. Now using definition 4.2, it can be shown more elegantly. In fact, to check the divergence property, we will need to evaluate the expression

$$(\text{DIV}\,\vec{W}, I)_{HC}\,,$$

where $I \in HC$ is a function where $I_{i,j} \equiv 1$. Using definition 4.2, we get

$$(\text{DIV}\,\vec{W}, I)_{HC} = \qquad (4.7)$$
$$\sum_{i=1}^{M-1}\sum_{j=1}^{N-1}\left(\frac{1}{V_{i,j}}\frac{dV_{i,j}}{dt}\right)V_{i,j} = \frac{d}{dt}\sum_{i=1}^{M-1}\sum_{j=1}^{N-1}V_{i,j} = \frac{d}{dt}V\,,$$

where V is the total volume. Since V depends only on the values of velocity on the boundary, the previous equations express the divergence property of the discrete operator DIV. Now, if all the nodes move with constant velocity, then volumes $V_{i,j}$ do not change, and consequently, operator DIV on the constant vector function is equal to zero. In the next section, we will show that from this property of DIV we can obtain the divergence property for GRAD.

As we know from the case of the cell-valued discretisation for scalar functions, the prime operator B is operator DIV in the internal cells and the approximation for the normal component of the vector on the boundary. Therefore, to complete the definition for the prime operator, we will need to determine the approximation for $(\vec{n}, \vec{W})$ on the boundary. We will use the formulas in 3.15 from Chapter 3 that we used for the approximation of

Robin boundary conditions for elliptic equations. For the bottom boundary $j = 1; i = 1, \ldots, N - 1$, we have

$$(\vec{n}, \vec{W})_{i,1} \approx \tag{4.8}$$

$$-\left(\frac{WX_{i,1} + WX_{i+1,1}}{2} \frac{y_{i+1,1} - y_{i,1}}{l\xi_{i,1}} - \right.$$

$$\left. \frac{WY_{i,1} + WY_{i+1,1}}{2} \frac{x_{i+1,1} - x_{i,1}}{l\xi_{i,1}} \right).$$

This expression is the approximation of a normal component of velocity in the middle of the boundary segment.

4.2.2 The Derived Operator

The derived operator GRAD in internal nodes $i = 2, \ldots, M - 1; j = 2, \ldots, N - 1$, has the following form (see formula 3.16 from Chapter 3):

$$GX_{ij} = \left(\frac{y_{i,j+1} - y_{i+1,j}}{2} u_{ij} + \frac{y_{i-1,j} - y_{i,j+1}}{2} u_{i-1,j} + \right.$$

$$\left. \frac{y_{i,j-1} - y_{i-1,j}}{2} u_{i-1,j-1} + \frac{y_{i+1,j} - y_{i,j-1}}{2} u_{i,j-1} \right) / VN_{ij},$$

$$\tag{4.9}$$

$$GY_{ij} = -\left(\frac{x_{i,j+1} - x_{i+1,j}}{2} u_{ij} + \frac{x_{i-1,j} - x_{i,j+1}}{2} u_{i-1,j} + \right.$$

$$\left. \frac{x_{i,j-1} - x_{i-1,j}}{2} u_{i-1,j-1} + \frac{x_{i+1,j} - x_{i,j-1}}{2} u_{i,j-1} \right) / VN_{ij}.$$

As we mentioned in the chapter related to elliptic equations, formulas for all nodes have the form 4.9, if we introduce the following fictitious nodes: for $j = 1, \ldots, N$

$$(x_{0,j} = x_{1,j}, y_{0,j} = y_{1,j}), (x_{M+1,j} = x_{M,j}, y_{M+1,j} = y_{M,j}),$$

and for $i = 1, \ldots, M$

$$(x_{i,0} = x_{i,1}, y_{i,0} = y_{i,1}), (x_{i,N+1} = x_{i,N}, y_{i,N+1} = y_{i,N}).$$

It is sometimes useful to use another form of the formulas in 4.9 that is similar to the formula in 4.4 for operator DIV, such as the following:

$$GX_{i,j} = -\frac{1}{VN_{i,j}} \sum_{k,l=0}^{1} \frac{\partial V_{i-k,j-l}}{\partial x_{i,j}} u_{i-k,j-l},$$

$$\tag{4.10}$$

$$GY_{i,j} = -\frac{1}{VN_{i,j}} \sum_{k,l=0}^{1} \frac{\partial V_{i-k,j-l}}{\partial y_{i,j}} u_{i-k,j-l}.$$

Let us show that operator GRAD satisfies the divergence property. Let us recall that operators $\mathcal{B}$ and $-$GRAD are adjoint to each other:

$$(\vec{W}, \mathrm{GRAD}\,p)_{\mathcal{H}N} = (\mathcal{B}\,\vec{W}, p)_{HC}\,.$$

If we take $\vec{W} = \vec{I} = (1, 0)$ (the constant vector), DIV $\vec{I} \equiv 0$ and the right-hand side of the previous equation will contain only the term related to the boundary, and we get

$$(\vec{I}, \mathrm{GRAD}\,u)_{\mathcal{H}N} = R \approx \oint_S (\vec{I}, \vec{n})\,p\,ds\,. \qquad (4.11)$$

Similar equations can be obtained for any constant vector, and therefore, we have proven that operator GRAD satisfies the divergence property.

4.2.3   Boundary Conditions and Discretisations

We will consider two types of boundary conditions. The first type of boundary condition is the *free boundary*, where we have pressure p as a given function of the coordinate and time on the boundary. The second type of boundary conditions is *impermeability boundary conditions*, where we have the normal component of velocity as a given function on the boundary. The latter boundary condition type, for example, corresponds to the physical problems where media is bounded by some impermeable walls that are moved with a given velocity.

In the case of free boundary conditions, the conditions are similar to the case of Dirichlet boundary conditions for elliptic equations. We introduce the values of pressure on the boundary, and these values are associated with the sides of cells that form the boundary (see Figure 4.20). For this type of boundary condition, pressure on the boundary is given, and these given values participate in the formulas for GRAD in boundary nodes.

In the case where the normal component of the velocity vector is given on the boundary, we will use the formulas in 4.8 to define WX and WY on the boundary (see details later). Let us note that we have the natural approximation for the normal component of the velocity vector, and for each boundary segment, we have a unique vector of normal.

4.3   The Semi-Discrete Finite-Difference Scheme in 2-D

First, let us consider the *semi-discrete* finite-difference scheme, where time is still continuous and only discretisation in space is made.

For the continuity equation, we have the following approximation:

$$\rho_{i,j}(t) = \frac{m^C_{i,j}}{V_{i,j}(t)}\,, \qquad (4.12)$$

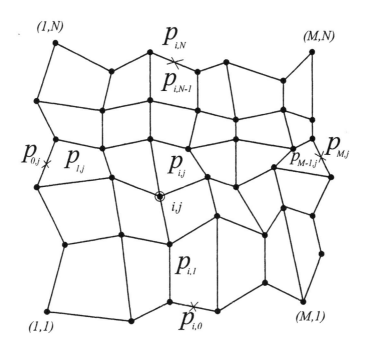

Figure 4.20: *Pressure discretisation in 2-D.*

and our definition of DIV is the same as

$$\frac{d\rho_{i,j}}{dt} + \rho_{i,j} \left(\text{DIV } \vec{W} \right)_{i,j} = 0. \tag{4.13}$$

Let us note that we do not have the problem with multiplication in the second term in the previous equation because ρ and DIV $\vec{W}$ belong to the same discrete space, HC.

The discrete equation for specific internal energy is as follows:

$$\rho_{i,j} \frac{d\varepsilon_{i,j}}{dt} = -p_{i,j} \left(\text{DIV } \vec{W} \right)_{i,j}. \tag{4.14}$$

There is also no problem here with multiplication because all quantities belong to space HC.

When approximating the momentum equation, we have the problem of how to approximate the multiplication of ρ and $d\vec{W}/dt$ because these quantities belong to different spaces. Since the right-hand side in the momentum equation is GRAD p, we need to have a projection operator from space HC to space HN. Let us denote this operator as M:

$$M : HC \rightarrow HN,$$

then, the discrete momentum equation will be as follows:

$$(M\,\rho)_{i,j} \frac{d\vec{W}_{i,j}}{dt} = -(\text{GRAD}\,p)_{i,j}. \tag{4.15}$$

Let us consider what requirements we have for operator M. First, the previous discrete equation must express conservation of momentum. Recall that conservation of momentum in the differential case follows from the equation

$$\rho \frac{d\vec{W}}{dt} = -\text{grad}\,p$$

by its multiplication by the elementary volume and then integration over the domain. In addition, it is taken into account that the mass of the fluid particle $\rho\,dV$ does not depend on time, and it is possible to factor differentiation with respect to time outside the integral sign. In the discrete case, conservation of momentum must be obtained in a similar way. Therefore, the expression

$$(M\,\rho)_{i,j}\,VN_{i,j}, \tag{4.16}$$

which is the analog of $\rho\,dV$, must not be dependent on time. The expression in 4.16 has the meaning of mass and is related to the node. We have only one quantity which has the meaning of mass and does not depend on time, and that is the mass of cell $m_{i,j}^C = \rho_{i,j}\, V_{i,j}$. Hence, it is natural to use the following relation to determine operator M:

$$(M\,\rho)_{i,j}\, V N_{i,j} = \frac{1}{4}\left(m_{i,j}^C + m_{i-1,j}^C + m_{i,j-1}^C + m_{i-1,j-1}^C\right) = \frac{1}{4}\sum_{k,l=0}^{1} m_{i-k,j-l}^C\,.$$

In the right-hand side of the last equation, we have a mass of cells that have node (i,j) as a vertex. From the previous equation, we get the following definition of operator M:

$$(M\,\rho)_{i,j} = \frac{1}{4\,V N_{i,j}} \sum_{k,l=0}^{1} \rho_{i-k,j-l}\, V_{i-k,j-l}\,. \tag{4.17}$$

From this definition of operator M, the discrete analog for the conservation of momentum has the following form:

$$\frac{d}{dt}\left((M\rho)\,\vec{W},\vec{I}\right)_{\mathcal{HN}} = -(\mathrm{GRAD}\,p,\vec{I})_{\mathcal{HN}} \approx \oint_S (\vec{n},\vec{I})\,p\,dS\,. \tag{4.18}$$

Let us now consider what is the discrete analog of the conservation of total energy. Using the same procedure as for the differential case, we get

$$\left(M\rho,\frac{d}{dt}\left(\frac{\vec{W}^2}{2}\right)\right)_{\mathcal{HN}} + \left(\rho,\frac{d\varepsilon}{dt}\right)_{HC} = \tag{4.19}$$
$$-(\mathrm{GRAD}\,p,\vec{W})_{\mathcal{HN}} - (p,\mathrm{DIV}\,\vec{W})_{HC}\,.$$

The notation $(\bullet,\bullet)_{HC}$ means that we make a summation only over the interior cells, and it is not a real inner product when p does not equal to zero on the boundary. From the definition of operator GRAD, we can conclude that the expression in the right-hand side is divergent. Now, using the definition of the adjoint operator for transformation of the first term in the left-hand side, we get

$$\left(\rho,\frac{d}{dt}\left[M^*\left(\frac{\vec{W}^2}{2}\right)+\varepsilon\right]\right)_{H\!C} \approx -\oint_S (\vec{W},\vec{n})\,p\,dS\,. \tag{4.20}$$

When deriving this equation, we are taking into account that

$$(M^*\,\varphi)_{i,j} = \frac{1}{4}\sum_{k,l=0}^{1} \varphi_{i+k,j+l}\,. \tag{4.21}$$

In other words, the coefficients of this operator do not depend on time and we can put operator M^* under time differentiation.

4.4 The Finite-Difference Algorithm in 2-D

4.4.1 The Fully Discrete, Explicit, Computational Algorithm

Initial Conditions

At the initial moment of time, we need to generate a grid in our domain. For our algorithm, it must be a logically rectangular, boundary fitted grid, that is, the segments of the grid approximate the boundary of the domain. There are many publications devoted to different aspects of the generation of logically rectangular grids. We do not have enough space here to discuss this problem and refer to book [70], where readers can find the necessary references.

The grid divides the domain into quadrangular meshes with vertices at the intersection of the grid lines. These meshes identify as Lagrangian (fluid) particles, and we will observe the behavior of all parameters related to these particles. Strictly speaking, the mesh which has the form of the quadrangle in the initial moment, on the next time moment, must have a curvilinear boundary. Since we have discrete equations only for the position of the nodes of the grid, the vertices of the meshes move with the fluid, and at each time moment, we connect these vertices by segments of straight lines. Therefore, for each time moment, we have a logically rectangular grid which is formed by rectangles.

The computational domain can contain different materials with different equations of state; therefore, we have to construct a grid so that each mesh will contain only one material. This means that we have an additional requirement for the grid generation procedure, namely, the grid lines must also coincide with the interface between different materials.

Since we use the Lagrangian description for media, we need to associate some mass $m_{i,j}^C$ with each fluid volume $V_{i,j}$, that will not be changed in time. The usual way to determine this mass is as follows. We have an initial condition for density, that is, we have density at the initial time moment as a function of space variables $\rho(x, y, 0)$, and we have the positions of the grid nodes at the initial time moment. Therefore, we can compute all volumes $V_{i,j}$, which is the volume of the rectangles, and then determine mass $m_{i,j}^C$ as follows:

$$m_{i,j}^C = \rho(x_{i,j}^*, y_{i,j}^*, 0)\, V_{i,j}\,.$$

From the statement of the original differential problem, we have the initial conditions for velocity as $\vec{W}$, ρ, ε, and the equation of state $p = P(\rho, \varepsilon)$ for pressure, or more generally, $F(p, \rho, \varepsilon) = 0$. Hence, we can then assume that for any point, we also have given values for pressure p. The computational domain can contain different materials with different equations of state; therefore, function P or F can be dependent on the fluid particle.

The initial conditions for the velocity vector are

$$WX_{i,j}^0 = wx^0(x_{i,j}, y_{i,j})\,, \quad WY_{i,j}^0 = wy^0(x_{i,j}, y_{i,j})\,. \tag{4.22}$$

Initial conditions for density and specific internal energy are

$$\rho^0_{i,j} = \rho^0(x^*_{i,j}, y^*_{i,j}, 0),$$ (4.23)

$$\varepsilon^0_{i,j} = \varepsilon^0(x^*_{i,j}, y^*_{i,j}, 0).$$ (4.24)

Values of pressure at the initial time moment can be computed from the equation of state:

$$p^0_{i,j} = P(\rho^0_{i,j}, \varepsilon^0_{i,j}).$$ (4.25)

New Time Step

The sequence of computation for the explicit finite-difference scheme is as follows. Suppose we know all the quantities on the n-th time level and we need to compute all the quantities on the next $n + 1$ time level.

First, we compute the new velocity field from the equation

$$(M\,\rho)^n_{i,j} \frac{\vec{W}^{n+1}_{i,j} - \vec{W}^n_{i,j}}{\Delta t} = -(\text{GRAD}\,p^n)_{i,j}.$$ (4.26)

It is important to note that the coefficients of operator GRAD depend on the coordinates of the grid point and, in the previous formula, these coordinates are taken from the n-th time level. In particular, volume $VN_{i,j}$, which is the denominator in the formula for GRAD, is taken from the n-th time level, or $VN^n_{i,j}$.

Next, we compute the new values for specific internal energy:

$$\rho^n_{i,j} \frac{\varepsilon^{n+1}_{i,j} - \varepsilon^n_{i,j}}{\Delta t} = -p^n_{i,j} \left(\text{DIV}\,\vec{W}^n\right)_{i,j}.$$ (4.27)

We can now compute the new positions for the grid points:

$$\frac{x^{n+1}_{i,j} - x^n_{i,j}}{\Delta t} = WX^n_{i,j},$$

$$\frac{y^{n+1}_{i,j} - y^n_{i,j}}{\Delta t} = WY^n_{i,j}.$$ (4.28)

Using these new coordinates, we can compute the volumes of the cells on the new time level:

$$V^{n+1}_{i,j} = F_V\left(x^{n+1}_{i+k,j+l}, y^{n+1}_{i+k,j+l}; k, l = 0, 1\right),$$ (4.29)

where F_V is given by the formula 4.5, and arguments for this function are the coordinates of the vertices of the cell.

If we know the volumes of the cells, we can compute the density on the new time level:

$$\rho^{n+1}_{i,j} = \frac{m^C_{i,j}}{V^{n+1}_{i,j}}.$$ (4.30)

Let us recall that the mass of the cell is constant and does not depend on time, and therefore, does not have a time index.

Finally, we can compute pressure on the new time level from the equation of state:

$$p_{i,j}^{n+1} = P_{i,j}(\rho_{i,j}^{n+1}, \varepsilon_{i,j}^{n+1}).$$ (4.31)

In the case where the equation of state is given in the implicit form,

$$F_{i,j}(p_{i,j}^{n+1}, \rho_{i,j}^{n+1}, \varepsilon_{i,j}^{n+1}) = 0,$$ (4.32)

we will need to solve one non-linear equation of one unknown in each cell to find the pressure. Functions $P_{i,j}$, $F_{i,j}$ have indices (i, j), because for each Lagrangian particle, the equation of state, in principle, can be different.

This is the purely explicit scheme, which has some restrictions on the ratio of time and space steps which ensure stability of the finite-difference scheme.

Let us consider what would happen with the conservation laws. First, we consider the law of variation for volume. Let us note that because the volume of the cell is bilinear, the function of the coordinates of its vertices and the coordinates of these vertices is a linear function of the corresponding component of the velocities (formula 4.28), and then the following formula is valid:

$$(\text{DIV}\vec{W}^n)_{i,j} = \frac{1}{V_{i,j}^n} \frac{V_{i,j}^{n+1} - V_{i,j}^n}{\Delta t}.$$

And in general,

$$(\text{DIV}\vec{W}^\sigma)_{i,j} = \frac{1}{V_{i,j}^n} \frac{V_{i,j}^{n+1} - V_{i,j}^n}{\Delta t}$$ (4.33)

if we use velocity $\vec{W}^\sigma = \sigma \vec{W}^{n+1} + \vec{W}^n$, instead of velocity $\vec{W}^n$ in the formulas in 4.28. If we rewrite the previous equation as:

$$\frac{V_{i,j}^{n+1} - V_{i,j}^n}{\Delta t} = V_{i,j}^n (\text{DIV}\vec{W}^\sigma)_{i,j},$$ (4.34)

then it is the same as the analog of the differential equation 2.9

Now let us show that using 4.33, we can rewrite the continuity equation in the form which is analogous to the differential case. The discrete analog of the continuity equation is

$$\frac{\rho_{i,j}^{n+1} - \rho_{i,j}^n}{\Delta t} = \frac{1}{\Delta t}\left(\frac{m_{i,j}^C}{V_{i,j}^{n+1}} - \frac{m_{i,j}^C}{V_{i,j}^n}\right) =$$ (4.35)

$$-\frac{m_{i,j}^C}{V_{i,j}^{n+1}} \frac{1}{\Delta t}\left[\frac{V_{i,j}^{n+1} - V_{i,j}^n}{V_{i,j}^n}\right] =$$

$$-\rho_{i,j}^{n+1} \text{DIV}\vec{W}^\sigma.$$

Let us now consider the law of conservation for momentum. Let us recall that the integral form for the conservation of momentum is

$$\int_{V(t_2)} \rho(x,y,t_2)\, \vec{W}(x,y,t_2)\, dV - \int_{V(t_1)} \rho(x,y,t_1)\, \vec{W}(x,y,t_1)\, dV =$$

$$-\int_{t_1}^{t_2} \oint_{\partial V(t)} p\vec{n}\, dS\, dt \approx -\Delta t \oint_{S(t^n)} p\vec{n}\, dS\, dt\,.$$

In the discrete case, the conservation of momentum follows from equation 4.26 as follows. If we multiply it by volume $VN_{i,j}^n$ of the nodes and take the summation of it over all nodes, we get the following:

$$\frac{\sum_{i,j}(M\rho)_{i,j}^n\, \vec{W}_{i,j}^{n+1}\, VN_{i,j}^n - \sum_{i,j}(M\rho)_{i,j}^n\, \vec{W}_{i,j}^n\, VN_{i,j}^n}{\Delta t} = \quad (4.36)$$

$$-\sum_{i,j}(\mathrm{GRAD}\, p^n)_{i,j}\, VN_{i,j} \approx \oint_{S(t^n)} \vec{n}\, p^n\, dS\,.$$

Since $(M\,\rho)_{i,j}\, VN_{i,j}$ does not depend on time, then

$$(M\,\rho)_{i,j}^n\, VN_{i,j}^n = (M\,\rho)_{i,j}^{n+1}\, VN_{i,j}^{n+1}$$

and we can replace the time index n at $(M\,\rho)_{i,j}$ and $VN_{i,j}$ in the first term in equation 4.36 by $n+1$. After this substitution, equation 4.36 looks similar to the differential case,

$$\sum_{i,j}(M\rho)_{i,j}^{n+1}\, \vec{W}_{i,j}^{n+1}\, VN_{i,j}^{n+1} - \sum_{i,j}(M\rho)_{i,j}^n\, \vec{W}_{i,j}^n\, VN_{i,j}^n = \quad (4.37)$$

$$-\Delta t \sum_{i,j}(\mathrm{GRA\dot D}\, p^n)_{i,j}\, VN_{i,j} \approx \oint_{S(t)} \vec{n}\, p^n\, dS\,.$$

It is easy to see that for the constructed finite-difference scheme, the discrete analog for the conservation of full energy is not satisfied. Let us recall that we first need to obtain the balance equation for kinetic energy, and this can be done by multiplication of the left-hand side of equation 4.26 by

$$\vec{W}^{n+0.5} = \frac{\vec{W}_{i,j}^{n+1} + \vec{W}_{i,j}^n}{2}$$

and then taking the summation over the nodes. That is:

$$\sum_{i,j}(M\,\rho)_{i,j}^n\, \frac{\dfrac{(\vec{W}_{i,j}^{n+1})^2 - (\vec{W}_{i,j}^n)^2}{2}\, VN_{i,j}^n}{\Delta t} = \quad (4.38)$$

$$\left(\mathrm{GRAD}\, p^n, \vec{W}^{n+0.5}\right)_{\mathcal{H}N}\,.$$

If we make the same consideration about $(M\,\rho)_{i,j}\,VN_{i,j}$ as for the momentum equation, we can rewrite the previous equation in the following form:

$$\frac{\sum_{i,j}(M\,\rho)_{i,j}^{n+1}\frac{\left(\vec{W}_{i,j}^{n+1}\right)^2 VN_{i,j}^{n+1}}{2} - \sum_{i,j}(M\,\rho)_{i,j}^{n}\frac{\left(\vec{W}_{i,j}^{n}\right)^2 VN_{i,j}^{n}}{2}}{\Delta t} =$$

$$\left(\text{GRAD}\,p^n, \vec{W}^{n+0.5}\right)_{\mathcal{HN}} . \tag{4.39}$$

Similar to the differential case, we can write the following balance equation for specific internal energy:

$$\sum_{i,j}\rho_{i,j}^{n}\frac{\epsilon_{i,j}^{n+1} - \epsilon_{i,j}^{n}}{\Delta t}VC_{i,j}^{n} = \tag{4.40}$$

$$-\sum_{i,j}p_{i,j}^{n}\left(\text{DIV}\vec{W}^{n}\right)_{i,j}VC_{i,j}^{n} = -\left(p^n, \text{DIV}\vec{W}^n\right)_{HC} ,$$

where the summation is over the cells and again $\rho_{i,j}^{n}\,VC_{i,j}^{n}$ is a mass of the cell and does not depend on time. Therefore, we can replace the time index n with $n+1$, or in other words, we can put this quantity under time differentiation. The conservation law for full energy will be satisfied if the sum of the right-hand sides of equations 4.39 and 4.40 can be reduced to the analog of the boundary integral. Similarly, to the 1-D case this is possible when the discrete functions p and $\vec{W}$ in both equations are the same (taken at the same time moment). When only function p is the same in both equations, the conservation law for full energy is not satisfied.

Thus, to preserve the conservation law for full energy for schemes that are discrete in space and time, we need to make additional considerations. One possible way to construct a finite-difference scheme which will satisfy the conservation law for full energy, is to take the velocity in the equation for internal energy as half of the sum of the values from the n and $n+1$ time step. That is, instead of equation 4.27, consider the following equation:

$$\rho_{i,j}^{n}\frac{\epsilon_{i,j}^{n+1} - \epsilon_{i,j}^{n}}{\Delta t} = -p_{i,j}^{n}\left(\text{DIV}\,\vec{W}^{n+0.5}\right)_{i,j} . \tag{4.41}$$

It is important to note that the finite-difference scheme still remains explicit in the sense that when we compute the specific internal energy from equation 4.41, we already know the velocity $\vec{W}^{n+1}$, and all the quantities in the right-hand side of equation 4.41 are also known.

The natural conservative explicit finite-difference scheme is

$$(M\,\rho)_{i,j}^{n}\frac{\vec{W}_{i,j}^{n+1} - \vec{W}_{i,j}^{n}}{\Delta t} = -(\text{GRAD}\,p^n)_{i,j} , \tag{4.42}$$

$$\rho_{i,j}^{n}\frac{\epsilon_{i,j}^{n+1} - \epsilon_{i,j}^{n}}{\Delta t} = -p_{i,j}^{n}\left(\text{DIV}\,\vec{W}^{n+0.5}\right)_{i,j} , \tag{4.43}$$

$$\frac{x_{i,j}^{n+1} - x_{i,j}^n}{\Delta t} = WX_{i,j}^{n+0.5},$$

$$\frac{y_{i,j}^{n+1} - y_{i,j}^n}{\Delta t} = WY_{i,j}^{n+0.5}, \tag{4.44}$$

$$V_{i,j}^{n+1} = F_V\left(x_{i+k,j+l}^{n+1}, y_{i+k,j+l}^{n+1}; k, l = 0, 1\right), \tag{4.45}$$

$$\rho_{i,j}^{n+1} = \frac{m_{i,j}}{V_{i,j}^{n+1}}, \tag{4.46}$$

$$p_{i,j}^{n+1} = P_{i,j}(\rho_{i,j}^{n+1}, \varepsilon_{i,j}^{n+1}). \tag{4.47}$$

These equations are presented in the sequence in which the actual computations are made.

Realization of Boundary Conditions

First, we consider the case of "free boundary" when pressure on the boundary is given as a function of time and position on the boundary. This is the simplest case because it can be considered in the same way as the Dirichlet boundary conditions for elliptic equations. Namely, we will just take the pressure on the boundary segments equal to the given pressure. In practice, it is convenient to consider the fictitious cells and take the pressure in these fictitious cells equal to the given pressure.

Now let us consider the condition of impermeability. First, we will consider the case of a straight wall. Assume that this wall is parallel to the y axis, which corresponds to grid line $i = 1$ (see Figure 4.21) and moves in an x direction with given velocity $\vec{W} = (w_x, 0)$. We will take the normal component of velocity, which in our case is $WX_{1,j}$, equal to w_x, which corresponds to the boundary conditions.

Now we need to compute $WY_{1,j}$, which is a tangential component of the velocity vector. As in the case for free boundary conditions, we introduce fictitious cells $(0, j)$. Let us show that the tangential component of the velocity vector does not depend on the pressure on the boundary. In fact, general formula 4.9 gives

$$GY_{1j} = -\left(\frac{x_{1,j+1} - x_{2,j}}{2} p_{1,j} + \frac{x_{1,j} - x_{1,j+1}}{2} p_{0,j} + \right.$$

$$\left. \frac{x_{1,j-1} - x_{1,j}}{2} p_{0,j-1} + \frac{x_{2,j} - x_{1,j-1}}{2} p_{1,j-1}\right) / VN_{ij}.$$

And now because we consider the case of a straight wall we have

$$x_{1,j} - x_{1,j+1} = 0, x_{1,j-1} - x_{1,j} = 0,$$

and therefore $GY_{1,j}$ does not contain pressure on the boundary:

$$GY_{1j} = -\left(\frac{x_{1,j+1} - x_{2,j}}{2} p_{1,j} + \frac{x_{2,j} - x_{1,j-1}}{2} p_{1,j-1}\right) / VN_{ij}.$$

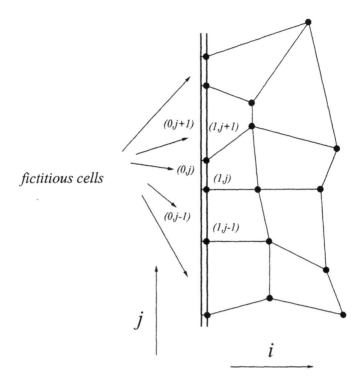

Figure 4.21: *Impermeability conditions on a plane wall.*

Therefore, we can compute the tangential component of velocity (in our case, it is simply $WY_{1,j}$) by using the general equation 4.42, where we can take pressure on the boundary arbitrarily.

Now let us consider a case where the wall is inclined to the x axis at angle φ (see Figure 4.22). Actually, the case of the inclined boundary, can be reduced to the case of the straight boundary by rotation of the coordinate system.

There is another approach which we will describe on the example of a general curvilinear boundary which is moving with a given velocity (see Figure 4.23). This procedure is based on introducing values of pressure in fictitious cells and using general formulas 4.42. For each boundary segment (i, N), we determine pressure from the discrete analog of the equation for the normal component of velocity:

$$\rho \frac{d}{dt}(\vec{W}, \vec{n}) = -\frac{dp}{dn}. \qquad (4.48)$$

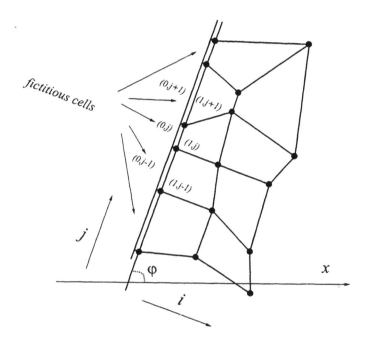

Figure 4.22: *Impermeability conditions on a plane inclined wall.*

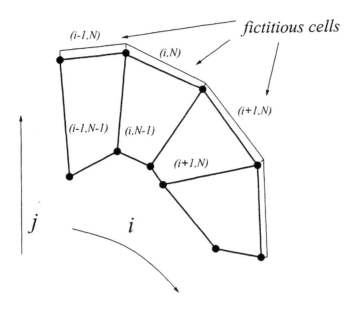

Figure 4.23: *Impermeability conditions on a curvilinear wall.*

For example, we can use following formula:

$$p_{i,N} = p_{i,N-1} - \left(\rho_{i,N-1} \frac{d}{dt}(\vec{W}, \vec{n}) \right) \, ln_{i,N-1} \, ,$$

where $\frac{d}{dt}(\vec{W}, \vec{n})$ is known from the boundary conditions, and $ln_{i,N-1}$ is the distance from the center of cell $(i, N-1)$ to the boundary. We can then compute both components of velocity, or $WX_{i,N}, WY_{i,N}$, by using general formulas.

Let us note that for the case of a curvilinear wall, we cannot claim that the tangential component of velocity vector does not depend on pressure on the boundary as it does for a straight wall. However, we can show that the coefficient near the pressure in the expression for a tangential component is of the order $O(h)$ for smooth enough grids.

4.5   Implicit Finite-Difference Scheme

New Time Step

We will consider the following general form of an implicit finite-difference scheme:

$$(M \, \rho)^n_{i,j} \, \frac{\vec{W}^{n+1}_{i,j} - \vec{W}^n_{i,j}}{\Delta t} = - \left(\mathrm{GRAD} \, p^{(\sigma_1)} \right)_{i,j} , \qquad (4.49)$$

$$\rho_{i,j}^n \frac{\varepsilon_{i,j}^{n+1} - \varepsilon_{i,j}^n}{\Delta t} = -p_{i,j}^{(\sigma_1)} \left(\text{DIV } \vec{W}^{(\sigma_3)} \right)_{i,j}, \tag{4.50}$$

$$\frac{x_{i,j}^{n+1} - x_{i,j}^n}{\Delta t} = WX_{i,j}^{(\sigma_2)},$$

$$\tag{4.51}$$

$$\frac{y_{i,j}^{n+1} - y_{i,j}^n}{\Delta t} = WY_{i,j}^{(\sigma_2)},$$

$$V_{i,j} = F_V \left(x_{i+k,j+l}^{n+1}, y_{i+k,j+l}^{n+1}; k, l = 0, 1 \right), \tag{4.52}$$

$$\rho_{i,j}^{n+1} = \frac{m_{i,j}}{V_{i,j}^{n+1}}, \tag{4.53}$$

$$p_{i,j}^{n+1} = P(\rho_{i,j}^{n+1}, \varepsilon_{i,j}^{n+1}). \tag{4.54}$$

The full set of conservation laws will be valid for these equations when $\sigma_3 = 0.5$; σ_1 and σ_2 can be chosen arbitrarily.

For our future consideration and from a computational point of view, it is useful to introduce the so-called mass of node and determine it to be as follows:

$$m_{i,j}^N = (M\,\rho)_{i,j}\, VN_{i,j} = \frac{m_{i,j}^C + m_{i-1,j}^C + m_{i,j-1}^C + m_{i-1,j-1}^C}{4}. \tag{4.55}$$

By definition, mass of node does not depend on time.

Now using the definition of $m_{i,j}^C$ and $m_{i,j}^N$ and the formulas for DIV and GRAD, 4.4 and 4.10, we can rewrite equations 4.49 and 4.50 in the following form:

$$m_{i,j}^N \frac{WX_{i,j}^{n+1} - WX_{i,j}^n}{\Delta t} = \sum_{k,l=0}^{1} \left(\frac{\partial V_{i-k,j-l}}{\partial x_{i,j}} \right)^n p_{i-k,j-l}^{(\sigma_1)},$$

$$\tag{4.56}$$

$$m_{i,j}^N \frac{WY_{i,j}^{n+1} - WY_{i,j}^n}{\Delta t} = \sum_{k,l=0}^{1} \left(\frac{\partial V_{i-k,j-l}}{\partial y_{i,j}} \right)^n p_{i-k,j-l}^{(\sigma_1)},$$

$$m_{i,j}^C \frac{\varepsilon_{i,j}^{n+1} - \varepsilon_{i,j}^n}{\Delta t} = \tag{4.57}$$

$$-p_{i,j}^{(\sigma_1)} \sum_{k,l=0}^{1} \left(\left(\frac{\partial V_{i,j}}{\partial x_{i+k,j+l}} \right)^n WX_{i+k,j+l}^{(\sigma_3)} + \left(\frac{\partial V_{i,j}}{\partial y_{i+k,j+l}} \right)^n WY_{i+k,j+l}^{(\sigma_3)} \right).$$

Let us recall that in both equations, the derivatives of volume with respect to the coordinates are computed on the n-th time level, and to emphasize this fact, we use the following notation:

$$\left(\frac{\partial V_{i,j}}{\partial x_{i+k,j+l}}\right)^n .$$

Let us analyze the structure for the system of non-linear equations 4.56, 4.57, 4.51, 4.52, and 4.53. Originally, it was the system where the role of unknowns were played by all the quantities: pressure, velocity, specific internal energy, coordinates of nodes, and density. But different variables play different roles in this system. To understand the structure of this system, let us suppose that we have found pressure $p_{i,j}^{n+1}$ on a new time level. Then, using equation 4.56, we can determine the velocity vector $\vec{W}_{i,j}^{n+1}$ on the new time level. Using this velocity, we can compute values for specific internal energy on the new time level from equation 4.57. Next, we compute new coordinates from equation 4.52 and using these coordinates, we compute the new volume and density from equation 4.53. These considerations mean that all variables can be considered as a composite function of pressure, and consequently, it is possible to eliminate all variables except pressure from the original system of equations and obtain the new system of non-linear equations which will contain only pressure.

To show the structure of this new system of equations, let us introduce some stencils. The first stencil, which we will call $St_p(i,j)$, is a stencil for values of pressure which participate in the approximation of the equation of motion in node (i,j). It contains four cells (i,j), $(i-1,j)$, $(i-1,j-1)$, $(i,j-1)$ and it is shown in Figure 4.24. The second stencil, which we will denote as St_ε, is a stencil for values of the velocity vector which participate in the approximation of the equation for specific internal energy in cell (i,j). It contains four nodes (i,j), $(i+1,j)$, $(i+1,j+1)$, $(i,j+1)$. The coordinates from the same stencil participate in the expression for volume of the cell (i,j).

Using these stencils, the structure of dependence in each equation can be written in the following form. For the equation of motion we have:

$$WX_{i,j}^{n+1}(p_{s,t}^{n+1} : (s,t) \in St_p(i,j)) = 0,$$
$$WY_{i,j}^{n+1}(p_{s,t}^{n+1} : (s,t) \in St_p(i,j)) = 0.$$

For the energy equation, we have:

$$\varepsilon_{i,j}^{n+1}(p_{i,j}^{n+1}, ((WX_{k,l}^{n+1}(p_{s,t}^{n+1} : (s,t) \in St_p(k,l))),$$
$$(WY_{k,l}^{n+1}(p_{s,t}^{n+1} : (s,t) \in St_p(k,l))) : (k,l) \in St_\varepsilon(i,j))) = 0.$$

The equation for the new coordinates is as follows:

$$x_{i,j}^{n+1}(WX_{i,j}^{n+1}) = 0,$$
$$y_{i,j}^{n+1}(WY_{i,j}^{n+1}) = 0.$$

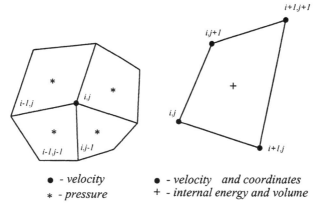

• - *velocity* • - *velocity   and coordinates*
* - *pressure* + - *internal energy and volume*

Stensil St$_p$(i,j) for equation of motion.   Stencil St$_\varepsilon$(i,j) for volume and internal energy.

Figure 4.24: *Stencils St$_p$ and St$_\varepsilon$.*

The dependence of volume on the coordinates is:

$$V_{i,j}^{n+1}\left(x_{k,l}^{n+1}, y_{k,l}^{n+1} : (k,l) \in St_\varepsilon(i,j)\right) = 0. \tag{4.58}$$

The equation for density is:

$$\rho_{i,j}^{n+1}(V_{i,j}^{n+1}) = 0. \tag{4.59}$$

And finally, for the equation of state we have:

$$F_{i,j}(p_{i,j}^{n+1}, \rho_{i,j}^{n+1}, \varepsilon_{i,j}^{n+1}) = 0. \tag{4.60}$$

Using these dependencies in the system of non-linear equations, we can express all variables in terms of pressure and substitute them into the equation of state 4.32. Finally, the system of non-linear equations, which contains only pressure, can be written in schematic form as follows:

$$\Phi_{i,j}\left(p_{k,l} : (k,l) \in St(i,j)\right) = \tag{4.61}$$
$$F_{i,j}\left(p_{i,j}, \rho_{i,j}(V_{i,j}(x_{k,l}(WX_{k,l}(p_{s,t} : (s,t) \in St_p(k,l)))),\right.$$
$$(y_{k,l}(WY_{k,l}(p_{s,t} : (s,t) \in St_p(k,l)))) : (k,l) \in St_\varepsilon(i,j)),$$
$$\varepsilon_{i,j}(p_{i,j}, ((WX_{k,l}(p_{s,t} : (s,t) \in St_p(k,l))),$$
$$\left.(WY_{k,l}(p_{s,t} : (s,t) \in St_p(k,l))) : (k,l) \in St_\varepsilon(i,j)))\right) = 0.$$

Here, we have omitted index $n + 1$ for pressure, because all values are taken from the new time step. The resulting stencil $St(i, j)$ for pressure for

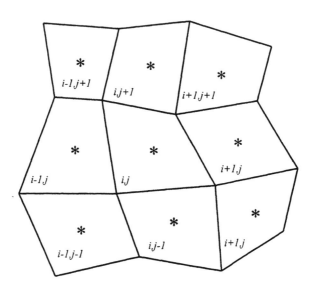

Figure 4.25: *Resulting stencil for pressure – $St(i, j)$.*

this system of nonlinear equations contains nine cells for each (i, j) and is presented in Figure 4.25.

We can now use Newton's method and some of its modifications to solve the system of non-linear equations 4.61. If some approximation $p_{i,j}^{(s)}$ for $p_{i,j}^{n+1}$ is given (s is used here for the iteration number), then the increment $\delta p_{(s+1)} = p_{i,j}^{(s+1)} - p_{i,j}^{(s)}$ must be determined from the linear system of equations:

$$\sum_{(k,l)\in St(i,j)} \left(\frac{\partial \Phi_{i,j}}{\partial p_{k,l}}\right)^{(s)} \delta p_{k,l}^{(s+1)} = -\Phi_{i,j}^{(s)}. \qquad (4.62)$$

Let us note that index (s) at derivatives $\frac{\partial \Phi_{i,j}}{\partial p_{k,l}}$ means that it must be computed by using values of its arguments on the previous (s) iteration.

It is now necessary to find the formulas for derivatives $\frac{\partial \Phi_{i,j}}{\partial p_{k,l}}$ and investigate the properties of the system of linear equations.

To obtain formulas for the derivatives, we will use the structure of dependence of function $\Phi_{i,j}$ on corresponding pressures:

$$\frac{\partial \Phi_{i,j}}{\partial p_{i,j}} = \tag{4.63}$$

$$\frac{\partial F_{i,j}}{\partial p_{i,j}} +$$

$$\frac{\partial F_{i,j}}{\partial \rho_{i,j}} \frac{\partial \rho_{i,j}}{\partial V_{i,j}} \sum_{\alpha,\beta=0}^{1} \left(\frac{\partial V_{i,j}}{\partial x_{i+\alpha,j+\beta}} \frac{\partial x_{i+\alpha,j+\beta}}{\partial WX_{i+\alpha,j+\beta}} \frac{\partial WX_{i+\alpha,j+\beta}}{\partial p_{i,j}} + \right.$$

$$\left. \frac{\partial V_{i,j}}{\partial y_{i+\alpha,j+\beta}} \frac{\partial y_{i+\alpha,j+\beta}}{\partial WY_{i+\alpha,j+\beta}} \frac{\partial WY_{i+\alpha,j+\beta}}{\partial p_{i,j}} \right) +$$

$$\frac{\partial F_{i,j}}{\partial \epsilon_{i,j}} \left[\frac{\partial \epsilon_{i,j}}{\partial p_{i,j}} + \sum_{\alpha,\beta=0}^{1} \left(\frac{\partial \epsilon_{i,j}}{\partial WX_{i+\alpha,j+\beta}} \frac{\partial WX_{i+\alpha,j+\beta}}{\partial p_{i,j}} + \right. \right.$$

$$\left. \left. \frac{\partial \epsilon_{i,j}}{\partial WY_{i+\alpha,j+\beta}} \frac{\partial WY_{i+\alpha,j+\beta}}{\partial p_{i,j}} \right) \right] ,$$

$$\frac{\partial \Phi_{i,j}}{\partial p_{i+1,j}} = \tag{4.64}$$

$$\frac{\partial F_{i,j}}{\partial \rho_{i,j}} \frac{\partial \rho_{i,j}}{\partial V_{i,j}} \sum_{\beta=0}^{1} \left(\frac{\partial V_{i,j}}{\partial x_{i+1,j+\beta}} \frac{\partial x_{i+1,j+\beta}}{\partial WX_{i+1,j+\beta}} \frac{\partial WX_{i+1,j+\beta}}{\partial p_{i+1,j}} + \right.$$

$$\left. \frac{\partial V_{i,j}}{\partial y_{i+1,j+\beta}} \frac{\partial y_{i+1,j+\beta}}{\partial WY_{i+1,j+\beta}} \frac{\partial WY_{i+1,j+\beta}}{\partial p_{i+1,j}} \right) +$$

$$\frac{\partial F_{i,j}}{\partial \epsilon_{i,j}} \sum_{\beta=0}^{1} \left(\frac{\partial \epsilon_{i,j}}{\partial WX_{i+1,j+\beta}} \frac{\partial WX_{i+1,j+\beta}}{\partial p_{i+1,j}} + \frac{\partial \epsilon_{i,j}}{\partial WY_{i+1,j+\beta}} \frac{\partial WY_{i+1,j+\beta}}{\partial p_{i+1,j}} \right) ,$$

$$\frac{\partial \Phi_{i,j}}{\partial p_{i+1,j+1}} = \tag{4.65}$$

$$\frac{\partial F_{i,j}}{\partial \rho_{i,j}} \frac{\partial \rho_{i,j}}{\partial V_{i,j}} \left(\frac{\partial V_{i,j}}{\partial x_{i+1,j+1}} \frac{\partial x_{i+1,j+1}}{\partial WX_{i+1,j+1}} \frac{\partial WX_{i+1,j+1}}{\partial p_{i+1,j+1}} + \right.$$

$$\left. \frac{\partial V_{i,j}}{\partial y_{i+1,j+1}} \frac{\partial y_{i+1,j+1}}{\partial WY_{i+1,j+1}} \frac{\partial WY_{i+1,j+1}}{\partial p_{i+1,j+1}} \right) +$$

$$\frac{\partial F_{i,j}}{\partial \epsilon_{i,j}} \left(\frac{\partial \epsilon_{i,j}}{\partial WX_{i+1,j+1}} \frac{\partial WX_{i+1,j+1}}{\partial p_{i+1,j+1}} + \frac{\partial \epsilon_{i,j}}{\partial WY_{i+1,j+1}} \frac{\partial WY_{i+1,j+1}}{\partial p_{i+1,j+1}} \right) .$$

The expressions for the remaining derivatives $\partial \Phi_{i,j}/\partial p_{k,l}$ can be obtained in a similar way. Again, let us note that all derivatives here are taken from the (s) iteration.

The formulas for the derivatives which participate in equations 4.63, 4.64, and 4.65 follow from the equations for finite-difference schemes 4.56, 4.57, 4.51, 4.52, 4.53 and are as follows:

$$\frac{\partial \rho_{i,j}}{\partial V_{i,j}} = -\frac{m_{i,j}^C}{V_{i,j}^2}, \tag{4.66}$$

$$\frac{\partial x_{i+\alpha,j+\beta}}{\partial W X_{i+\alpha,j+\beta}} = \Delta t\, \sigma_2, \tag{4.67}$$

$$\frac{\partial y_{i+\alpha,j+\beta}}{\partial W Y_{i+\alpha,j+\beta}} = \Delta t\, \sigma_2, \tag{4.68}$$

$$\frac{\partial W X_{i+\alpha,j+\beta}}{\partial p_{i,j}} = \frac{\sigma_1 \Delta t}{m_{i+\alpha,j+\beta}^N} \left(\frac{\partial V_{i,j}}{\partial x_{i+\alpha,j+\beta}}\right)^n, \tag{4.69}$$

$$\frac{\partial W Y_{i+\alpha,j+\beta}}{\partial p_{i,j}} = \frac{\sigma_1 \Delta t}{m_{i+\alpha,j+\beta}^N} \left(\frac{\partial V_{i,j}}{\partial y_{i+\alpha,j+\beta}}\right)^n, \tag{4.70}$$

$$\frac{\partial \varepsilon_{i,j}}{\partial p_{i,j}} = -\frac{\sigma_1 \Delta t}{m_{i,j}^C} \sum_{\alpha,\beta=0}^{1} \left(\left(\frac{\partial V_{i,j}}{\partial x_{i+\alpha,j+\beta}}\right)^n W X_{i+\alpha,j+\beta}^{(\sigma_3)} + \right.$$

$$\left. \left(\frac{\partial V_{i,j}}{\partial y_{i+\alpha,j+\beta}}\right)^n W Y_{i+\alpha,j+\beta}^{(\sigma_3)} \right), \tag{4.71}$$

$$\frac{\partial \varepsilon_{i,j}}{\partial W X_{i+\alpha,j+\beta}} = -\frac{\sigma_3\, p_{i,j}^{(\sigma_1)} \Delta t}{m_{i,j}^C} \left(\frac{\partial V_{i,j}}{\partial x_{i+\alpha,j+\beta}}\right)^n, \tag{4.72}$$

$$\frac{\partial \varepsilon_{i,j}}{\partial W Y_{i+\alpha,j+\beta}} = -\frac{\sigma_3\, p_{i,j}^{(\sigma_1)} \Delta t}{m_{i,j}^C} \left(\frac{\partial V_{i,j}}{\partial y_{i+\alpha,j+\beta}}\right)^n. \tag{4.73}$$

Using these expressions, we can obtain the following formulas for derivatives $\partial \Phi_{i,j}/\partial p_{k,l}$:

$$\frac{\partial \Phi_{i,j}}{\partial p_{i,j}} = \tag{4.74}$$

$$\frac{\partial F_{i,j}}{\partial p_{i,j}} -$$

$$\frac{m_{i,j}^C}{V_{i,j}^2} \frac{\partial F_{i,j}}{\partial \rho_{i,j}} \sigma_1 \sigma_2 (\Delta t)^2 \times$$

$$\sum_{\alpha,\beta=0}^{1} \left[\left(\frac{\partial V_{i,j}}{\partial x_{i+\alpha,j+\beta}}\right)^{(s)} \left(\frac{\partial V_{i,j}}{\partial x_{i+\alpha,j+\beta}}\right)^n \Big/ m_{i+\alpha,j+\beta}^N + \right.$$

$$\left. \left(\frac{\partial V_{i,j}}{\partial y_{i+\alpha,j+\beta}}\right)^{(s)} \left(\frac{\partial V_{i,j}}{\partial y_{i+\alpha,j+\beta}}\right)^n \Big/ m_{i+\alpha,j+\beta}^N \right] -$$

$$\frac{\partial F_{i,j}}{\partial \varepsilon_{i,j}} \frac{\sigma_1 \Delta t}{m_{i,j}^C} \left[\sum_{\alpha,\beta=0}^{1} \left(\left(\frac{\partial V_{i,j}}{\partial x_{i+\alpha,j+\beta}} \right)^n WX_{i+\alpha,j+\beta}^{(\sigma_3)} + \right. \right.$$

$$\left. \left(\frac{\partial V_{i,j}}{\partial y_{i+\alpha,j+\beta}} \right)^n WY_{i+\alpha,j+\beta}^{(\sigma_3)} \right) +$$

$$\left. \sigma_3 \Delta t \, p_{i,j}^{(\sigma_1)} \sum_{\alpha,\beta=0}^{1} \left(\frac{\left(\left(\frac{\partial V_{i,j}}{\partial x_{i+\alpha,j+\beta}} \right)^n \right)^2 + \left(\left(\frac{\partial V_{i,j}}{\partial y_{i+\alpha,j+\beta}} \right)^n \right)^2}{m_{i+\alpha,j+\beta}^N} \right) \right],$$

$$\frac{\partial \Phi_{i,j}}{\partial p_{i+1,j}} = \tag{4.75}$$

$$- \frac{m_{i,j}^C}{V_{i,j}^2} \frac{\partial F_{i,j}}{\partial \rho_{i,j}} \sigma_1 \sigma_2 (\Delta t)^2 \times$$

$$\sum_{\beta=0}^{1} \frac{\left(\frac{\partial V_{i,j}}{\partial x_{i+1,j+\beta}} \right)^{(s)} \left(\frac{\partial V_{i+1,j}}{\partial x_{i+1,j+\beta}} \right)^n + \left(\frac{\partial V_{i,j}}{\partial y_{i+1,j+\beta}} \right)^{(s)} \left(\frac{\partial V_{i+1,j}}{\partial y_{i+1,j+\beta}} \right)^n}{m_{i+1,j+\beta}^N} -$$

$$\frac{\partial F_{i,j}}{\partial \varepsilon_{i,j}} \frac{\sigma_1 \sigma_3}{m_{i,j}^C} (\Delta t)^2 \, p_{i,j}^{(\sigma_1)} \times$$

$$\sum_{\beta=0}^{1} \frac{\left(\frac{\partial V_{i,j}}{\partial x_{i+1,j+\beta}} \right)^{(s)} \left(\frac{\partial V_{i+1,j}}{\partial x_{i+1,j+\beta}} \right)^n + \left(\frac{\partial V_{i,j}}{\partial y_{i+1,j+\beta}} \right)^{(s)} \left(\frac{\partial V_{i+1,j}}{\partial y_{i+1,j+\beta}} \right)^n}{m_{i+1,j+\beta}^N},$$

$$\frac{\partial \Phi_{i,j}}{\partial p_{i+1,j+1}} = \tag{4.76}$$

$$- \frac{m_{i,j}^C}{V_{i,j}^2} \frac{\partial F_{i,j}}{\partial \rho_{i,j}} \sigma_1 \sigma_2 (\Delta t)^2 \times$$

$$\frac{\left(\frac{\partial V_{i,j}}{\partial x_{i+1,j+1}} \right)^{(s)} \left(\frac{\partial V_{i+1,j+1}}{\partial x_{i+1,j+1}} \right)^n + \left(\frac{\partial V_{i,j}}{\partial y_{i+1,j+1}} \right)^{(s)} \left(\frac{\partial V_{i+1,j+1}}{\partial y_{i+1,j+1}} \right)^n}{m_{i+1,j+1}^N} -$$

$$\frac{\partial F_{i,j}}{\partial \varepsilon_{i,j}} \frac{\sigma_1 \sigma_3 (\Delta t)^2}{m_{i,j}^C} \, p_{i,j}^{(\sigma_1)} \times$$

$$\frac{\left(\frac{\partial V_{i,j}}{\partial x_{i+1,j+1}} \right)^n \left(\frac{\partial V_{i+1,j+1}}{\partial x_{i+1,j+1}} \right)^n + \left(\frac{\partial V_{i,j}}{\partial y_{i+1,j+1}} \right)^n \left(\frac{\partial V_{i+1,j+1}}{\partial y_{i+1,j+1}} \right)^n}{m_{i+1,j+1}^N}.$$

To find $\delta p_{i,j}^{(s+1)}$, it is necessary to solve the nine-point linear system of equations at each iteration of Newton's method, whose coefficients are calculated by the previous formulas from the values of the quantities at the preceding s-iteration. We note the difficulties arising in the realization of Newton's classical method for our equations. First, it is necessary to recalculate the coefficients of the linear system at each iteration, which requires an extensive amount of computer time. Second, the operator of the linear system is not, in general, self-adjoint and positive definite, which makes it difficult to use efficient iterative methods and also requires more computer memory. The solution to a system of linear equations, for which the properties of the operator are not known in advance, is a complex problem [111]. These difficulties can be avoided by using the method of Newton-Kantorovich, or by the method of parallel chords [94], where, instead of using matrix $||\partial \Phi_{i,j}^{(s)}/\partial p_{k,l}||$, a matrix similar to this is used. We chose this matrix to be independent of the iteration number and self-adjoint and positive definite.

There are different ways to construct implicit finite-difference schemes and solve the related system of non-linear equations. Interested readers can find more information in [59], [3], [40], and [99].

4.5.1   The Method of Parallel Chords

Similar to the 1-D case, at first, we take all the quantities in formulas 4.74, 4.75, and 4.76 from the previous time step. In other words, instead of values from the (s) iteration, we will take the values from time step n. Second, we divide equation 4.62 by the following:

$$- \left(\frac{m_{i,j}^C}{V_{i,j}^2} \frac{\partial F_{i,j}}{\partial \rho_{i,j}} \sigma_1 \sigma_2 (\Delta t)^2 + \frac{\partial F_{i,j}}{\partial \varepsilon_{i,j}} \frac{p_{i,j}}{m_{i,j}^C} \sigma_1 \sigma_3 (\Delta t)^2 \right) .$$

Finally, we obtain the following equations in the internal nodes:

$$\sum_{\alpha,\beta=0}^{1} A_{i+\alpha,j+\beta}^{i,j} \, \delta p_{i+\alpha,j+\beta}^{(s+1)} = -\tilde{\Phi}_{i,j} \,, \tag{4.77}$$

where

$$A_{i,j}^{i,j} = \left[-\frac{\partial F_{i,j}}{\partial p_{i,j}} + \frac{\partial F_{i,j}}{\partial \varepsilon_{i,j}} \frac{\sigma_1 \Delta t}{m_{i,j}^C} \sum_{\alpha,\beta=0}^{1} \left(\frac{\partial V_{i,j}}{\partial x_{i+\alpha,j+\beta}} W X_{i+\alpha,j+\beta} + \right. \right.$$

$$\left. \left. \frac{\partial V_{i,j}}{\partial y_{i+\alpha,j+\beta}} W Y_{i+\alpha,j+\beta} \right) \right] \Big/$$

$$\left(\frac{m_{i,j}^C}{V_{i,j}^2} \frac{\partial F_{i,j}}{\partial \rho_{i,j}} \sigma_1 \sigma_2 (\Delta t)^2 + \frac{\partial F_{i,j}}{\partial \varepsilon_{i,j}} \frac{p_{i,j}}{m_{i,j}^C} \sigma_1 \sigma_3 (\Delta t)^2 \right) +$$

$$\sum_{\alpha,\beta=0}^{1} \frac{\left(\frac{\partial V_{i,j}}{\partial x_{i+\alpha,j+\beta}} \right)^2 + \left(\frac{\partial V_{i,j}}{\partial y_{i+\alpha,j+\beta}} \right)^2}{m_{i+\alpha,j+\beta}^N} \,,$$

$$A_{i+1,j}^{i,j} = \sum_{\beta=0} \frac{\dfrac{\partial V_{i,j}}{\partial x_{i+1,j+\beta}}\dfrac{\partial V_{i+1,j}}{\partial x_{i+1j+\beta}} + \dfrac{\partial V_{i,j}}{\partial y_{i+1,j+\beta}}\dfrac{\partial V_{i+1,j}}{\partial y_{i+1,j+\beta}} +}{m_{i+1,j+\beta}^N},$$

$$A_{i+1,j+1}^{i,j} = \frac{\dfrac{\partial V_{i,j}}{\partial x_{i+1,j+1}}\dfrac{\partial V_{i+1,j+1}}{\partial x_{i+1,j+1}} + \dfrac{\partial V_{i,j}}{\partial y_{i+1,j+1}}\dfrac{\partial V_{i+1,j+1}}{\partial y_{i+1,j+1}}}{m_{i+1,j+1}^N},$$

$$A_{i,j+1}^{i,j} = \sum_{\alpha=0} \frac{\dfrac{\partial V_{i,j}}{\partial x_{i+\alpha,j+1}}\dfrac{\partial V_{i,j+1}}{\partial x_{i+\alpha,j+1}} + \dfrac{\partial V_{i,j}}{\partial y_{i+\alpha,j+1}}\dfrac{\partial V_{i,j+1}}{\partial y_{i+\alpha,j}} +}{m_{i+\alpha,j+1}^N},$$

$$A_{i-1,j+1}^{i,j} = \frac{\dfrac{\partial V_{i,j}}{\partial x_{i,j+1}}\dfrac{\partial V_{i-1,j+1}}{\partial x_{i,j+1}} + \dfrac{\partial V_{i,j}}{\partial y_{i,j+1}}\dfrac{\partial V_{i-1,j+1}}{\partial y_{i,j+1}}}{m_{i,j+1}^N},$$

$$A_{i-1,j}^{i,j} = \sum_{\beta=0} \frac{\dfrac{\partial V_{i,j}}{\partial x_{i,j+\beta}}\dfrac{\partial V_{i-1,j}}{\partial x_{i,j+\beta}} + \dfrac{\partial V_{i,j}}{\partial y_{i,j+\beta}}\dfrac{\partial V_{i-1,j}}{\partial y_{i,j+\beta}} +}{m_{i,j+\beta}^N},$$

$$A_{i-1,j-1}^{i,j} = \frac{\dfrac{\partial V_{i,j}}{\partial x_{i,j}}\dfrac{\partial V_{i-1,j-1}}{\partial x_{i,j+1}} + \dfrac{\partial V_{i,j}}{\partial y_{i,j}}\dfrac{\partial V_{i-1,j-1}}{\partial y_{i,j}}}{m_{i,j}^N},$$

$$A_{i,j-1}^{i,j} = \sum_{\alpha=0} \frac{\dfrac{\partial V_{i,j}}{\partial x_{i+\alpha,j}}\dfrac{\partial V_{i,j-1}}{\partial x_{i+\alpha,j}} + \dfrac{\partial V_{i,j}}{\partial y_{i+\alpha,j}}\dfrac{\partial V_{i,j-1}}{\partial y_{i+\alpha,j}} +}{m_{i+\alpha,j}^N},$$

$$A_{i+1,j-1}^{i,j} = \frac{\dfrac{\partial V_{i,j}}{\partial x_{i+1,j}}\dfrac{\partial V_{i+1,j-1}}{\partial x_{i+1,j}} + \dfrac{\partial V_{i,j}}{\partial y_{i+1,j}}\dfrac{\partial V_{i+1,j-1}}{\partial y_{i+1,j}}}{m_{i+1,j}^N},$$

$$\tilde{\Phi}_{i,j} = \Phi_{i,j} \left/ \left(\frac{m_{i,j}^C}{V_{i,j}^2} \frac{\partial F_{i,j}}{\partial \rho_{i,j}} \sigma_1 \sigma_2 (\Delta t)^2 \frac{\partial F_{i,j}}{\partial \varepsilon_{i,j}} \frac{p_{i,j}}{m_{i,j}^C} \sigma_1 \sigma_3 (\Delta t)^2 \right) \right. .$$

4.5.2 Boundary Conditions

When calculating the matrices of the coefficients, the following relations were used:

$$\frac{\partial W X_{i+\alpha,j+\beta}}{\partial p_{i,j}} = \frac{\sigma_1 \Delta t}{m_{i+\alpha,j+\beta}^N} \left(\frac{\partial V_{i,j}}{\partial x_{i+\alpha,j+\beta}} \right)^n, \qquad (4.78)$$

$$\frac{\partial W Y_{i+\alpha,j+\beta}}{\partial p_{i,j}} = \frac{\sigma_1 \Delta t}{m_{i+\alpha,j+\beta}^N} \left(\frac{\partial V_{i,j}}{\partial y_{i+\alpha,j+\beta}} \right)^n,$$

which follow from the equations of motion and are valid for the internal nodes. In the case of a free boundary, these formulas remain true, because using fictitious points, the motion of the boundary nodes and the internal nodes can be calculated by the same formulas.

Let us now consider impermeability conditions. In this case, the normal component of the velocity vector is given on the boundary. For definitiveness, let us consider a boundary where $j = N$ (see Figure 4.23). If $W X_{i,N}^{n+1}$

and $WY_{i,N}^{n+1}$ are the velocities of the boundary nodes, then the boundary condition for the boundary segment $(i, N) - (i + 1, N)$ is as follows:

$$WX_{i,N}^{n+1} \, nx_{i,N} + WY_{i,N}^{n+1} \, ny_{i,N} = \nu_{i,N}^{n+1} , \qquad (4.79)$$

where $nx_{i,N}$, $ny_{i,N}$ are the components of the outward normal to the boundary in node i, N. The definition of the normal vector to a broken line is not trivial, and was discussed in previous chapters. For now, let us assume that nx and ny are given.

For the case of impermeability boundary conditions, we also want to use general formulas for computing the velocity in the boundary nodes. However, we need to introduce pressure into the fictitious cells. We can define these values similar to explicit algorithms. It is important that the error in the value of tangential velocity, which is related to the the values in fictitious cells, is order h, for a straight wall this error equal to zero.

Now we will describe the procedure for computing the velocities in the boundary nodes. Assume we know the right values of the pressure in all the real cells on the new time level, and let us take values of the pressure in the fictitious cells as in explicit algorithms. Let us denote by $\tilde{WX}_{i,N}^{n+1}$ and $\tilde{WY}_{i,N}^{n+1}$ velocities of the boundary nodes, which are obtained by general formulas. If we compute the tangential component of the velocity vector by using these values, we must obtain the values, which are very close to right values. Therefore, we can use the following equation, which gives us the relation between velocities WX, WY, and $\tilde{WX}$, $\tilde{WY}$:

$$- WX_{i,N}^{n+1} \, ny_{i,N} + WY_{i,N}^{n+1} \, nx_{i,N} = -\tilde{WX}_{i,N}^{n+1} \, ny_{i,N} + \tilde{WY}_{i,N}^{n+1} \, nx_{i,N} , \qquad (4.80)$$

taking into account that the unit tangential vector is $(-ny_{i,N}, nx_{i,N})$.

Now to compute the actual velocities $WX_{i,j}^{n+1}$, $WY_{i,j}^{n+1}$, we have the system of two equations with two unknowns 4.79 and 4.80, which gives us the following expressions:

$$WX_{i,N}^{n+1} = -ny_{i,N} \left(-\tilde{WX}_{i,N}^{n+1} \, ny_{i,N} + \tilde{WY}_{i,N}^{n+1} \, nx_{i,N} \right) - nx_{i,N} \, \nu_{i,N}^{n+1} , \qquad (4.81)$$

$$WY_{i,N}^{n+1} = nx_{i,N} \left(-\tilde{WX}_{i,N}^{n+1} \, ny_{i,N} + \tilde{WY}_{i,N}^{n+1} \, nx_{i,N} \right) + nx_{i,N} \, \nu_{i,N}^{n+1} .$$

Now we can use these velocities to compute the new positions of the boundary nodes.

Therefore, if we know the pressure in the internal cells, we have a two-stage procedure to compute the actual velocities of the boundary nodes. In the first stage, we compute some intermediate values using general formulas and then compute the actual values using the formulas in 4.81.

We now need to describe how to compute the derivatives of $\partial \Phi_{i,j} / \partial p_{k,l}$. We first need to make some changes in the formulas where the boundary nodes participate. Let us note that because of the condition for velocities in 4.79, the coordinates $x_{i,N}^{n+1}$, $y_{i,N}^{n+1}$ are not independent and satisfy the following condition:

$$x_{i,N}^{n+1} = x_{i,N}^n + \Delta t \frac{\nu_{i,N}^{n+1}}{nx_{i,N}} + y_{i,N}^n \frac{ny_{i,N}}{nx_{i,N}} - y_{i,N}^{n+1} \frac{ny_{i,N}}{nx_{i,N}}. \tag{4.82}$$

If $nx_{i,N}$ is equal to zero, then we can rewrite the previous equation and express the coordinate $y_{i,N}^{n+1}$. Now we can consider the problem of computation of the derivatives. The first problem is how to compute the derivatives $\partial V_{i,j}/\partial p_{k,l}$ if the vertices of $V_{i,j}$ belong to the boundary. For the case of the free boundary, we get the following terms:

$$\frac{\partial V_{i,j}}{\partial x_{i+\alpha,j+\beta}} \frac{\partial x_{i+\alpha,j+\beta}}{\partial WX_{i+\alpha,j+\beta}} \frac{\partial WX_{i+\alpha,j+\beta}}{\partial p_{k,l}} + \tag{4.83}$$

$$\frac{\partial V_{i,j}}{\partial y_{i+\alpha,j+\beta}} \frac{\partial y_{i+\alpha,j+\beta}}{\partial WY_{i+\alpha,j+\beta}} \frac{\partial WY_{i+\alpha,j+\beta}}{\partial p_{k,l}}. \tag{4.84}$$

If node $(x_{i+\alpha,j+\beta}, y_{i+\alpha,j+\beta})$ belongs to the boundary, then it will be necessary to use the following expression:

$$\left(\frac{\partial V_{i,j}}{\partial x_{i+\alpha,j+\beta}} + \frac{\partial V_{i,j}}{\partial y_{i+\alpha,j+\beta}} \frac{\partial y_{i+\alpha,j+\beta}}{\partial x_{i+\alpha,j+\beta}} \right) \frac{\partial x_{i+\alpha,j+\beta}}{\partial WX_{i+\alpha,j+\beta}} \frac{\partial WX_{i+\alpha,j+\beta}}{\partial p_{k,l}}. \tag{4.85}$$

For the derivative $\frac{\partial y_{i+\alpha,j+\beta}}{\partial x_{i+\alpha,j+\beta}}$ in the previous equation, we have the following expression, which follows from 4.82:

$$\frac{\partial y_{i+\alpha,j+\beta}}{\partial x_{i+\alpha,j+\beta}} = -\frac{nx_{i+\alpha,j+\beta}}{ny_{i+\alpha,j+\beta}}. \tag{4.86}$$

The expression for the derivative $\partial WX_{i+\alpha,j+\beta}/\partial p_{k,l}$ follows from equation 4.81:

$$\frac{\partial WX_{i+\alpha,j+\beta}}{\partial p_{k,l}} = \tag{4.87}$$

$$-ny_{i+\alpha,j+\beta} \left(-\frac{\partial \widetilde{WX}_{i+\alpha,j+\beta}}{\partial p_{k,l}} ny_{i+\alpha,j+\beta} + \right.$$

$$\left. \frac{\partial \widetilde{WY}_{i+\alpha,j+\beta}}{\partial p_{k,l}} nx_{i+\alpha,j+\beta} \right) =$$

$$-\frac{ny_{i+\alpha,j+\beta}}{m_{i+\alpha,j+\beta}^N} \sigma_1 \Delta t \left(-\frac{\partial V_{k,l}}{\partial x_{i+\alpha,j+\beta}} ny_{i+\alpha,j+\beta} + \frac{\partial V_{k,l}}{\partial y_{i+\alpha,j+\beta}} nx_{i+\alpha,j+\beta} \right).$$

Using these formulas, we can transform 4.85 into the following form:

$$\sigma_1\,\sigma_2\,(\Delta t)^2\,\left(-\frac{\partial V_{i,j}}{\partial x_{i+\alpha,j+\beta}}\,ny_{i+\alpha,j+\beta}+\frac{\partial V_{i,j}}{\partial y_{i+\alpha,j+\beta}}\,nx_{i+\alpha,j+\beta}\right)\,\times$$

$$\frac{1}{m^N_{i+\alpha,j+\beta}}\,\left(-\frac{\partial V_{k,l}}{\partial x_{i+\alpha,j+\beta}}\,ny_{i+\alpha,j+\beta}+\frac{\partial V_{k,l}}{\partial y_{i+\alpha,j+\beta}}\,nx_{i+\alpha,j+\beta}\right).\,(4.88)$$

Let us now consider how to compute the derivative $\frac{\partial \varepsilon_{i,j}}{\partial p_{k,l}}$. Since $\varepsilon_{i,j}$ depends on $p_{k,l}$ through the velocities, and the components of the velocities satisfy the condition in 4.79, then similar to the previous case, we get

$$\frac{\partial \varepsilon_{i,j}}{\partial p_{k,l}} = \qquad\qquad\qquad\qquad\qquad\qquad\qquad\qquad (4.89)$$

$$\left(\frac{\partial \varepsilon_{i,j}}{\partial WX_{i+\alpha,j+\beta}}+\frac{\partial \varepsilon_{i,j}}{\partial WY_{i+\alpha,j+\beta}}\,\frac{\partial WY_{i+\alpha,j+\beta}}{\partial WX_{i+\alpha,j+\beta}}\right)\frac{\partial WX_{i+\alpha,j+\beta}}{\partial p_{k,l}} =$$

$$\frac{\sigma_1\,\sigma_3\,p^{\sigma_1}_{i,j}\,(\Delta t)^2}{m^C_{i,j}}\,\left(-\frac{\partial V_{i,j}}{\partial x_{i+\alpha,j+\beta}}\,ny_{i+\alpha,j+\beta}+\frac{\partial V_{i,j}}{\partial y_{i+\alpha,j+\beta}}\,nx_{i+\alpha,j+\beta}\right)\,\times$$

$$\frac{1}{m^N_{i+\alpha,j+\beta}}\,\left(-\frac{\partial V_{k,l}}{\partial x_{i+\alpha,j+\beta}}\,ny_{i+\alpha,j+\beta}+\frac{\partial V_{k,l}}{\partial y_{i+\alpha,j+\beta}}\,nx_{i+\alpha,j+\beta}\right).$$

Therefore, we know how to compute all the derivatives in the case of impermeability boundary conditions. In the next section, we will show that for the case of free boundary conditions and for the case of impermeability boundary conditions, the matrix of the system of linear equations for δp is symmetric and positive definite.

We will now consider another approach for managing impermeability boundary conditions, which gives better results for a very curved boundary.

Let us recall that in the discrete case, the boundary is a broken line and the unique definition of the normal vector can only be done for segments of the boundary and cannot be done for nodes. This is the reason that for a very curved boundary, we need to take the average of the velocity from the nodes to the center of the boundary segment and write the boundary condition for this segment.

$$\frac{WX^{n+1}_{i,N}+WX^{n+1}_{i+1,N}}{2}\,\left(-\frac{y^n_{i+1,N}-y^n_{i,N}}{l\xi_{i,N}}\right)+$$

$$\frac{WY^{n+1}_{i,N}+WY^{n+1}_{i+1,N}}{2}\,\left(\frac{x^n_{i+1,N}-x^n_{i,N}}{l\xi_{i,N}}\right)=\nu^{n+1}_{i,N}.\qquad (4.90)$$

Similar to the previous approach, we will compute the velocities on the boundary by using general formulas. We will not recompute the actual velocities using boundary conditions, but will try to find the pressure in the fictitious cells which will give us the velocities that satisfy the conditions in 4.90. Thus, we will introduce pressures $p_{i,N}^{n+1}$ in the fictitious cells which will be additional unknowns. The number of these unknowns is equal to the number of boundary segments where we have impermeability conditions. The role of the additional equations will be played by the conditions in 4.90, and in actuality, because we will use the general formulas for computing the velocities on the boundary, the conditions in 4.90 are the equations with respect to pressure.

In the differential case, such an approach can be roughly interpreted as follows. If we divide the equation of motion by density and then take the inner product with the vector of normal to the boundary, we get

$$\frac{d(\vec{W}, \vec{n})}{dt} = \frac{1}{\rho}(\vec{n}, \mathrm{grad}p).\tag{4.91}$$

Hence, the expression in the left-hand side is given from the boundary conditions and

$$(\vec{n}, \mathrm{grad}p) = \frac{dp}{dn}$$

is the directional derivative of p. Therefore, equation 4.91 can be rewritten as follows:

$$\frac{1}{\rho}\frac{dp}{dn} = f(x, y, t),\tag{4.92}$$

where function f is given. The equation 4.92 plays the role of the boundary condition for pressure. Therefore, in some sense, the conditions of impermeability are equivalent to the specifications of the normal derivatives of pressure on the boundary.

Boundary conditions can be written in the following form:

$$F_{i,N}(WX_{i,N}^{n+1}, WX_{i+1,N}^{n+1}; WY_{i,N}^{n+1}, WY_{i+1,N}^{n+1}) =\tag{4.93}$$

$$\frac{WX_{i,N}^{n+1} + WX_{i+1,N}^{n+1}}{2}\left(-\frac{y_{i+1,N}^n - y_{i,N}^n}{l\xi_{i,N}}\right) +$$

$$\frac{WY_{i,N}^{n+1} + WY_{i+1,N}^{n+1}}{2}\left(\frac{x_{i+1,N}^n - x_{i,N}^n}{l\xi_{i,N}}\right) - \nu_{i,N}^{n+1} = 0.\tag{4.94}$$

If we consider function F as a composite function of the values of pressure, then we can write

$$\Phi_{i,N}\left(p_{k,l}^{n+1} : (k, l) \in St(i, N)\right) =\tag{4.95}$$

$$F_{i,N}\left(WX_{i,N}^{n+1}(p_{i-1,N-1}^{n+1}, p_{i-1,N}^{n+1}, p_{i,N-1}^{n+1}, p_{i,N}^{n+1}),\right.$$

$$WX_{i+1,N}^{n+1}(p_{i,N-1}^{n+1}, p_{i,N}^{n+1}, p_{i+1,N-1}^{n+1}, p_{i+1,N}^{n+1});$$

$$WY_{i,N}^{n+1}(p_{i-1,N-1}^{n+1}, p_{i-1,N}^{n+1}, p_{i,N-1}^{n+1}, p_{i,N}^{n+1}),$$

$$\left.WY_{i+1,N}^{n+1}(p_{i,N-1}^{n+1}, p_{i,N}^{n+1}, p_{i+1,N-1}^{n+1}, p_{i+1,N}^{n+1})\right),$$

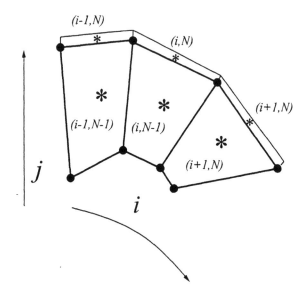

Figure 4.26: *Stencil $St(i, N)$ for pressure for impermeability boundary conditions.*

where the stencil

$$St(i, N) =$$
$$\{(i, N - 1), (i - 1, N - 1), (i + 1, N - 1),$$
$$(i, N), (i - 1, N), (i + 1, N)\}$$

is shown in Figure 4.26, and contains six values of pressure, three of which are in fictitious cells.

Theoretically, it is clear how to compute the derivatives $d\Phi_{i,N}/dp_{k,l}$, therefore, we will not go into further details. However, it is important to note that it is also possible to make some slight modifications to these derivatives that will provide symmetry and positive definite properties to the resulting matrix of the system of linear equations.

4.5.3   Properties of the System of Linear Equations

Let us first consider the properties of the matrix A of the system of linear equations for the case of the free boundary conditions. In this case, we have the following at any cell:

$$(A\,z)_{i,j} = \tag{4.96}$$

$$\left\{ \left[-\frac{\partial F_{i,j}}{\partial p_{i,j}} + \frac{\partial F_{i,j}}{\partial \varepsilon_{i,j}} \frac{\sigma_1 \Delta t}{m_{i,j}^C} \times \right. \right.$$

$$\sum_{\alpha,\beta=0}^{1} \left(\frac{\partial V_{i,j}}{\partial x_{i+\alpha,j+\beta}} W X_{i+\alpha,j+\beta} + \frac{\partial V_{i,j}}{\partial y_{i+\alpha,j+\beta}} W Y_{i+\alpha,j+\beta} \right) \bigg] \bigg/$$

$$\left. \left(\frac{m_{i,j}^C}{V_{i,j}^2} \frac{\partial F_{i,j}}{\partial \rho_{i,j}} \sigma_1 \sigma_2 \left(\Delta t \right)^2 + \frac{\partial F_{i,j}}{\partial \varepsilon_{i,j}} \frac{\sigma_1 \sigma_3}{m_{i,j}^C} \left(\Delta t \right)^2 p_{i,j} \right) \right\} z_{i,j} +$$

$$\sum_{\alpha,\beta=0}^{1} \left(\frac{1}{m_{i+\alpha,j+\beta}^N} \frac{\partial V_{i,j}}{\partial x_{i+\alpha,j+\beta}} \sum_{k,l=0}^{1} \frac{\partial V_{i+\alpha-k,j+\beta-l}}{\partial x_{i+\alpha,j+\beta}} z_{i-k,j-l} + \right.$$

$$\left. \frac{1}{m_{i+\alpha,j+\beta}^N} \frac{\partial V_{i,j}}{\partial y_{i+\alpha,j+\beta}} \sum_{k,l=0}^{1} \frac{\partial V_{i+\alpha-k,j+\beta-l}}{\partial y_{i+\alpha,j+\beta}} z_{i-k,j-l} \right).$$

From this equation, we can conclude that matrix A can be represented in the following form:

$$A = D_x \, \overline{M} \, G_x + D_y \, \overline{M} \, G_y + R, \tag{4.97}$$

where

$$(D_x \, z)_{i,j} = \sum_{\alpha,\beta=0,1} \frac{\partial V_{i,j}}{\partial x_{i+\alpha,j+\beta}} z_{i+\alpha,j+\beta}, \tag{4.98}$$

$$(D_y \, z)_{i,j} = \sum_{\alpha,\beta=0,1} \frac{\partial V_{i,j}}{\partial y_{i+\alpha,j+\beta}} z_{i+\alpha,j+\beta}, \tag{4.99}$$

$$(G_x \, z)_{i,j} = \sum_{\alpha,\beta=0,1} \frac{\partial V_{i-\alpha,j-\beta}}{\partial x_{i,j}} z_{i-\alpha,j-\beta}, \tag{4.100}$$

$$(G_y \, z)_{i,j} = \sum_{\alpha,\beta=0,1} \frac{\partial V_{i-\alpha,j-\beta}}{\partial y_{i,j}} z_{i-\alpha,j-\beta}, \tag{4.101}$$

$$(\overline{M} \, z)_{i,j} = \frac{z_{i,j}}{m_{i,j}^N}, \tag{4.102}$$

$$(R\,z)_{i,j} = \qquad\qquad\qquad\qquad\qquad\qquad\qquad\qquad (4.103)$$

$$\left\{\left[-\frac{\partial F_{i,j}}{\partial p_{i,j}} + \frac{\partial F_{i,j}}{\partial \varepsilon_{i,j}}\frac{\sigma_1\,\Delta t}{m_{i,j}^C}\times\right.\right.$$

$$\sum_{\alpha,\beta=0}^{1}\left(\frac{\partial V_{i,j}}{\partial x_{i+\alpha,j+\beta}}\,WX_{i+\alpha,j+\beta} + \frac{\partial V_{i,j}}{\partial y_{i+\alpha,j+\beta}}\,WY_{i+\alpha,j+\beta}\right)\Bigg]\Bigg/$$

$$\left.\left(\frac{m_{i,j}^C}{V_{i,j}^2}\frac{\partial F_{i,j}}{\partial \rho_{i,j}}\sigma_1\sigma_2\,(\Delta t)^2 + \frac{\partial F_{i,j}}{\partial \varepsilon_{i,j}}\frac{\sigma_1\sigma_3}{m_{i,j}^C}(\Delta t)^2\,p_{i,j}\right)\right\}\,z_{i,j}\,.$$

It is obvious that $\overline{M} = \overline{M}^* > 0$. If the natural requirements

$$\frac{\partial F_{i,j}}{\partial p_{i,j}} > 0,\ \frac{\partial F_{i,j}}{\partial \rho_{i,j}} < 0,\ \frac{\partial F_{i,j}}{\partial \varepsilon_{i,j}} < 0, \qquad\qquad (4.104)$$

are satisfied, then the diagonal operator R is also positive for reasonable constraints on the time step. It is also easy to check that

$$G_x = (D_x)^*,\ G_y = (D_y)^* \qquad\qquad\qquad (4.105)$$

in the matrix sense. Therefore,

$$A = D_x\,\overline{M}\,(D_x)^* + D_y\,\overline{M}\,(D_y)^* + R \qquad\qquad (4.106)$$

which implies that $A = A^* > 0$.

Let us now consider the case of impermeability boundary conditions. As we mentioned, in this case, instead of using the expression

$$\frac{\partial V_{i,j}}{\partial x_{k,l}}\frac{\partial V_{p,q}}{\partial x_{k,l}} + \frac{\partial V_{i,j}}{\partial y_{k,l}}\frac{\partial V_{p,q}}{\partial y_{k,l}}\,, \qquad\qquad (4.107)$$

we have to take

$$\left(-ny_{k,l}\frac{\partial V_{i,j}}{\partial x_{k,l}} + nx_{k,l}\frac{\partial V_{i,j}}{\partial x_{k,l}}\right)\left(-ny_{k,l}\frac{\partial V_{p,q}}{\partial x_{k,l}} + nx_{k,l}\frac{\partial V_{p,q}}{\partial x_{k,l}}\right). \quad (4.108)$$

Now we will show that by the corresponding redefinition of quantities $\partial V/\partial x$, $\partial V/\partial y$ on the boundary nodes, we can obtain the expression 4.108 from formula 4.107. To do this, we define these quantities as follows:

$$\overline{\frac{\partial V i,j}{\partial x_{k,l}}} = -ny_{k,l}\left(-ny_{k,l}\frac{\partial V_{i,j}}{\partial x_{k,l}} + nx_{k,l}\frac{\partial V_{i,j}}{\partial x_{k,l}}\right),$$

$$\qquad\qquad\qquad\qquad\qquad\qquad\qquad\qquad\qquad (4.109)$$

$$\overline{\frac{\partial V i,j}{\partial y_{k,l}}} = nx_{k,l}\left(-ny_{k,l}\frac{\partial V_{i,j}}{\partial x_{k,l}} + nx_{k,l}\frac{\partial V_{i,j}}{\partial x_{k,l}}\right).$$

Then, we get

$$\overline{\frac{\partial V_{i,j}}{\partial x_{k,l}} \frac{\partial V_{p,q}}{\partial x_{k,l}}} + \overline{\frac{\partial V_{i,j}}{\partial y_{k,l}} \frac{\partial V_{p,q}}{\partial y_{k,l}}} = \tag{4.110}$$

$$(nx_{k,l}^2 + ny_{k,l}^2) \left(-ny_{k,l} \frac{\partial V_{i,j}}{\partial x_{k,l}} + nx_{k,l} \frac{\partial V_{i,j}}{\partial x_{k,l}} \right) \times$$

$$\left(-ny_{k,l} \frac{\partial V_{p,q}}{\partial x_{k,l}} + nx_{k,l} \frac{\partial V_{p,q}}{\partial x_{k,l}} \right) =$$

$$\left(-ny_{k,l} \frac{\partial V_{i,j}}{\partial x_{k,l}} + nx_{k,l} \frac{\partial V_{i,j}}{\partial x_{k,l}} \right) \left(-ny_{k,l} \frac{\partial V_{p,q}}{\partial x_{k,l}} + nx_{k,l} \frac{\partial V_{p,q}}{\partial x_{k,l}} \right).$$

Therefore, in the case of the impermeability conditions, the coefficients of the linear system can be calculated by formulas for the case of free boundary conditions and only quantities $\partial V/\partial x$, $\partial V/\partial y$ must be redefined at the boundary nodes. Hence, all the resulting properties of the matrix of the system of linear equations in this case also remain valid.

4.5.4   Some Properties of Algorithm

Conservation Laws and Iteration method

In the previous chapter we considered the question of how iteration methods can violate conservation properties of FDS for the heat equation. Here we consider this question for gas dynamics equations.

Suppose we have some criterion for the termination of parallel chord iterations. Usually, this criterion looks as follows:

$$\max_{i,j} |F_{i,j} \left(p_{i,j}^{(s)}, \rho_{i,j}^{(s)}, \varepsilon_{i,j}^{(s)} \right)| < \delta, \tag{4.111}$$

where δ is the given tolerance. If this criterion is satisfied, then we take

$$p_{i,j}^{n+1} = p_{i,j}^{(s)}.$$

In our iteration method all quantities are the functions of pressure, and consequently all equations, except the equation of state, are satisfied *exactly*. It is important to note that it is true for any accuracy of the iteration method, that is, for any δ in 4.111. Therefore, quantities obtained by the iteration method satisfy the same equations (which related to conservation laws) as the exact solution of FDS. Hence if the original FDS is conservative, then our iteration method will not violate its conservation properties.

The Case of Incompressible Fluid

This is the case where, from a formal point of view, the equation of state is

$$\rho_{i,j}^{n+1} = const. \tag{4.112}$$

In terms of function $F_{i,j}$, which we were using previously, this means that

$$F_{i,j}(p_{i,j}, \rho_{i,j}, \varepsilon_{i,j}) = \rho_{i,j} - const.\qquad(4.113)$$

In our consideration, this means that

$$\frac{\partial F_{i,j}}{\partial p_{i,j}} \equiv 0,\qquad(4.114)$$

$$\frac{\partial F_{i,j}}{\partial \rho_{i,j}} \equiv 1,\qquad(4.115)$$

$$\frac{\partial F_{i,j}}{\partial \varepsilon_{i,j}} \equiv 0,\qquad(4.116)$$

and all considerations are still valid. That is, the algorithm will work and the properties of the system of linear equations will be the same.

4.6   Artificial Viscosity in 2-D

Similar to 1-D, we will consider only scalar artificial viscosity, which was introduced in the finite-difference scheme as an addition to pressure, and consequently, is defined in the cell.

Here, we will consider two types of viscosity, linear and quadratic viscosity, which are generalizations of 1-D results in a 2-D case.

The differential form of linear viscosity is

$$\omega = -\nu \operatorname{div} \vec{W}.\qquad(4.117)$$

Similar to the 1-D case, using a dimensional consideration, we can choose the form of ν:

$$\nu = \nu_* L \rho U,\qquad(4.118)$$

where ν_* is a dimensionless parameter, L characterizes local linear size, and U is a characteristic velocity, which is analogous to the 1-D case and we will choose as the local adiabatic speed of sound. Thus,

$$\omega = -\nu_* L \rho C_A \operatorname{div} \vec{W}.\qquad(4.119)$$

Since $\omega \in HC$ in the discrete case, then for the approximation of div, we can use the same discrete operator as in the energy equation (see 4.14). This will give us the approximation that is consistent with the differential case. In particular, the additional term in the equation of motion will be:

$$\operatorname{GRAD} \nu \operatorname{DIV} \vec{W}$$

and the operator which acts on the velocity is sign definite and symmetric. Also, the additional term in the energy equation is

$$\nu \, (\operatorname{DIV} \vec{W})^2,$$

which is positive. This means that work of the artificial viscosity forces leads to the transfer of kinetic energy to internal energy and that internal energy increases. It is correct from a physical point of view.

In the discrete case, we can take the characteristic density is equal to the density in the cell. The role of C_A can be played by the maximal value of adiabatic speed of sound over some neighborhood of the given cell.

The main problem is how to choose the parameter L. Let us describe one approach for choosing L [28]. The parameter L shows us the order of width of the shock front. If we know the direction of the front propagation, then it is natural to require that a given cell be the zone of the shock front. This means that we need to choose L as the linear projection of a given cell to the normal direction of the shock front.

The direction that is normal to the shock front is the direction of maximal compression, which coincides with one of the principal axes of the strain velocity tensor.

The angles φ of these axes with the x-coordinate are:

$$\tan \varphi_r = \qquad\qquad\qquad\qquad (4.120)$$

$$\frac{\partial WY/\partial y - \partial WX/\partial x}{\partial WX/\partial y + \partial WY/\partial x} \pm$$

$$\frac{\sqrt{(\partial WY/\partial y - \partial WX/\partial x)^2 + (\partial WX/\partial y - \partial WY/\partial x)^2}}{\partial WX/\partial y + \partial WY/\partial x},$$

$$r = 1, 2.$$

This expression follows from the condition

$$\frac{d}{d\varphi}\left[\left(\frac{d\vec{W_l}}{dl}\right)^2\right] = 0,$$

where $\vec{W_l}$ is the projection of vector $\vec{W}$ on the direction $\vec{l}$ and $d\vec{W_l}/dl$ is the directional derivative on this direction:

$$\frac{d\vec{W_l}}{dl} = \frac{\partial WX}{\partial x}\cos^2\varphi + \left(\frac{\partial WX}{\partial y} + \frac{\partial WY}{\partial x}\right)\cos\varphi\sin\varphi + \frac{\partial WY}{\partial y}\sin^2\varphi.$$

If we know the directions that are given by formula 4.120, then we can find the values of $\left(\vec{W_l}\right)_r$, $r = 1, 2$ in the vertices of the cell. And finally,

$$L = \max_{r=1,2}\left\{\frac{\left(\vec{W_l}\right)_r^{min} - \left(\vec{W_l}\right)_r^{max}}{\left(\frac{\vec{W_l}}{dl}\right)_r}, 0\right\}. \qquad (4.121)$$

The details can be found in [28].

4.7 The Numerical Example

We will consider, as an example of numerical computation, the same problem as for the 1-D scheme. That is, a problem of the formation of a strong shock wave by a piston. The 2-D statement of the problem is the following. The problem consists of a rectangular box in which the walls form reflective boundaries and the left-hand side wall acts as a piston, initially driving a strong shock wave towards the right. Since the upper and lower boundaries are reflective (the normal component of the velocity is equal to zero), and the initial conditions are independent of the vertical coordinate direction, the problem is expected to be one dimensional, independent of the width of the box. The working fluid is assumed to be an ideal gas with $\gamma = 5/3$, compressed by a piston (the left wall of the box) moving to the right with a velocity of 1. The initial conditions involve a stationary gas which occupies the box $[0, 1] \times [0, 0.1]$, with a density of 1 and an internal energy of 10^{-4}. The right wall of the box is a stationary wall. Then, the initial conditions are the same as for the 1-D problem, which was considered in section 3.9. Therefore, the analytical solution has to depend only on x, and for a given y, the profiles of all functions have to be the same as in 1-D. That is, the post shock conditions are given by a pressure of 1.333, density of 4, internal energy of 0.5, and a shock speed of 1.333.

Computations were performed using $M = 101$ and $N = 11$ and linear artificial viscosity with coefficient $\nu_* = 1.5$. For an explicit finite-difference scheme, we use $dt = 0.00025$, and for an implicit one, we use $dt = 0.001$.

For the solution of the system of linear equations for dp in the framework of an implicit scheme, we have used the Gauss-Seidel method, where the tolerance for the linear and non-linear iterations was 10^{-4}.

The numerical solutions for explicit and implicit schemes are practically the same, and we present here only the numerical results for the explicit scheme.

This problem was solved using two types of the initial grid. The first grid is the tensor product grid or rectangular grid, where we have used 101 nodes in the x direction and 11 nodes in the y direction. In this case, for each given y (line $j = const$), the solution is the same and coincides with the solution of the 1-D problem.

In Figure 4.27, we present a grid at time $t = 0.5$. Pressure for the same time moment is presented in Figure 4.28. From these figures, we can conclude that for the initial rectangular grid, the 2-D finite-difference scheme reproduces the 1-D solution.

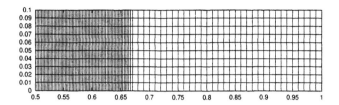

Figure 4.27: *Grid, explicit algorithm, $t = 0.5$.*

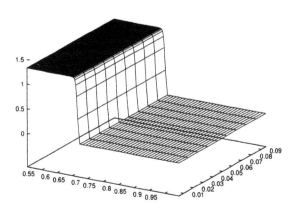

Figure 4.28: *Pressure, explicit algorithm, $t = 0.5$.*

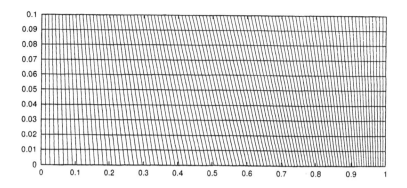

Figure 4.29: *Initial non-uniform grid, Saltzman test problem.*

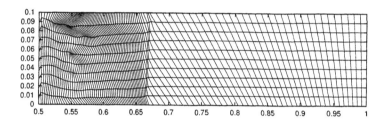

Figure 4.30: *Grid, explicit algorithm, $t = 0.5$.*

The second grid is constructed using the following formulas:

$$x_{i,j} = (i - 1) * dx + (11 - j) * dy * \sin \frac{\pi (i - 1)}{100} \, ,$$
$$y_{i,j} = (j - 1) * dy \, ,$$
$$i = 1, 2, \ldots, 101; \quad j = 1, 2, \ldots, 11 \, ,$$

where $dx = dy = 0.01$. This initial grid is displayed in Figure 4.29. This type of grid is used in the Saltzman test problem [105], which tests the ability of the code to retain a one-dimensional solution to a one-dimensional problem on a non-uniform, two-dimensional grid. There are many publications where different algorithms are tested on this problem (see, for example, [25] and [88]). For our purpose, it is just an example of a 2-D computation, where we know the analytical solution. The one-dimensional symmetry is broken by the mesh.

In Figure 4.30, we present the Lagrangian grid at $t = 0.5$.

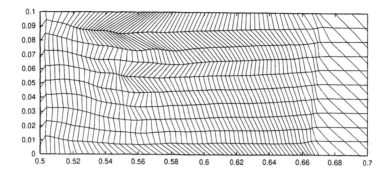

Figure 4.31: *Fragment of the grid, explicit algorithm, $t = 0.5$.*

The fragment of the grid, which is already involved in movement, is shown in Figure 4.31.

It is obvious that the solution is not one-dimensional. The reasons for this phenomena and some approaches to reduce distortions of the grid, are investigated in [25], [88], and a few other publications. Here, we note that such a distortion is usual for most Lagrangian codes, and in order to avoid it, some special types of artificial viscosity [88], or more complicated procedures can be used, which involve estimations and subtracting spurious numerical vorticity [25].

In Figures 4.32, 4.33, and 4.34, we present 2-D profiles for pressure, density, and internal energy, respectively.

The velocity field for part media, which is involved in movement, is presented in Figure 4.35.

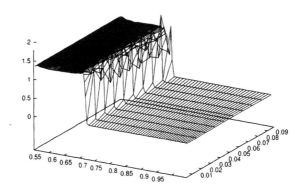

Figure 4.32: *Pressure, explicit algorithm, t = 0.5.*

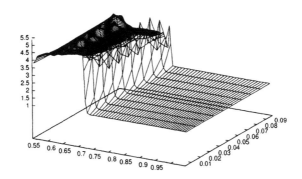

Figure 4.33: *Density, explicit algorithm, t = 0.5.*

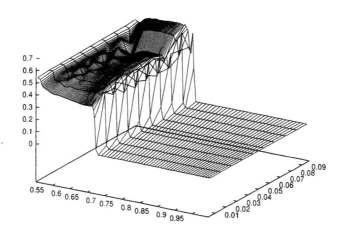

Figure 4.34: *Energy, explicit algorithm,* $t = 0.5$.

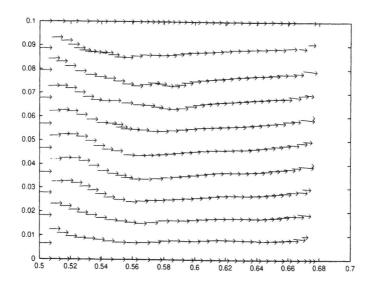

Figure 4.35: *Velocity field, explicit algorithm,* $t = 0.5$.

Chapter 6

Conclusion

In this book we described the construction of finite-difference schemes for elliptic, parabolic, and hyperbolic equations on logically rectangular grids by means of the support-operators method. This method produces FDSs, which are conservative and mimic the main properties of original differential problems. The numerical experiments shows that constructed FDSs are second-order accurate on smooth grids, and first-order accurate on rough grids, which confirms their robustness. We have constructed discrete analogs of the operators div and grad which then can be used to approximate differential equations which contain these operators.

This book can be considered as an introduction to the method of support operators. Here we enumerate some references where the method of support-operators was used in more complicated situations. Application to the equations of magnetic field diffusion, which involve the operator curl, can be found in [139], while the case of anisotropic conductivity was considered in [29]. Finite-difference schemes for Maxwell equations are constructed in [24]. Approximation of the general equations of motion, which involve operations on tensor objects, such as the divergence of a tensor and the gradient of vectors, is considered in [113, 134]. Applications to the equations of gas dynamics in Eulerian form can be found in [27].

Using the local basis system of components for discrete vector fields is considered in [138], [31], [139], [115]. For example, for the heat equation, component of fluxes, which is normal to sides of the cells, is used because these components are continuous on interface. Using such components improves the accuracy, and is especially useful for problems with discontinuous coefficients. For magnetic field diffusion, for the same reason, we use normal components of magnetic field, and tangential components of electric field. Construction of finite-difference schemes in the curvilinear orthogonal coordinate system is considered in [66], [67].

The method of support-operators also was applied to construct finite-difference schemes on different types of grids: triangular grids [47], Voronoi grids [126], [127], [128], [129], [2], and grids with local refinement [125].

In a recent publication [14] the freedom in approximating integrals in integral identities is used to construct high-order schemes.

The properties of the discrete operators, which follow from the construction of FDSs by the support-operators method, can be used to prove convergence theorems for linear and non-linear problems [20, 21, 4, 5, 6], and to understand the stability of FDSs [49], [104]. These properties also help in the construction of effective iteration methods for solving the non-linear system of discrete equations, which arise in the application of the SOM to hydrodynamics and magneto-hydrodynamics [40, 44, 3].

An analysis of procedures for constructing FDSs by the method of support-operators shows that this process can be formalized. Namely, the process can be reduced to operations such as the construction of adjoint operators, superposition of operators, and so on. Therefore, a computer algebra system can be used to symbolically compute the formulas needed for the construction of computer codes implementing the support-operators method. The symbolic computations can be used for the construction and investigation of FDSs [141, 39, 36, 142, 123, 38, 37, 143], which eliminates most of the errors resulting from hand implementation of the schemes. Computer algebra systems can also be used to analyze the numerical method.

Examples of practical applications include simulations of: controlled laser fusion [147]; the collapse of a quasi-spherical target in a hard cone [30]; the Rayleigh-Taylor instability for incompressible flows [41]; the compression of a toroidal plasma by a quasi-spherical liner [43, 42]; an over-compressed detonation wave in a conic channel [69]; a magnetic field in a spiral band reel [8]; the magnetic field of a toroidal spiral with a screen [9]; the flows of a viscous incompressible liquid with a free surface [22]; a microwave plasma generator [82].

Appendix A

Fortran Code Directory

This appendix provides a description of the codes that are on the disk that comes with the book. The disk is formatted for IBM MS-DOS computers. Initially all programs were prepared and tested on the SUN workstation using the f 77 compiler. All files which contain programs in Fortran have the extension ".f"; files which have the extension "**output**", contain the test output, and can be used for checking correctness of programs. That is, if you run the programs on your computer without changing them, you will have the same output results that are in the files on the disk.

The first section describes the general structure of the disk. In the second section, we describe the contents of the directory related to elliptic equations. The next section provides a description of programs related to the heat conduction equation, and in the last section, we describe the programs for Lagrangian gas dynamics.

1 The Structure of Directories on the Disk

The disk contains three main directories: **ELEQ**, **HEATEQ** and **GAS-DEQ**, as shown in Figure 1.1.

The directory **ELEQ** contains programs related to the elliptic equations, and has sub-directories **E1D** and **E2D**, which contain programs for cases on one- and two-space dimensions, respectively. In turn, each of these sub-directories, **E1D** and **E2D**, contains sub-directories **DIRICH** and **ROBIN**, for cases of Dirichlet and Robin boundary conditions. In each of these sub-directories, there are sub-directories **NODAL** and **CELL**, where programs for nodal and cell-centered discretisations are placed. The structure of directory **ELEQ** is shown in Figure 1.2.

The directory **HEATEQ** contains programs related to the heat conduction equation, and have sub-directories **H1D** and **H2D**, which contain programs for cases on one and two space dimensions respectively. In turn, each

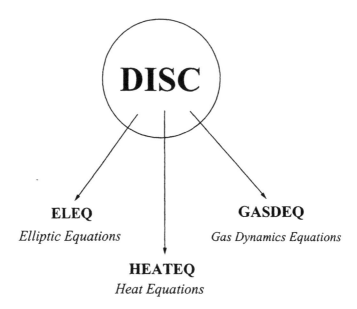

Figure 1.1: *Disk structure.*

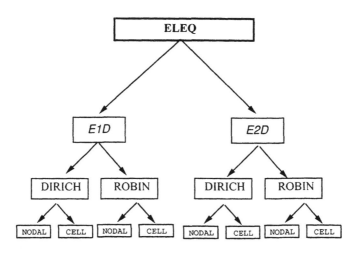

Figure 1.2: *ELEQ directory structure.*

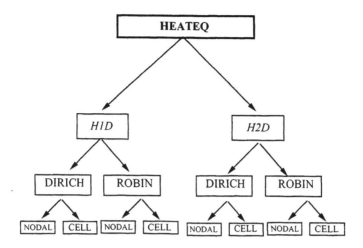

Figure 1.3: *HEATEQ directory structure.*

of these sub-directories, **H1D** and **H2D**, contains sub-directories **DIRICH** and **ROBIN**, where programs for Dirichlet and Robin boundary conditions are placed. In each of these sub-directories, there are sub-directories **NODAL** and **CELL**, where programs for nodal and cell-centered discretisations are placed. The structure of directory **HEATEQ** is shown in Figure 1.3.

The directory **GASDEQ** contains programs related to the gas dynamics equations and have sub-directories **G1D** and **G2D**, which contain programs for cases on one- and two-space dimensions respectively. In turn, each of these sub-directories, **G1D** and **G2D**, contains sub-directories **EXP** and **IMP**, where programs for explicit and implicit finite-difference schemes are placed. The structure of directory **GASDEQ** is shown in figure 1.4.

2 Programs for Elliptic Equations

In this section we describe in detail the contents of directory **ELEQ**, which contains programs for elliptic equations.

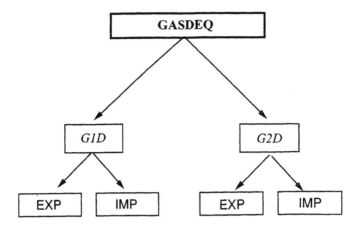

Figure 1.4: *GASDEQ directory structure.*

2.1 Programs for 1-D Equations

2.1.1 Dirichlet Boundary Conditions

The programs for 1-D equations and Dirichlet boundary conditions, and nodal discretisation of scalar functions, are placed in directory

/ELEQ/E1D/DIRICH/NODAL.

This directory contains only two files: file **p1ddnc.f**, which contains all Fortran programs related to this case, and file **p1ddnc.output**, which corresponds to the test problem that is described in Section 2.1.10 from Chapter 3. The programs in file **p1ddnc.f** are:

- **p1ddnc** - the main program,

- **CFDS** - subroutine for computation of coefficients and the right-hand side of the finite-difference scheme,

- **seidel** - subroutine for the Gauss-Seidel iteration method,

- **mkgrid** - subroutine which generates a computational grid,

- **exact** - subroutine which computes an exact solution,

- **output** - subroutine which prepares file **p1ddnc.output**,

- **error** - subroutine which computes errors,

- **bcon** - subroutine which determines boundary conditions,

- **coeff** - subroutine which computes values of the coefficient of a differential equation in a cell,

- **fk** - real function which determines the analytical expression for coefficients of a differential equation,

- **right** - subroutine which computes values of the right-hand side of the original differential equation in nodes,

- **fr** - real function which determines the values of the right-hand side of the original differential equation.

The programs for 1-D equations and Dirichlet boundary conditions, and cell-valued discretisation of scalar functions, are placed in directory

/ELEQ/E1D/DIRICH/CELL .

This directory contains only two files: file **p1ddcn.f**, which contains all Fortran programs related to this case, and file **p1ddcn.output**, which corresponds to the test problem that is described in Section 2.3 from Chapter 3. The programs in file **p1ddcn.f** are:

- **p1ddcn** - the main program,

- **CFDS** - subroutine for the computation of coefficients and the right-hand side of a finite-difference scheme,

- **seidel** - subroutine for the Gauss-Seidel iteration method,

- **mkgrid** - subroutine which generates a computational grid,

- **exact** - subroutine which computes an exact solution,

- **output** - subroutine which prepares file **p1ddnc.output**,

- **error** - subroutine which computes errors,

- **bcon** - subroutine which determines boundary conditions,

- **coeff** - subroutine which computes the values of the coefficient of the differential equation in a node,

- **fk** - real function which determines the analytical expression for coefficients of the differential equation,

- **right** - subroutine which computes values of the right-hand side of the original differential equation in cells,

- **fr** - real function which determines values of the right-hand side of the original differential equation.

2.1.2   Robin Boundary Conditions

The programs for 1-D equations and Robin boundary conditions, and nodal discretisation of scalar functions, are placed in directory

/ELEQ/E1D/ROBIN/NODAL.

This directory contains only two files: file **p1drnc.f**, which contains all Fortran programs related to this case, and file **p1drnc.output**, which corresponds to the test problems that are described in Section 2.1.10 from Chapter 3. The programs in file **p1drnc.f** are:

- **p1drnc** - the main program,

- **CFDS** - subroutine for the computation of coefficients and the right-hand side of finite-difference scheme with first- and second-order approximation of boundary conditions. This program has the parameter **IAPP**, where "**IAPP** is equal to 1" corresponds to the first-order approximations of boundary conditions and "**IAPP** is equal to 2" corresponds to the second-order approximations of boundary conditions,

- **seidel** - subroutine for the Gauss-Seidel iteration method,

- **mkgrid** - subroutine which generates a computational grid,

- **exact** - subroutine which computes an exact solution,

- **output** - subroutine which prepares file **p1drnc.output**,

- **error** - subroutine which computes errors,

- **bcon** - subroutine which determines boundary conditions,

- **coeff** - subroutine which computes values of the coefficient of the differential equation in cells,

- **fk** - real function which determines the analytical expression for coefficients of a differential equation,

- **right** - subroutine which computes values of the right-hand side of the original differential equation in nodes,

- **fr** - real function which determines values of the right-hand side of the original differential equation.

The programs for 1-D equations and Robin boundary conditions, and cell-valued discretisation of scalar functions, are placed in directory

/ELEQ/E1D/ROBIN/CELL.

This directory contains only two files: file **p1drcn.f**, which contains all Fortran programs related to this case, and file **p1drcn.output**, which corresponds to the test problems that are described in Section 2.3 from Chapter 3. The programs in file **p1drcn.f** are:

- **p1drcn** - the main program,

- **CFDS** - subroutine for the computation of coefficients and the right-hand side of a finite-difference scheme,

- **seidel** - subroutine for the Gauss-Seidel iteration method,

- **mkgrid** - subroutine which generates a computational grid,

- **exact** - subroutine which computes an exact solution,

- **output** - subroutine which prepares file **p1drcn.output**,

- **error** - subroutine which computes errors,

- **bcon** - subroutine which determines boundary conditions,

- **coeff** - subroutine which computes values of the coefficient of the differential equation in nodes,

- **fk** - real function which determines the analytical expression for coefficients of a differential equation,

- **right** - subroutine which computes values of the right-hand side of the original differential equation in cells,

- **fr** - real function which determines values of the right-hand side of the original differential equation.

2.2   Programs for 2-D Equations

2.2.1   Dirichlet Boundary Conditions

The programs for 2-D equations and Dirichlet boundary conditions, and nodal discretisation of scalar functions, are placed in directory

/ELEQ/E2D/DIRICH/NODAL .

This directory contains only two files: file **p2ddnc.f**, which contains all Fortran programs related to this case, and file **p2ddnc.output**, which corresponds to the test problem that is described in Section 3.3.1 from Chapter 3. The programs in file **p2ddnc.f** are:

- **p2ddnc** - the main program,

- **CFDS** - subroutine for computation of the coefficients and the right-hand side of a finite-difference scheme,

- **seidel** - subroutine for the Gauss-Seidel iteration method,

- **mkgrid** - subroutine which generates a computational grid,

- **exact** - subroutine which computes an exact solution,

- **output** - subroutine which prepares file **p2ddnc.output**,

- **error** - subroutine which computes errors,

- **bcon** - subroutine which determines boundary conditions,

- **fpsin** - real function which determines the Dirichlet boundary conditions on the north (top) boundary,

- **fpsis** - real function which determines the Dirichlet boundary conditions on the south (bottom) boundary,

- **fpsie** - real function which determines the Dirichlet boundary conditions on the east (right) boundary,

- **fpsiw** - real function which determines the Dirichlet boundary conditions on the west (left) boundary,

- **coeff** - subroutine which computes values of the coefficients of the differential equation in cells,

- **fkxx** - real function which determines the analytical expression for coefficient KXX of a differential equation,

- **fkxy** - real function which determines the analytical expression for coefficient KXY of a differential equation,

- **fkyy** - real function which determines the analytical expression for coefficient KYY of a differential equation,

- **right** - subroutine which computes values of the right-hand side of the original differential equation in nodes,

- **ff** - real function which determines values of the right-hand side of the original differential equation.

The programs for 2-D equations and Dirichlet boundary conditions, and cell-valued discretisation of scalar functions, are placed in directory

/ELEQ/E2D/DIRICH/CELL .

This directory contains only two files: file **p2ddcn.f**, which contains all Fortran programs related to this case, and file **p2ddcn.output**, which corresponds to the test problem that is described in Section 3.3.1 from Chapter 3. The programs in file **p2ddcn.f** are:

- **p2ddcn** - the main program,

- **CFDS** - subroutine for the computation of coefficients and the right-hand side of a finite-difference scheme,

- **seidel** - subroutine for the Gauss-Seidel iteration method,

- **mkgrid** - subroutine which generates a computational grid,

- **exact** - subroutine which computes an exact solution,

- **output** - subroutine which prepares file **p2ddcn.output**,

- **error** - subroutine which computes errors,

- **bcon** - subroutine which determines boundary conditions,

- **fpsin** - real function which determines the Dirichlet boundary conditions on the north (top) boundary,

- **fpsis** - real function which determines the Dirichlet boundary conditions on the south (bottom) boundary,

- **fpsie** - real function which determines the Dirichlet boundary conditions on the east (right) boundary,

- **fpsiw** - real function which determines the Dirichlet boundary conditions on the west (left) boundary,

- **coeff** - subroutine which computes the values of coefficients of the differential equation in nodes,

- **fkxx** - real function which determines the analytical expression for coefficient KXX of a differential equation,

- **fkxy** - real function which determines the analytical expression for coefficient KXY of a differential equation,

- **fkyy** - real function which determines the analytical expression for coefficient KYY of a differential equation,

- **right** - subroutine which computes the values of the right-hand side of the original differential equation in cells,

- **ff** - real function which determines the values of the right-hand side of the original differential equation.

2.2.2   Robin Boundary Conditions

The programs for 2-D equations and Robin boundary conditions, and nodal discretisation of scalar functions, are placed in directory

/ELEQ/E2D/ROBIN/NODAL .

This directory contains only two files: file **p2drnc.f**, which contains all Fortran programs related to this case, and file **p2drnc.output**, which corresponds to test problems that are described in Section 3.3.2 from Chapter 3. The programs in file **p2drnc.f** are:

- **p2drnc** - the main program,

- **CFDS** - subroutine for the computation of coefficients and the right-hand side of a finite-difference scheme,

- **seidel** - subroutine for the Gauss-Seidel iteration method,

- **mkgrid** - subroutine which generates a computational grid,

- **exact** - subroutine which computes an exact solution,

- **output** - subroutine which prepares file **p2drnc.output**,

- **error** - subroutine which computes errors,

- **bcon** - subroutine which determines boundary conditions,

- **falphan** - real function which determines α in Robin boundary conditions on the north (top) boundary,

- **falphas** - real function which determines α in Robin boundary conditions on the south (bottom) boundary,

- **falphae** - real function which determines α in Robin boundary conditions on the east (right) boundary,

- **falphaw** - real function which determines α in Robin boundary conditions on the west (left) boundary,

- **fgamman** - real function which determines γ in Robin boundary conditions on the north (top) boundary,

- **fgammas** - real function which determines γ in Robin boundary conditions on the south (bottom) boundary,

- **fgammae** - real function which determines γ in Robin boundary conditions on the east (right) boundary,

- **fgammaw** - real function which determines γ in Robin boundary conditions on the west (left) boundary,

- **coeff** - subroutine which computes values of the coefficients of the differential equation in cells,

- **fkxx** - real function which determines the analytical expression for coefficient KXX of a differential equation,

- **fkxy** - real function which determines the analytical expression for coefficient KXY of a differential equation,

- **fkyy** - real function which determines the analytical expression for coefficient KYY of a differential equation,

- **right** - subroutine which computes values of the right-hand side of the original differential equation in nodes,

- **ff** - real function which determines values of the right-hand side of the original differential equation.

The programs for 2-D equations and Robin boundary conditions, and cell-valued discretisation of scalar functions, are placed in directory

/ELEQ/E2D/ROBIN/CELL

This directory contains only two files: file **p2drcn.f**, which contains all Fortran programs related to this case, and file **p2drcn.output**, which corresponds to test problems that are described in Section 3.3.2 from Chapter 3. The programs in file **p2drcn.f** are:

- **p2drcn** - the main program,

- **CFDS** - subroutine for the computation of coefficients and the right-hand side of a finite-difference scheme,

- **seidel** - subroutine for the Gauss-Seidel iteration method,

- **mkgrid** - subroutine which generates a computational grid,

- **exact** - subroutine which computes an exact solution,

- **output** - subroutine which prepares file **p2drnc.output**,

- **error** - subroutine which computes errors,

- **bcon** - subroutine which determines boundary conditions,

- **falphan** - real function which determines α in Robin boundary conditions on the north (top) boundary,

- **falphas** - real function which determines α in Robin boundary conditions on the south (bottom) boundary,

- **falphae** - real function which determines α in Robin boundary conditions on the east (right) boundary,

- **falphaw** - real function which determines α in Robin boundary conditions on the west (left) boundary,

- **fgamman** - real function which determines γ in Robin boundary conditions on the north (top) boundary,

- **fgammas** - real function which determines γ in Robin boundary conditions on the south (bottom) boundary,

- **fgammae** - real function which determines γ in Robin boundary conditions on the east (right) boundary,

- **fgammaw** - real function which determines γ in Robin boundary conditions on the west (left) boundary,

- **coeff** - subroutine which computes values of the coefficients of the differential equation in nodes,

- **fkxx** - real function which determines the analytical expression for coefficient KXX of a differential equation,

- **fkxy** - real function which determines the analytical expression for coefficient KXY of a differential equation,

- **fkyy** - real function which determines the analytical expression for coefficient KYY of a differential equation,

- **right** - subroutine which computes values of the right-hand side of the original differential equation in cells,

- **ff** - real function which determines values of the right-hand side of the original differential equation.

3 Programs for Heat Equations

In this section, we describe in detail, the contents of directory **HEATEQ**, which contains programs for the heat equation.

3.1 Programs for 1-D Equations

3.1.1 Dirichlet Boundary Conditions

The programs for 1-D equations and Dirichlet boundary conditions, and nodal discretisation of scalar functions, are placed in directory

/HEATEQ/H1D/DIRICH/NODAL .

This directory contains only two files: file **h1ddnc.f**, which contains all Fortran programs related to this case, and file **h1ddnc.output**, which corresponds to the test problem that is described in Section 2.2.1 from Chapter 4. The programs in file **h1ddnc.f** are:

- **h1ddnc** - the main program,

- **ellcoef** - subroutine for the computation of coefficients of discrete approximations of an elliptic operator,

- **seidel** - subroutine for the Gauss-Seidel iteration method,

- **mkgrid** - subroutine which generates a computational grid,

- **exact** - subroutine which computes an exact solution,

- **output** - subroutine which prepares file **h1ddnc.output**,

- **error** - subroutine which computes errors,

- **bcon** - subroutine which determines boundary conditions,

- **coeff** - subroutine which computes values of the coefficient of the differential equation in cells,

- **fk** - real function which determines the analytical expression for coefficients of a differential equation,

- **right** - subroutine which computes values of the right-hand side of the original differential equation in nodes,

- **fr** - real function which determines values of the right-hand side of the original differential equation,

- **initial** - subroutine which computes the initial values in the nodes,

- **step** - subroutine that makes one time step by an explicit finite-difference scheme,

- **fmu0** - real function which computes the Dirichlet boundary condition on the right end of a segment as a function of time,

- **fmu1** - real function which computes the Dirichlet boundary condition on the the left end of a segment as a function of time,

- **geometry** - subroutine which computes the volumes of cells and nodes,

- **hcoeff** - subroutine which computes the coefficients of implicit finite-difference schemes for the heat equation,

- **sphi** - subroutine, right-hand side of a finite-difference scheme.

The programs for 1-D equations and Dirichlet boundary conditions, and cell-valued discretisation of scalar functions, are placed in directory

/HEATEQ/H1D/DIRICH/CELL .

This directory contains only two files: file **h1ddcn.f**, which contains all Fortran programs related to this case, and file **h1ddcn.output**, which corresponds to the test problem that is described in Section 2.3 from Chapter 4. The programs in file **h1ddcn.f** are:

- **h1ddcn** - the main program,

- **ellcoef** - subroutine for the computation of coefficients of discrete approximations of an elliptic operator,

- **siedel** - subroutine for the Gauss-Seidel iteration method,

- **mkgrid** - subroutine which generates a computational grid,

- **exact** - subroutine which computes an exact solution,

- **output** - subroutine which prepares file **h1ddcn.output**,

- **error** - subroutine which computes errors,

- **bcon** - subroutine which determines boundary conditions,

- **coeff** - subroutine which computes values of the coefficient of the differential equation in nodes,

- **fk** - real function which determines the analytical expression for co-efficients of a differential equation,

- **right** - subroutine which computes values of the right-hand side of the original differential equation in nodes,

- **fr** - real function which determines values of the right-hand side of the original differential equation,

- **initial** - subroutine which computes the initial values in the cells,

- **step** - subroutine which makes one time step by an explicit finite-difference scheme,

- **fmu0** - real function which computes the Dirichlet boundary condition on the right end of a segment as a function of time,

- **fmu1** - real function which computes the Dirichlet boundary condition on the left end of a segment as a function of time,

- **geometry** - subroutine which computes volumes of cells and nodes,

- **hcoeff** - subroutine which computes the coefficients of implicit finite-difference schemes for the heat equation,

- **sphi** - subroutine, right-hand side of finite-difference scheme.

3.1.2 Robin Boundary Conditions

The programs for 1-D equations and Robin boundary conditions, and nodal discretisation of scalar functions, are placed in directory

$$/\text{HEATEQ}/\text{H1D}/\text{ROBIN}/\text{NODAL}.$$

This directory contains only two files: file **h1drnc.f**, which contains all Fortran programs related to this case, and file **h1drnc.output**, which corresponds to test problems that are described in Section 2.2.1 from Chapter 4. The programs in file **h1drnc.f** are:

- **h1drnc** - the main program,

- **ellcoef** - subroutine for the computation of coefficients of discrete approximations of an elliptic operator,

- **siedel** - subroutine for the Gauss-Seidel iteration method,

- **mkgrid** - subroutine which generates a computational grid,

- **exact** - subroutine which computes an exact solution,

- **output** - subroutine which prepares file **h1drnc.output**,

- **error** - subroutine which computes errors,

- **bcon** - subroutine which determines boundary conditions,

- **coeff** - subroutine computes values of the coefficient of a differential equation in cells,

- **fk** - real function which determines the analytical expression for co-efficients of a differential equation,

- **right** - subroutine which computes values of the right-hand side of the original differential equation in nodes,

- **fr** - real function which determines values of the right-hand side of the original differential equation,

- **initial** - subroutine which computes the initial values in the nodes,

- **step(IAPP)** - subroutine which makes one time step by an explicit finite-difference scheme, **IAPP** order of approximation of boundary conditions,

- **geometry** - subroutine which computes volumes of cells and nodes,

- **falpa** - real function which computes coefficient α in Robin boundary conditions at the right end of a segment,

- **falpb** - real function which computes coefficient α in Robin boundary conditions at the left end of a segment,

- **fpsia** - real function which computes the right-hand side ψ in Robin boundary conditions at the right end of a segment,

- **fpsib** - real function which computes the right-hand side ψ in Robin boundary conditions at the left end of a segment,

- **hcoeff(IAPP)** - subroutine which computes coefficients of implicit finite-difference schemes for the heat equation, **IAPP** order of approximation of boundary conditions,

- **sphi(IAPP)** - subroutine, right-hand side of finite-difference scheme, **IAPP** order of approximation of boundary conditions.

The programs for 1-D equations and Robin boundary conditions, and cell-valued discretisation of scalar functions, are placed in directory

$$/\text{HEATEQ}/\text{H1D}/\text{ROBIN}/\text{CELL}.$$

This directory contains only two files: file **h1drcn.f**, which contains all Fortran programs related to this case, and file **h1drcn.output**, which corresponds to test problems that are described in Section 2.3 from Chapter 4. The programs in file **h1drcn.f** are:

- **h1drcn** - the main program,

- **ellcoef** - subroutine for the computation of coefficients of discrete approximations of an elliptic operator,

- **siedel** - subroutine for the Gauss-Seidel iteration method,

- **mkgrid** - subroutine which generates a computational grid,

- **exact** - subroutine which computes an exact solution,

- **output** - subroutine which prepares file **h1drcn.output**,

- **error** - subroutine which computes errors,

- **bcon** - subroutine which determines boundary conditions,

- **coeff** - subroutine which computes values of the coefficient of the differential equation in nodes,

- **fk** - real function which determines the analytical expression for coefficients of a differential equation,

- **right** - subroutine which computes values of the right-hand side of the original differential equation in cells,

- **fr** - real function which determines values of the right-hand side of the original differential equation,

- **initial** - subroutine which computes the initial values in the cells,

- **step** - subroutine which makes one time step by an explicit finite-difference scheme,

- **geometry** - subroutine which computes volumes of cells and nodes,

- **falpa** - real function which computes coefficient α in Robin boundary conditions at the right end of a segment,

- **falpb** - real function which computes coefficient α in Robin boundary conditions at the left end of a segment,

- **fpsia** - real function which computes the right-hand side ψ in Robin boundary conditions at the right end of a segment,

- **fpsib** - real function which computes the right-hand side ψ in Robin boundary conditions at the left end of a segment,

- **hcoeff** - subroutine which computes coefficients of implicit finite-difference schemes for the heat equation,

- **sphi** - subroutine, the right-hand side of a finite-difference scheme.

3.2   Programs for 2-D Equations

3.2.1   Dirichlet Boundary Conditions

The programs for 2-D equations and Dirichlet boundary conditions, and nodal discretisation of scalar functions, are placed in directory

/HEATEQ/H2D/DIRICH/NODAL,

which contains only two files: file **h2ddnc.f**, which contains all Fortran programs related to this case, and file **h2ddnc.output**, which corresponds to the test problem that is described in Section 3.3 from Chapter 4. The programs in file **h2ddnc.f** are:

- **h2ddnc** - the main program,

- **ellcoef** - subroutine for the computation of coefficients of the discrete analog of an elliptic operator,

- **siedel** - subroutine for the Gauss-Seidel iteration method,

- **mkgrid** - subroutine which generates a computational grid,

- **exact** - subroutine which computes an exact solution,

- **output** - subroutine which prepares file **h2ddnc.output**,

- **error** - subroutine which computes errors,

- **bcon** - subroutine which determines boundary conditions,

- **fpsin** - real function which determines Dirichlet boundary conditions on the north (top) boundary,

- **fpsis** - real function which determines Dirichlet boundary conditions on the south (bottom) boundary,

- **fpsie** - real function which determines Dirichlet boundary conditions on the east (right) boundary,

- **fpsiw** - real function which determines Dirichlet boundary conditions on the west (left) boundary,

- **coeff** - subroutine which computes values of coefficients of differential equations in cells,

- **fkxx** - real function which determines the analytical expression for coefficient KXX of a differential equation,

- **fkxy** - real function which determines the analytical expression for coefficient KXY of a differential equation,

- **fkyy** - real function which determines the analytical expression for coefficient KYY of a differential equation,

- **right** - subroutine which computes values of the right-hand side of the original differential equation in nodes,

- **ff** - real function which determines values of the right-hand side of the original differential equation,

- **geometry** - subroutine which computes volumes of cells and nodes,

- **initial** - subroutine which computes the initial values for temperature,

- **step** - subroutine, makes one time step by an explicit finite-difference scheme,

- **hcoeff** - subroutine which computes coefficients and the right-hand side of an implicit finite-difference scheme,

- **sphi** - subroutine which computes values of the right-hand side.

The programs for 2-D equations and Dirichlet boundary conditions, and cell-valued discretisation of scalar functions, are placed in directory

/HEATEQ/H2D/DIRICH/CELL .

This directory contains only two files: file **h2ddcn.f**, which contains all Fortran programs related to this case, and file **h2ddcn.output**, which corresponds to the test problem that is described in Section 3.3 from Chapter 4. The programs in file **h2ddcn.f** are:

- **h2ddcn** - the main program,

- **ellcoef** - subroutine for the computation of coefficients of the discrete analog of an elliptic operator,

- **siedel** - subroutine for the Gauss-Seidel iteration method,

- **mkgrid** - subroutine which generates a computational grid,

- **exact** - subroutine which computes an exact solution,

- **output** - subroutine which prepares file **h2ddcn.output**,

- **error** - subroutine which computes errors,

- **bcon** - subroutine which determines boundary conditions,

- **fpsin** - real function which determines Dirichlet boundary conditions on the north (top) boundary,

- **fpsis** - real function which determines Dirichlet boundary conditions on the south (bottom) boundary,

- **fpsie** - real function which determines Dirichlet boundary conditions on the east (right) boundary,

- **fpsiw** - real function which determines Dirichlet boundary conditions on the west (left) boundary,

- **coeff** - subroutine which computes values of coefficients of differential equations in nodes,

- **fkxx** - real function which determines the analytical expression for coefficient KXX of a differential equation,

- **fkxy** - real function which determines the analytical expression for coefficient KXY of a differential equation,

- **fkyy** - real function which determines the analytical expression for coefficient KYY of a differential equation,

- **right** - subroutine which computes values of the right-hand side of the original differential equation in cells,

- **ff** - real function which determines values of the right-hand side of the original differential equation,

- **geometry** - subroutine which computes volumes of cells and nodes,

- **initial** - subroutine which computes the initial values for temperature,

- **hcoeff** - subroutine which computes coefficients and the right-hand side of an implicit finite-difference scheme,

- **sphi** - subroutine which computes values of the right-hand side,

- **step** - subroutine which makes one step in time by an explicit finite-difference scheme.

3.2.2   Robin Boundary Conditions

The programs for 2-D equations and Robin boundary conditions, and nodal discretisation of scalar functions, are placed in directory

$$/HEATEQ/H2D/ROBIN/NODAL.$$

This directory contains only two files: file **h2drnc.f**, which contains all Fortran programs related to this case, and file **h2drnc.output**, which corresponds to test problems that are described in Section 3.3 from Chapter 4. The programs in file **h2drnc.f** are:

- **h2drnc** - the main program,

- **ellcoef** - subroutine for the computation of coefficients of the discrete analog of an elliptic operator,

- **siedel** - subroutine for the Gauss-Seidel iteration method,

- **mkgrid** - subroutine which generates a computational grid,

- **exact** - subroutine which computes an exact solution,

- **output** - subroutine which prepares file **h2drnc.output**,

- **error** - subroutine which computes errors,

- **bcon** - subroutine which determines boundary conditions,

- **falphan** - real function which determines function α in Robin conditions on the north (top) boundary,

- **falphas** - real function which determines function α in Robin conditions on the south (bottom) boundary,

- **falphae** - real function which determines function α in Robin conditions on the east (right) boundary,

- **falphaw** - real function which determines function α in Robin conditions on the west (left) boundary,

- **fgamman** - real function which determines function γ in Robin conditions on the north (top) boundary,

- **fgammas** - real function which determines function γ in Robin conditions on the south (bottom) boundary,

- **fgammae** - real function which determines function γ in Robin conditions on the east (right) boundary,

- **fgammaw** - real function which determines function γ in Robin conditions on the west (left) boundary,

- **coeff** - subroutine which computes values of coefficients of the differential equation in cells,

- **fkxx** - real function which determines the analytical expression for coefficient KXX of a differential equation,

- **fkxy** - real function which determines the analytical expression for coefficient KXY of a differential equation,

- **fkyy** - real function which determines the analytical expression for coefficient KYY of a differential equation,

- **right** - subroutine which computes values of the right-hand side of the original differential equation in nodes,

- **ff** - real function which determines values of the right-hand side of the original differential equation,

- **geometry** - subroutine which computes volumes of cells and nodes,

- **initial** - subroutine which computes the initial values for temperature,

- **hcoeff** - subroutine which computes coefficients and the right-hand side of an implicit finite-difference scheme,

- **sphi** - subroutine which computes values of the right-hand side.

The programs for 2-D equations and Robin boundary conditions, and cell-valued discretisation of scalar functions, are placed in directory

/HEATEQ/H2D/ROBIN/CELL.

This directory contains only two files: file **h2drcn.f**, which contains all Fortran programs related to this case, and file **h2drcn.output**, which corresponds to test problems that are described in Section 3.3 from Chapter 4. The programs in file **h2drcn.f** are:

- **h2drcn** - the main program,

- **ellcoef** - subroutine for the computation of coefficients of the discrete analog of an elliptic operator,

- **sphi** - subroutine which computes values of the right-hand side,

- **siedel** - subroutine for the Gauss-Seidel iteration method,

- **mkgrid** - subroutine which generates a computational grid,

- **exact** - subroutine which computes an exact solution,

- **output** - subroutine which prepares file **h2drcc.output**,

- **error** - subroutine which computes errors,

- **bcon** - subroutine which determines boundary conditions,

- **falphan** - real function which determines function α in Robin conditions on the north (top) boundary,

- **falphas** - real function which determines function α in Robin conditions on the south (bottom) boundary,

- **falphae** - real function which determines function α in Robin conditions on the east (right) boundary,

- **falphaw** - real function which determines function α in Robin conditions on the west (left) boundary,

- **fgamman** - real function which determines function γ in Robin conditions on the north (top) boundary,

- **fgammas** - real function which determines function γ in Robin conditions on the south (bottom) boundary,

- **fgammae** - real function which determines function γ in Robin conditions on the east (right) boundary,

- **fgammaw** - real function which determines function γ in Robin conditions on the west (left) boundary,

- **coeff** - subroutine which computes values of coefficients of the differential equation in nodes,

- **fkxx** - real function which determines the analytical expression for coefficient KXX of a differential equation,

- **fkxy** - real function which determines the analytical expression for coefficient KXY of a differential equation,

- **fkyy** - real function which determines the analytical expression for coefficient KYY of a differential equation,

- **right** - subroutine which computes values of the right-hand side of the original differential equation in cells,

- **ff** - real function which determines values of the right-hand side of the original differential equations,

- **geometry** - subroutine which computes volumes of cells and nodes,

- **initial** -subroutine which computes the initial values for temperature,

- **hcoeff** - subroutine which computes coefficients and the right-hand side of an implicit finite-difference scheme.

4 Programs for Gas Dynamics Equations

In this section, we describe in detail, the contents of directory **GASDEQ**, which contains programs for the solution of test problems from Sections 3.9 and 4.7 from Chapter 5.

4.1 Programs for 1-D Equations

4.1.1 Explicit Finite-Difference Scheme

The programs for 1-D equations for the solution of a 1-D test problem from Section 3.9 from Chapter 5, which correspond to an explicit finite-difference scheme, are placed in directory

/GASDEQ/G1D/EXP.

This directory contains only two files: file **g1de.f**, which contains all Fortran programs related to this case, and file **g1de.output**, which corresponds to the test problem that is described in Section 3.9. The programs in file **g1de.f** are:

- **gid1de** - the main program,

- **ini** - subroutine which computes the initial data,

- **fp** - real function which computes pressure from the equation of state,

- **fc** - real function which computes the adiabatic speed of sound,

- **frho** - real function which determines the initial profile for density,

- **fe** - real function which determines the initial profile for internal energy,

- **fw** - real function which determines the initial profile for velocity,

- **stepe** - subroutine which performs computations for one step in time for an explicit finite-difference scheme,

- **output** - output of flow parameters.

4.1.2 Implicit Finite-Difference Scheme

The programs for 1-D equations for the solution of a 1-D test problem from Section 3.9, which correspond to an implicit finite-difference scheme, are placed in directory

/GASDEQ/G1D/IMP.

This directory contains only two files: file **g1di.f**, which contains all Fortran programs related to this case, and file **g1di.output**, which corresponds to the test problem that is described in Section 3.9. The programs in file **g1di.f** are:

- **gid1di** - the main program,

- **ini** - subroutine which computes the initial data,

- **fp** - real function which computes pressure from the equation of state,

- **ff** - real function which determines the equation of state in the form $F(p, \rho, \varepsilon) = 0$,

- **fdfdrho** - real function which determines the derivative $\partial F/\partial \rho$,

- **fdfde** - real function which determines the derivative $\partial F/\partial \varepsilon$,

- **fc** - real function which computes the adiabatic speed of sound,

- **frho** - real function which determines the initial profile for density,

- **fe** - real function which determines the initial profile for internal energy,

- **fw** - real function which determines the initial profile for velocity,

- **stepi** - subroutine which performs the computations for one step in time for an explicit finite-difference scheme,

- **output** - output of flow parameters.

4.2 Programs for 2-D Equations

4.2.1 Explicit Finite-Difference Scheme.

The programs for 2-D equations for the solution of a 2-D test problem from Section 4.7, which correspond to an explicit finite-difference scheme, are placed in directory

/GASDEQ/G2D/EXP .

This directory contains only two files: file **g2de.f**, which contains all Fortran programs related to this case, and file **g2de.output**, which corresponds to the test problem that is described in Section 4.7 from Chapter 5. The programs in file **g2de.f** are

- **gid2de** - the main program,

- **ini** - subroutine which computes the initial data,

- **fp** - real function which computes pressure from the equation of state,

- **fc** - real function which computes the adiabatic speed of sound,

- **frho** - real function which determines the initial profile for density,

- **fe** - real function which determines the initial profile for internal energy,

- **fw** - real function which determines the initial profile for velocity,

- **fdvdx** - real function which computes the derivative of volume of the cell with respect to the x coordinate: $\partial V_{ij}/\partial x_{kl}$,

- **fdvdy** - real function which computes the derivative of volume of the cell with respect to the y coordinate: $\partial V_{ij}/\partial y_{kl}$,

- **stepe** - subroutine which performs computations for one step in time for an explicit finite-difference scheme,

- **output** - output of flow parameters.

4.2.2   Implicit Finite-Difference Scheme

The programs for 2-D equations for the solution of a 2-D test problem from Section 4.7, which correspond to an implicit finite-difference scheme, are placed in directory

/GASDEQ/G2D/IMP .

This directory contains only two files: file **g2di.f**, which contains all Fortran programs related to this case, and file **g2di.output**, which corresponds to the test problem that is described in Section 4.7. The programs in file **g2di.f** are:

- **gid2di** - the main program,

- **ini** - subroutine which computes the initial data,

- **fp** - real function which computes pressure from the equation of state,

- **ff** - real function which determines the equation of state in the form $F(p, \rho, \varepsilon) = 0$,

- **fdfdrho** - real function which determines the derivative $\partial F/\partial \rho$,

- **fdfde** - real function which determines the derivative $\partial F/\partial \varepsilon$,

- **fc** - real function which computes the adiabatic speed of sound,

- **frho** - real function which determines the initial profile for density,

- **fe** - real function which determines the initial profile for internal energy,

- **fw** - real function which determines the initial profile for velocity,

- **fdvdx** - real function which computes the derivative of volume of a cell with respect to the x coordinate: $\partial V_{ij}/\partial x_{kl}$,

- **fdvdy** - real function which computes the derivative of volume of a cell with respect to the y coordinate: $\partial V_{ij}/\partial y_{kl}$,

- **stepi** - subroutine which performs the computations for one step in time for an explicit finite-difference scheme,

- **output** - output of flow parameters.

Bibliography

[1] Ames, W.F. [1977]. *Numerical Methods for Partial Differential Equations*, Academic Press, New York.

[2] Apanovich, Yu.V., and Lymkis, E.D. [1988]. *Difference Schemes for the Navier-Stokes Equations on a Net Consisting of Dirichlet Cells*, USSR Comput. Math. and Math. Phys., [1] **28**, No. 2, 1988, pp. 57-63.

[3] Ardelyan, N.V. [1983]. *On the Use of Iterative Methods when Realizing Implicit Difference Schemes of Two-Dimensional Magnetohydrodynamics*, USSR Comput. Math. and Math. Phys., **23**, No. 6, 1983, pp. 84-90.

[4] Ardelyan, N.V. [1983]. *The Convergence of Difference Schemes for Two-Dimensional Equations of Acoustics and Maxwell's Equations*, USSR Comput. Math. and Math. Phys., **23**, No. 5, 1983, pp. 93-99.

[5] Ardelyan, N.V., and Chernigovskii, S.V. [1984]. *Convergence of Difference Schemes for Two-Dimensional Gas-Dynamic Equations in Acoustics Approximations with Gravitation Taken into Account*, Diff. Eqns. [2] **20**, No. 7, 1984, pp. 807-813.

[6] Ardelyan, N.V. [1987]. *Method of Investigating the Convergence of Nonlinear Finite-Difference Schemes*, Diff. Eqns. **23**, No. 7, 1987, pp. 737-745.

[7] Arfken, G. [1985]. *Mathematical Methods for Physicists*, Academic Press, Inc.

[8] Bakirova, M., Burdiashvili, M., Vo'tenko, D., Ivanov, A., Karpov, V., Kirov, A., Korshiaya, T., Krukovskii, A., Lubimov, B., Tishkin, V., Favorskii, A., and Shashkov, M. [1981]. *On Simulation of a Magnetic*

---

[1] U.S.S.R. Computational Mathematics and Mathematical Physics. - Pergamon Press, printed in Great Britain. Cover-to-cover English translation of Zhurnal vychislitel'noi matematiki i matematicheskoi fiziki.

[2] Differential Equations, (Differentsial'nye Uravneneia) Translated from Russian, by Consultants Bureau, New York, Plenum Publishing Corp.

Field in a Spiral Band Reel. - Preprint Keldysh Inst. of Appl. Math. the USSR Ac. of Sc., **110**, 1981, (In Russian).

[9] Burdiashvili, M., Vo'tenko, D., Ivanov, A., Kirov, A., Tishkin, V., Favorskii, A., and Shashkov M. [1984]. *A Magnetic Field of a Toroidal Spiral with a Screen.* - Preprint Keldysh Inst. of Appl. Math. the USSR Ac. of Sc., **63**, 1984, (in Russian).

[10] Bathelor, G.K. [1967]. *An Introduction to Fluid Dynamics*, Cambridge U.P.

[11] Bers, L., John, F., and Schecter, M. [1964]. *Partial Differential Equations*, Interscience, New York.

[12] Birkhoff, G., and Lynch, R.E. [1984]. *Numerical Solution of Elliptic Problems*, SIAM, Philadelphia.

[13] Castillo, J.E., ed. [1991]. *Mathematical Aspects of Numerical Grid Generation*, SIAM, Philidelphia.

[14] Castillo, J., Hyman, J.M., Shashkov, M.J., and Steinberg, S. [1995]. *The Sensitivity and Accuracy of Fourth Order Finite-Difference Schemes on Nonuniform Grids in One Dimension.* To appear in Journal of Computers and Applications.

[15] Courant, R., Isacson, E., and Rees, H. [1952]. *On the Solution of Nonlinear Hyperbolic Differential Equations by Finite Differences*, Comm. Pure Appl. Math., 1952, vol. 5, No. 3, pp. 243-254.

[16] Celia, M.A., and Gray, W.G. [1992]. *Numerical Methods for Differential Equations*, Prentice Hall, Engelwood Cliffs, NJ.

[17] Chu, W.H. [1971]. *Development of a General Finite Difference Approximation for a General Domain. Part 1: Machine Transformation*, J. Comp. Physics, **8**, (1971), 392-408.

[18] Crumpton, P.I., Shaw, G.I., and Ware, A.F. [1995]. *Discretisation and Multigrid Solution of Elliptic Equations with Mixed Derivative Terms and Strongly Discontinuous Coefficients*, J. Comp. Physics, **116**, (1995), 343-358.

[19] Davis, J.L. [1986]. *Finite Difference Methods in Dynamic of Continuous Media*, New York: Macmillan.

[20] Denisov, A.A., Koldoba, A.V., and Poveshchenko, Yu.A. [1989]. *The Convergence to Generalized Solutions of Difference Schemes of the Reference-Operator Method for Poisson's Equation*, USSR Comput. Math. and Math. Phys., **29**, No. 2, 1989, pp. 32-38.

[21] Koldoba, A.V., Poveshchenko, Yu.A., and Popov, Yu.P. [1983]. *The Approximation of Differential Operators on Nonorthogonal Meshes*, Diff. Equations, **19**, No. 7, 1983, pp. 919-927.

[22] Demin, A.V., Korobitsyn, V.A., Mazurenko, A.I., and Khe, A.I. [1988]. *Calculation of the Flows of a Viscous Incompressible Liquid with a Free Surface on Two Dimensional Lagrangian Nets*, U.S.S.R. Computational Mathematics and Mathematical Physics, 1988, **28**, No. 6, pp. 81-87.

[23] DeVries, P.L. [1994]. *A First Course in Computational Physics*, John Wiley & Sons, Inc.

[24] Dmitrieva, M.V., Ivanov, A.A., Tishkin, V.F., and Favorskii, A.P. [1985]. *Construction and Investigation of Support-Operators Finite-Difference Schemes for Maxwell Equations in Cylindrical Geometry*, Preprint Keldysh Inst. of Appl. Math. the USSR Ac. of Sc., **27**, 1985, (in Russian).

[25] Dukowicz, J.K., and Meltz, B.J.A. [1992]. *Vorticity Errors in Multidimensional Lagrangian Codes*, Journal of Computational Physics, **99**, (1992), pp. 115-134.

[26] Epstein, B. [1962]. *Partial Differential Equations, An Introduction*, Robert E. Krieger Pub. Co., Malarbar, Florida.

[27] Favorskii, A., Samarskii, A., Tishkin, V., and Shashkov, M. [1981]. *On Constructing Full Conservative Difference Schemes for Gas Dynamic Equations in Eulerian Form by the Method of Basic Operators.* - Preprint Keldysh Inst. of Appl. Math. the USSR Ac. of Sc., **63**, 1981, (in Russian).

[28] Favorskii, A.P. [1980]. *Variational-Discrete Models of Hydrodynamics Equations*, Diff. Equations, **16**, No. 7, 1980, pp. 834-845.

[29] Favorskii, A.P., Korshiya, T., Tishkin, V.F., and Shashkov, M. [1980]. *Difference Schemes for Equations of Electro-Magnetic Field Diffusion with Anisotropic Conductivity Coefficients.* - Preprint Keldysh Inst. of Appl. Math. the USSR Ac. of Sc., 4, 1980, (in Russian).

[30] Taran, M., Tishkin, V., Favorskii, A., Feoktistov, L., and Shashkov, M. [1980]. *On Simulation of the Collapse of a Quasi-Spherical Target in a Hard Cone.* - Preprint Keldysh Inst. of Appl. Math. the USSR Ac. of Sc., **127**, 1980, (in Russian).

[31] Favorskii, A., Tishkin, V., and Shashkov, M. [1979]. *Variational-Difference Schemes for the Heat Conduction Equation on Non-Regular Grids.* - Sov. Phys. Dokl. **24**, (6), p. 446, 1979, American Institute of Physics.

[32] Fletcher, C.A.J. [1988]. *Computational Techniques for Fluid Dynamics*, Springer-Verlag, Berlin.

[33] Fong, P. [1963]. *Foundation of Thermodynamics*, Oxford, 1963.

[34] Forsythe, G.E., and Wasow, W.R. [1960]. *Finite Difference Methods for Partial Differential Equations*, Wiley, New York.

[35] Fridrics, K.O. [1954]. *Symmetric Hyperbolic Differential Equations*, Comm. Pure Appl. Math., 1954, vol. 7, No. 2, pp. 345-392.

[36] Ganhga, V., Solov'ov, A., and Shashkov, M. [1991]. *Algorithms for Operation with Difference Operators and Grid Function in Symbolic Form,* IY International Conference on Computer Algebra in Physical Research, Dubna, USSR. Editors Shirikov, D.V., Rostovsev, V.A., and Gerdt, V.P., World Scientific Publishing Co. Pte. Ltd. 1991, pp. 175-181.

[37] Ganhga, V., Solov'ov, A., and Shashkov, M. [1993]. *Algorithms for Operation with Difference Operators and Grid Function in Symbolic Form*, CAAM-90, International Seminar on Computer Algebra and Application to Mechanics, Novosibirsk-Irkutsk, USSR. Proceedings of CAAM-90 in book "Computer Algebra and Its Application to Mechanics" (Edited by V.Ganzha, V.Rudenko and E.Voroztsov), NOVA Science Publisher, 1993, pp. 63-69.

[38] Ganhga, V., Solov'ov, A., and Shashkov, M. [1992]. *A New Algorithm for the Symbolic Computation of the Analysis of Local Approximation of Difference Operators*, Advances in Computer Methods for Partial Differential Equations VII, New Brunswick, New Jersey, USA. Proceedings in book "Advances in Computer Methods for Partial Differential Equations - VII " (Edited by R.Vichnevetsky, D.Knight and G. Richter), IMACS, 1992, pp. 280-286.

[39] Ganhga, V., and Shashkov, M. [1990]. *Local Approximation Study of Difference Operators by Means of REDUCE System*, ISSAC'90 - International Symposium on Symbolic and Algebraic Computation, Tokyo, Japan. Proceedings of ISSAC'90 - International Symposium on Symbolic and Algebraic Computation, S.Watanabe, and M.Nagata (eds.), Tokyo, Japan, 1990, pp. 185-192.

[40] Gasilov, V.A., Tishkin, V.F., Favorskii, A.P., and Shashkov, M.Yu. [1981]. *The Use of the Parallel-Chord Method to Solve Hydrodynamics Difference Equations*, U.S.S.R. Computational Mathematics and Mathematical Physics, **21**, No. 3, 1981, pp. 178-192.

[41] Gasilov, V., Goloviznin, V., Taran, M., Tishkin, V., Favorskii, A., and Shashkov, M. [1979a]. *Numerical Simulation of the Rayleigh-Taylor*

Instability for Incompressible Flows.- Preprint Keldysh Inst. of Appl. Math. the USSR Ac. of Sc., **70**, 1979, (in Russian).

[42] Gasilov, V., Goloviznin, V., Kurtmullaev, R., Semenov, V., Sosnin, N., Tishkin, V., Favorskii, A., and Shashkov, M. [1979b]. *Numerical Simulation of the Compression of Toroidal Plasma by Quasi-Spherical Liner.* - Preprint Keldysh Inst. of Appl. Math. the USSR Ac. of Sc., **71**, 1979 (In Russian).

[43] Gasilov, V., Goloviznin, V., Kurtmullaev, R., Semenov, V., Favorskii, A., and Shashkov, M. [1979c]. *The Numerical Simulation of the Quasi Spherical Metal Liner Dynamics*, In Proceedings of Second International Conference on Megagauss Magnetic Field Generation and Related Topics, Washington, D.C., USA, July 1979.

[44] Gasilov, V., Korshiya, T., Lubimov, V., Tishkin, V., Favorskii, A., and Shashkov, M. [1983]. *Application of the Parallel-Chord Method to Solve Implicit Difference Equations of Magneto-Hydrodynamics.* - U.S.S.R. Comput. Maths. Math. Phys., **23**, No. 4, 1983, pp. 104-111.

[45] Gilbarg, D., and Trudinger, N.S. [1983]. *Elliptic Partial Differential Equations of Second Order*, Berlin, New York, Springer-Verlag.

[46] Godunov, S.K., and Ryabenkii, V.S. [1987]. *Difference Schemes. An Introduction to the Underlying Theory*, Studies in Mathematics and Its Applications. Volume 19. Editors: Lions, J.L., Papanicolaou, C., Fujita, H., Keller, H.B., Elsevier Science Publishers B.V.

[47] Goloviznin, V.M., Korshunov, V.K., and Samarskii, A.A. [1982]. *Two-Dimensional Difference Schemes of Magneto-Hydrodynamics on Triangle Lagrange Meshes*, USSR Comput. Math. and Math. Phys., **22**, No. 4 , 1982, pp. 160-178 .

[48] Goloviznin, V.M., Kanyukova, V.D., and Samarskaya, E.A. [1983]. *Highly Implicit Finite-Difference Schemes for Gas-Dynamics*, Diff. Equations, **19**, No. 7, (1983), pp. 878-887.

[49] Goloviznin, V.M., Korshunov, V.K., Sabitova, A., and Samarskaya, E.A. [1984]. *Stability of Variational-Difference Schemes in Gas-Dynamics*, Diff. Equations, **20**, No. 7, (1984), pp. 852-858.

[50] Golub, G.H., and Van Loan, C.F. [1989]. *Matrix Computation*, The Johns Hopkins University Press. Baltimore and London.

[51] Golub, G.H., and Ortega, J.M. [1991]. *Scientific Computing and Differential Equations*, Academic Press, Boston.

[52] Gurtin, M.E. [1981]. *An Introduction to Continuum Mechanics*, Academic Press, New York.

[53] Hackbusch, W. [1985]. *Multi Grid Methods and Applications*, Springer-Verlag, New York.

[54] Hageman, L.A., and Young, D.M. [1981]. *Applied Iterative Methods*, Academic Press, New York.

[55] Hall, C.A., and Porsching, T.A. [1990]. *Numerical Analysis of Partial Differential Equations*, Prentice Hall, Englewood Cliffs, NJ.

[56] Hairer, E., Norsett, S.P., and Wanner, G. [1993]. *Solving Ordinary Differential Equations I*, Springer Series in Computational Mathematics. Springer-Verlag.

[57] Harten, A. [1978]. *The Artificial Compression Method for Computation of Shocks and Contact Discontinuities: III. Self-Adjusting Hybrid Schemes*, Math. of Comput. **32**, No. 142, (1978), pp. 363-389.

[58] Hirt, C.W. [1968]. *Heuristic Stability Theory for Finite Difference Equations*, J. Comp. Phys., **2**, (1968), pp. 339-355.

[59] Hirt, C.W., Amsden, A.A., and Cook, J.L. [1974]. *An Arbitrary Lagrangian Computing Method for All Flow Speeds*, Journ. Computational Phys., **14**, (1974), pp. 227-253.

[60] Harten, A. [1978]. *The Artificial Compression Method for computation of Shocks and Contact Discontinuities: II. Self-Adjusting Hybrid Schemes*, Math. of Comp., 1978, **32**, No. 142, pp. 363-389.

[61] Heinrich, B. [1987]. *Finite Difference Methods on Irregular Networks, A Generalized Approach to Second Order Elliptic Problems*, Birkäuser, Basel.

[62] Hoffman, K.A. [1989]. *Computational Fluid Dynamics for Engineers*, Application of Engineering Education System, Texas. USA.

[63] Horn, R.A., and Johnson, C.R. [1985]. *Matrix Analysis*, Cambridge Univ. Press, Cambridge.

[64] Hyman, J.M. [1979]. *A Method of Lines Approach to Numerical Solution of the Conservation Laws*, Advances in Computer Methods for Partial Differential Equations - III, Editors, Vichnevetsky R., and Stepleman R.S., IMACS Publication, 1979, pp. 313-321.

[65] Kantorovich, L.V., and Akilov, G.P. [1978]. *Functional Analysis*, Academic Press, New York.

[66] Korobitsin, V.A. [1989]. *Axisymmetric Difference Operators in an Orthogonal Coordinate System*, USSR Comput. Math. and Math. Phys., **29**, No. 6, 1989, pp. 13-21.

[67] Korobitsin, V.A. [1990]. *Basic Operators Method for Construction of Difference Schemes in Curvilinear Orthogonal Coordinate System*, Mathematical Modeling, 1990, **2**, No. 6, pp. 110-117.

[68] Kershaw, D.S. [1980]. *Differencing of the Diffusion Equation in Lagrangian Hydrodynamic Codes*, Journal of Computational Physics, **39**, (1981), pp. 375-395.

[69] Kirpichenko, P., Sokolov, V., Tarasov, J., Tishkin, V., Turina, N., Favorskii, A., and Shashkov M. [1984]. *A Numerical Simulation of Over- Compressed Detonation Wave in a Conic Canal*, - Preprint Keldysh Inst. of Appl. Math. the USSR Ac. of Sc., **82**, 1984, (in Russian).

[70] Knupp, P.M., and Steinberg, S. [1993]. *The Fundamentals of Grid Generation*, CRC Press, Boca Raton, 1993.

[71] Kreiss, H.-O. [1978]. *Numerical Methods for Solving Time-Dependent Problem for Partial Differential Equations*, University of Montreal Press, Montreal.

[72] Kreiss, H.-O., Manteuffel, T., Swartz, B., Wendroff, B., and White, A. [1986]. *Supra-Convergent Schemes on Irregular Grids*, Math. Comput. **47**, (1986), pp. 537-554.

[73] Ladyzhenskaia, O.A. [1968a]. *Linear and Quasi-Linear Elliptic Equations*, Academic Press, New York.

[74] Ladyzhenskaia, O.A., Solonnikov, V.A., and Uralceva, N.N. [1968b]. *Linear and Quasi-Linear Equations of Parabolic Type*, American Mathematical Society, Providence.

[75] Lambert, J.D. [1991]. *Numerical Methods for Ordinary Differential Systems*, John Wiley and Sons.

[76] Landshoff, R. [1955]. *A Numerical Method for Treating Fluid Flow in the Presence of Shocks*, LASL Rept. No. LA-1930, Los Alamos, New Mexixo.

[77] Latter, R. [1955]. *Similarity Solution for a Spherical Shock Wave*, J. Appl. Phys., **26**, pp. 954-960.

[78] LeVeque, R.J. [1987]. *High Resolution Finite-Volume Methods on Arbitrary Grids via Wave Propagation*, Hampton, Va.: NASA Langley Research Center.

[79] LeVeque, R.J. [1990]. *Numerical Methods for Conservation Laws*, Basel, Burkhauser Verlag.

[80] Lees, M. [1960]. *A Priori Estimates for the Solution of Difference Approximations to Parabolic Partial Differential Equations*, Duke Math. J. , **27**, 3, (1960), pp. 297-311.

[81] Maenchen, G., and Sack, S. [1964]. *The Tensor Code*, in Methods in Computational Physics, Editors B.Adler, S.Fernbach, and M.Rotenberg, vol. 3, Fundamental Methods in Hydrodynamics, Academic Press, 1964, pp. 181-210.

[82] Maikov, A. R., Sveshnikov, A. G., and Yakunin, S. A. [1985]. *Mathematical Modelling of Microwave Plasma Generator*, USSR Comput. Math. and Math. Phys., **25**, No. 3, 1985, pp. 149-157.

[83] Margolin, L.G., and Nichols, B.D. [1983]. *Momentum Control Volumes for Finite Difference codes*, LA-UR-83-524, Los Alamos National Laboratory, NM, 1983.

[84] Margolin, L.G., and Adams, T.F [1985]. *Spatial Differencing for Finite Difference Codes*, LA-UR-85-10249, Los Alamos National Laboratory, Los Alamos, NM, 1985.

[85] Margolin, L.G., Ruppel, H.M and Demuth, R.B. [1985]. *Gradient Scaling for Nonuniform Meshes*, Proceedings of Fourth International Conference on "Numerical Methods in Laminar and Turbulent Flow", The University of Wales, Swansea, UK, July 9-12, p. 1477, 1985.

[86] Margolin, L.G., and Tarwater, A.E. [1987]. *A Diffusion Operator for Lagrangian Codes*, UCRL-95652, Lawrence Livermore National Laboratory, Livermore, CA, 1987, and Proceeding of the 5th International Conference on Numerical Methods in Thermal Transfer, Motreal, Canada, Eds. R.W.Lewis, K.Morgan, and W.G.Habashi, p. 1252, June 1987.

[87] Margolin, L.G., and Pyun, J.J. [1987]. *A Method for Treating Hourglassing Patterns*, LA-UR-87-439, Los Alamos National Laboratory, Los Alamos, NM, 1987, and Proceeding of Fourth International Conference on "Numerical Methods in Laminar and Turbulent Flow", Montreal, Quebec, Canada, July 6-10, 1987.

[88] Margolin, L.G. [1988]. *A Centered Artificial Viscosity for Cells with Large Aspect Ratio*, UCRL-53882, Lawrence Livermore National Laboratory, Livermore, CA, 1988.

[89] McCormic, S.F. (ed.) [1988]. *Multigrid Methods: Theory,Applications and Supercomputing*, Marcel Dekker, New York.

[90] Mitchell, A.R., and Griffith, D.F. [1980] *The Finite Difference Method in Partial Differential Equations*, Willey, New York.

[91] Morel, J.E., Dendy, J.E., Jr., Hall, M.L., and White, S. W. [1992]. *A Cell-Centered Lagrangian-Mesh Diffusion Differencing Scheme*, Journal of Computational Physics, **103**, (1992), pp. 286-299.

[92] Mukhin, S.I., Popov, S.B., and Popov, Yu.P. [1983]. *On Difference Schemes with Artificial Dispersion*, USSR Comput. Math. and Math. Phys., **23**, No. 6, 1983, pp. 45-53.

[93] Neumann, J. von. and Richtmyer, R. [1950]. *A Method for the Numerical Calculation of Hydrodynamic Shocks*, J. Appl. Phys., 1950, vol. 21, No. 2, pp. 232-257.

[94] Ortega, J.M., and Rheinboldt, W.C. [1970] *Iterative Solution of Nonlinear Equations in Several Variables*, Academic Press, New York.

[95] Patankar, S.V. [1980]. *Numerical Heat Transfer and Fluid Flow*, Hemisphere Publishing Corporation.

[96] Paterson, A.R. [1983]. *A First Course in Fluid Dynamics*, New York, Cambridge University Press.

[97] Peyret, R., and Taylor, T.D. [1983]. *Computational Methods for Fluid Flow*, Springer-Verlag, New York.

[98] Popov, Yu.P., and Samarskii, A.A. [1976]. *Numerical Methods of Solving One-Dimensional Non-Stationary Gas-Dynamic Problem*, USSR Comput. Math. and Math. Phys., **16**, No. 6, 1976, pp. 120-136.

[99] Popov, Yu.P., and Samarskaya, E.A. [1977]. *Convergence of Newton's Iterative Method for Solving Gas-Dynamic Difference Equations*, USSR Comput. Math. and Math. Phys., **17**, No. 1, 1977, pp. 281-286.

[100] Richtmyer, R.D., and Morton, K.W. [1967]. *Difference Methods for Initial Value Problems*, Willey-Interscience. New York.

[101] Roache, P.J. [1976]. *Computational Fluid Dynamics*, Hermosa Publishers, Albuquerque.

[102] Rozdestvenskii, B.,L., and Janenko, N.N. [1983]. *Systems of Quasilinear Equations and Their Applications to Gas Dynamics*, American Mathematical Society, Providence.

[103] Rutherford, A. [1962]. *Vectors, Tensors and Basic Equations of Fluid Mechanics*, Prentice-Hall, Englewood Cliffs, NJ.

[104] Sabitova, A., and Samarskaya, E.A. [1985]. *Stability of Variational-Difference Schemes for the Problems of Gas-Dynamics with Heat Conduction*, Diff. Eqns., **21**, No. 7, 1985, 861-864.

[105] Saltzman, J., and Colella, P. [1985]. LA-UR-85-678, Los Alamos National Laboratory, Los Alamos, NM, 1985.

[106] Samarskii, A.A., and Arsenin, V.Ja. [1961]. *On the Numerical Solution of the Equations in Gas Dynamics with Various Types of Viscosity*, USSR Comput. Math. and Math. Phys., 2, 1962, pp. 382-387.

[107] Samarskii, A.A., and Gulin, A.V. [1973]. *Stability of Finite-Difference Schemes*, (In Russian), Nauka Publishers, Moscow.

[108] Samarskii, A.A., and Andreev V.B. [1976]. *Finite-Difference Methods for Elliptic Equations*, (In Russian), Nauka Publishers, Moscow.

[109] Samarskii, A.A. [1977]. *The Theory of Finite-Difference Schemes*, (In Russian), Nauka Publishers, Moscow.

[110] Samarskii, A.A., and Popov, Yu.P. [1980]. *The Finite-Difference Methods for Problems of Gas Dynamics*, (In Russian), Nauka Publishers, Moscow.

[111] Samarskii, A., and Nikolaev, E. [1989]. *Numerical Methods for Grid Equations*, vol. 1 Direct Methods, vol. 2 Iterative Methods. 1989. Birkhauser Verlag, Basel, Boston, Berlin.

[112] Samarskii, A.A., Tishkin, V.F., Favorskii, A.P., and Shashkov, M.Yu. [1981]. *Operational Finite-Difference Schemes*, Diff. Eqns., 17, No. 7, 854-862.

[113] Samarskii, A.A., Tishkin, V.F., Favorskii, A.P., and Shashkov, M.Yu. [1982]. *Employment of the Reference-Operator Method in the Construction of Finite Difference Analogs of Tensor Operations*, Diff. Eqns., 18, No. 7, 881-885.

[114] *Methods in Computational Physics*, Editors B.Adler, S.Fernbach, and M.Rotenberg, vol. 3, Fundamental Methods in Hydrodynamics, Academic Press, 1964, pp. 1-45.

[115] Shashkov, M., and Steinberg, S. [1995]. *Support-Operator Finite-Difference Algorithms for General Elliptic Problems*, Journal of Computational Physics, 118, (1995), pp. 131-151.

[116] Shestakov, A.I., Harte, J.A., and Kershaw, D.S. [1988]. *Solution of the Diffusion Equation by the Finite Elements in Lagrangian Hydrodynamic Codes*, Journal of Computational Physics, 76, (1988), pp. 385-413.

[117] Schulz, W.D. [1964]. *Two Dimensional Lagrangian Hydrodynamic Difference Equations*, in Methods in Computational Physics, Editors B.Adler, S.Fernbach, and M.Rotenberg, vol. 3, Fundamental Methods in Hydrodynamics, Academic Press, 1964, pp. 1-45.

[118] Schulz, W.D. [1964]. *Tensor Artificial Viscosity for Numerical Hydrodynamics*, J. Math. Phys., vol. 5, p. 133.

[119] Shyy, W. [1994]. *Computational Modeling for Fluid Flow and Interfacial Transport*, New York, Elsevier.

[120] Smith, G.D. [1992]. *Numerical Solution of Partial Differential Equations. Finite Difference Method*, Clarendon Press. Oxford.

[121] Sod, G.A. [1978]. *A Survey of Several Finite Difference Methods for System of Nonlinear Hyperbolic Conservation Laws*, J. Comp. Phys., **27**, (1978), pp. 1-31.

[122] Sod, G.A. [1985]. *Numerical Methods in Fluid Dynamics*, Cambridge University Press, Cambridge.

[123] Solov'ov, A., and Shashkov, M. [1991a]. *Algorithms for Operation with Difference Operators and Grid Function in Symbolic Form*, ICIAM'91 - Second International Conference on Industrial and Applied Mathematics, Washington, D.C., USA Abstract Book, Washington, D.C., USA, 1991, p.193.

[124] Solov'ov, A., and Shashkov, M. [1991b]. *On the One Technique for Free Boundary Simulation in the Dirichlet Particle Method*, ICIAM'91 - Second International Conference on Industrial and Applied Mathematics, Washington, D.C., USA Abstract Book, Washington, D.C., USA, 1991, p.194.

[125] Solov'ova E., and Shashkov, M. [1984]. *Application of the Basic Operator Method for Difference Scheme Construction on Un-Concordant Meshes*, - Preprint Keldysh Inst. of Appl. Math. the USSR Ac. of Sc., **6**, 1984, (in Russian).

[126] Solov'ev, A., Solov'eva, E., Tishkin, V., Favorskii, A., and Shashkov, M. [1986]. *Approximation of Finite-Difference Operators on a Mesh of Dirichlet Cells*, - Differential Equations, **22**, No. 7, 1986, pp. 863-872.

[127] Mikhailova, N., Tishkin, V., Turina, N., Favorskii, A., and Shashkov, M. [1986]. *Numerical Modelling of Two-Dimensional Gas-Dynamic Flows on a Variable-Structure Mesh*, - U.S.S.R. Comput. Maths. Math. Phys., 1986, **26**, No. 5, 1986, pp. 74-84.

[128] Solov'ev, A., and Shashkov, M. [1988]. *Difference Scheme for the "Dirichlet Particles" Method in Cylindrical Coordinates, Conserving Symmetry of Gas-Dynamical Flow*, - Differential Equations, **24**, No. 7, 1988, pp. 817-823.

[129] Shashkov, M., and Solov'ov, A. [1991]. *Numerical Simulation of Two-Dimensional Flows by the Free-Lagrangian Method.* - Report of Mathematisces Institut, der Techniscen Universitat Munchen - TUM-M9105, Mai 1991, 52 p.

[130] Steinberg, S., and Roache, P.J. [1985]. *Symbolic Manipulation and Computational Fluid Dynamics,* J. Comp. Physics, 57, 251-284.

[131] Strang, G., and Fix, G. [1973]. *An Analysis of the Finite Element Method,* Englewood Cliffs, NJ.: Prentice Hall.

[132] Strikwerda, J.C. [1989]. *Finite Difference Schemes and Partial Differential Equations,* Wadsworth & Brooks Cole, Pacific Grove.

[133] Tikhonov, A.N., and Samarskii, A.A. [1963]. *Equations of Mathematical Physics,* Macmillan, New York.

[134] Tishkin, V.F. [1985]. *Variational-Difference Schemes for the Dynamical Equations for Deformable Media,* Diff. Eqns., 21, No. 7, 1985, pp. 865-870.

[135] Thompson, J.F., Warsi, Z.U.A., and Mastin, C.W. [1985]. *Numerical Grid Generation: Foundations and Applications,* North-Holland, Elsevier, New York.

[136] Vinokur, M. [1974]. *Conservation Equations of Gas-Dynamics in Curvilinear Coordinate Systems,* J. Comp. Phys., 14, (1974), 105-125.

[137] Vinokur, M. [1989]. *An Analysis of Finite-Difference and Finite-Volume Formulations of Conservation Laws,* J. Comp. Phys., 81, (1989), 1-52.

[138] Favorskii, A., Korshiya, T., Shashkov, M., and Tishkin, V. [1980]. *A Variational Approach to the Construction of Difference Schemes on Curvilinear Meshes for Heat-Conduction Equation,* U.S.S.R. Computational Mathematics and Mathematical Physics, 20, No. 2, 1980, 135-155.

[139] Favorskii, A.P., Korshiya, T., Shashkov, M., and Tishkin, V. [1982]. *Variational Approach to the Construction of Finite-Difference Schemes for the Diffusion Equations for Magnetic Field,* Differential Equations, 18, No. 7, 1982, pp. 863-872.

[140] Shashkov, M. [1982]. *Violation of Conservation Laws when Solving Difference Equation by the Iteration Methods,* U.S.S.R. Computational Mathematics and Mathematical Physics, 22, No. 5, 1982, 131-139.

[141] Shashkov, M., and Shchenkov, I. [1986]. *The Use of Symbolic Transformation to Construct and Study Difference Schemes*, U.S.S.R. Computational Mathematics and Mathematical Physics, **26**, No. 3, 1986, 73-79.

[142] Liska, L., and Shashkov, M. [1991]. *Algorithms for Difference Schemes Construction on Non-Orthogonal Logically Rectangular Meshes*, IS-SAC'91 - International Symposium on Symbolic and Algebraic Computation, Bonn, Germany. Proceedings of ISSAC'91 - International Symposium on Symbolic and Algebraic Computation, Bonn, Germany, (S. Watt, Editor) ACM Press, 1991, pp. 419-429.

[143] Liska, R., Shashkov, M., and Solovjov, A. [1993]. *Support-Operators Method for PDE Discretisation: Symbolic Algorithms and Realization*, Mathematics and Computers in Simulation, **35**, (1993), 173-184.

[144] Stoer, J., and Bulirsch, R. [1980]. *Introduction to Numerical Analysis*, Springer-Verlag, New York.

[145] Strang, G. [1986]. *Introduction to Applied Mathematics*, Wellesley-Cambridge Press, Wellesley, Massachusetts.

[146] Varga, R.S. [1962]. *Matrix Iterative Analysis*, Prentice-Hall, Englewood Cliffs, NJ.

[147] Volkova, R., Kruglakova, L., Misheskaya, E., Tishkin, V., Turina, N., Favorskii, A., and Shashkov, M. [1985]. *SAFRA. Functional Filling. The Program for Solving 2-D Problems of the Controlled Laser Fusion. Manual.* - Keldysh Inst. of Appl. Math. the USSR Ac. of Sc., 1985, 63 p., (In Russian).

[148] Warsi, Z.U.A. [1993]. *Fluid Dynamics. Theoretical and Computational Approaches*, CRC Press, Boca Raton, Florida.

[149] Weinberger, H.F. [1965]. *A First Course in Partial Differential Equations*, John Willey, New York.

[150] Wilkins, M.L. [1964]. *Calculation of Elastic-Plastic Flow*, in Methods in Computational Physics, Editors B.Adler, S.Fernbach, and M.Rotenberg, vol. 3, Fundamental Methods in Hydrodynamics, Academic Press, 1964, pp. 211-264.

[151] Wilkins M.L. [1980]. *Use of Artificial Viscosity in Multidimensional Fluid Dynamic Calculations*, Journal of Computational Physics, **36**, (1980), pp. 281-303.

[152] Young, D.M. [1971]. *Iterative Solution of Large Linear Systems*, Academic Press. New York.

[153] Warming, R.F., and Hyett, B.J. [1974]. *The Modified Equation Approach to the Stability and Accuracy Analysis of Finite-Difference Methods*, J. Comp. Phys., **14**, (1974), 159-179.

[154] Yanenko, N.N., and Shokin, Yu.I. [1969]. *First Differential Approximation Method and Approximate Viscosity of Difference Schemes*, Phys. Fluids, suppl. II, vol. 12, 1969, pp. II-28-II-33.

Index

Accuracy, 12, 20
Acoustic approximation, 212
Adjoint operator, 56
Algebraic grid generation, 29
Angle of the cell, 31
Anisotropic conductivity, 311
Approximation properties,
    1-D, C-N discretisation, 99
    1-D, N-C discretisation, 79
Approximation viscosity, 263
Artificial viscosity, 229, 263, 302

Block difference operators, 36
Block Gauss-Seidel method, 185
Body sources of energy, 202
Boundary conditions
    in Lagrangian
    variables, 209

C-N Discretisation, 64, 237
Cell 7, 31
Cell-valued discretisation
    in 2-D, 32
Cell-valued discretisation, 11
Characteristic form of
    gas dynamics equations, 217
    of hyperbolic equations, 216
Characteristics, 216
    for acoustics, 214
    of gas dynamics
    equations, 217
Collapse of quasi-spherical target
    in a hard cone, 312
Compression of a toroidal plasma
    by quasi-spherical
    liner, 312

Computer algebra, 312
Conditional stability, 157, 254
Conservation law for heat
    equation, 149
Conservation laws and iteration
    method, 176, 301
Conservative FDSs, 47, 50
Consistency, 21, 22
Consistent discrete operators
    in 1-D, 53
    in 2-D, 57
Contact discontinuity, 223, 261
Continuity equation, 51
Continuous arguments, 6
Continuum elliptic problems
    with Dirichlet boundary
    conditions, 65
    with Robin boundary
    conditions, 68
Continuum hypothesis, 195
Controlled laser fusion, 312
Convergence theorems, 312
Convergence, 20, 22
Courant condition, 258
Criteria of stopping of iteration
    process, 83

Density of heat flux, 202
Density, 196
Derivative following the
    motion, 5
Derived operators, 49, 51
    GRAD for C-N
    discretisation, 64
    DIV for N-C
    discretisation, 73

Difference analogs
    of integral identity, 51
    of the second order differen-
        tial operation in 1-D, 56
Difference equations, 6
Difference operators in 2-D, 32,
    34
Dirichlet boundary conditions, 3,
    19, 65
Discontinuous solutions, 220
Discrete analog
    for the conservation of
        momentum, 241
    for the conservation of total
        energy, 241
    of norms, 12, 34
Discrete norms in 2-D, 34
Discrete scalar functions, 32
Discrete vector functions, 36
Divergence of a tensor, 311
Divergence property, 50, 51
    of operator divergence, 236
    of operator gradient, 235
Divergent form of discrete
        operators, 52
Domain of dependence, 220

Effective linear size, 181
Effective width of
        the shock front, 231
Eigenvalue, 215
Elliptic equations, 1, 65
Energy equation,
        different forms, 210
Entropy, 198
    for ideal gas, 211
    trace, 265
Equation of gas dynamic
    in Lagrangian form, 4
    in Eulerian form, 311
Equations of magnetic field
        diffusion, 311
Equations of state, 5, 199
Error, 12, 20

Eulerian Approach, 200
Explicit FDS for
        heat equation, 151
External bodies, 197
External parameters, 197

Fictitious cells, 239, 247, 253
Fictitious nodes, 239, 271
Finite-difference analogs of
        second derivatives, 18
Finite-difference operators, 13
    and the related matrix, 23
Finite-difference scheme, 6
    and related matrix, 24
First thermodynamics
        principle, 198
Fluid cut, 208
Fluid particle, 199
Fluid volume, 201
Flux form
    of Robin BC, 71
    of boundary value
        problem, 71
    of discrete equations, 179
Free boundary, 5, 239
Free boundary
        conditions 195, 272
Fully implicit FDS for
        heat equation, 152
Function of discrete arguments, 6
Functional space, 6

Gas constant, 199
Gas dynamic equations, 51
General theory of stability, 155
Gradient of vectors, 311
Green formula, 34, 59, 132
Grid function, 6, 7
Grid, 5
    in 1-D, 7
    in 2-D, 26
    regularity, 7, 9, 32
    with local refinement, 312

Half indices, 11

Heat conductivity, 202
Heat equation, 3
Heat flux, 50
High-order FDS, 312
Homogeneous FDS, 260, 263
Hugoniot adiabat, 224
Hugoniot relations, 222
Hyperbolic system of
        equations, 215

Ideal gas, 199
Imbalance in amount of heat, 178
Impermeability and free
        boundary
        conditions, 195
Impermeability boundary
        conditions in 2-D, 272
Implicit FDS
    for heat equation, 151
    for gas dynamics
        equations, 248
Improvement of approximation
    on the solution of
        PDE, 22, 82
Initial conditions, 149
Inner product
    for Robin BC, 68
    in discrete spaces, 56, 58, 72
Integer indices, 11
Integral average, 12
Integral identity, 50, 66
    in 1-D, 53
    in terms of first
        derivatives, 57
    in terms of inner products in
        2-D, 58
Invariant differential
        operators, 47
Isoentropic flows, 211
Isothermal flows, 211
Isothermal speed of sound, 214

Lagrange mass
        coordinate, 207, 209
Lagrangian approach, 200

Lagrangian variable, 4, 200
Landshoff viscosity, 263
Laplace equation, 1
Law of conservation of
        mass, 51, 201, 236
        momentum, 51, 52, 201, 235
        full energy, 51, 52, 201, 235
Law of heat conservation, 49
Law of variation of volume, 234
Left eigenvector, 215
Linear algebraic equations, 20, 37
Linear viscosity, 231, 263
Local basis systems, 311
Logical space, 9
Logically rectangular grid, 27

Magnetic field in
        a spiral band reel, 312
Mapping between nodal and cell-
        valued functions
    in 1-D, 240
    in 2-D, 274
Mapping, 29
Mass of cell, 237
Mass of node, 247
Mass speed of sound, 214
Matrix form of difference
        operators, 35
Matrix problem, 20
    for Dirichlet BC, 1-D, N-C
        discretisation, 77
    for Robin BC, 1-D, N-C
        discretisation, 78
Maximum principle, 159
Maxwell equations, 311
Method of finite-differences, 5
Method of frozen coefficients, 158
Method of harmonics, 256
Method of parallel chords, 292
Microwave plasma generator, 312

N-C discretisation in 2-D, 59
Neumann boundary conditions, 3
Neumann's method for the inves-
        tigation of stability, 155

Neumann and Richtmyer
        viscosity, 263
Newton method, 41, 249, 288
Newton-Kantorovich method, 42
Nodal discretisation, 9, 19
        in 2-D, 32
Node, 7
Non-conservative form of
        the energy equation, 233
Non-conservative iteration
        process, 176
Non-smooth grid, 8
        in 1-D, 7, 9
        in 2-D, 31
Non-uniform rectangular grid, 26
Norm, 6

Operator form
        of Dirichlet boundary value
        problem, 66
        of Robin boundary
        value problem, 68
Over-compressed detonation
        wave in conic
        channel, 312

Physical continuum, 195
Physical space, 9
Poisson adiabat, 227
Poisson equation, 2
Positive definite matrix, 39
Positive matrix, 24
Pressure, 196
Prime operator, 49, 51, 53
        DIV, 64
        GRAD, 59
        for Lagrangian
        gas dynamics, 236
Principle of frozen
        coefficients, 258
Projection operator, 12, 15
Properties of first-order
        operators, 67
Pseudoviscosity method, 193
Purely explicit FDS

in 1-D, 244
in 2-D, 278

Quadratic viscosity, 232, 263

Radius-vector, 196
Range of influence, 220
Rarefaction shocks, 228
Rayleigh-Taylor instability, 312
Realization of BC
        in 1-D explicit FDS, 247
        in 1-D implicit FDS, 253
        in 2-D, explicit FDS, 281
        in 2-D, implicit FDS, 293
Residual, 20
Residual on the solution of
        PDE, 21
Riemann's invariants, 218
Robin boundary conditions, 3, 19,
        65, 68

Saltzman test problem, 306
Scheme viscosity, 229
Second principle of
        thermodynamics, 198
Second-order approximation for
        Robin BC for heat
        equation, 164
Semi-discrete FDS, 240, 272
Shock waves, 224
        in an ideal gas, 224
Side of the cell, 31
Smooth grid, 8, 9
Smooth and non-smooth grid, 87
Smooth transformation, 17
Speed of sound, 214
Square grid, 27
Stability, 22
Stability conditions
        for 1-D acoustic
        equations, 259
        for heat equation in 2-D, 180
        of the FDS for heat
        equation, 154

with respect to initial
    data, 154
with respect to the
    right-hand side, 154
Stationary heat transfer, 2
Stencil of discrete operators, 18,
    13, 35
Strong discontinuity,
    shock wave, 261
Strong shock wave, 228
Structure of the shock front, 229
Substantive derivative, 5
Support-operators method, main
    stages, 48
Symmetric matrix, 24, 39
Symmetry preserving FDSs and
    iteration methods, 181

Temperature, 198
Tensor product grid, 26
The elliptic equations, 49
Thermodynamic equilibrium
    process, 198
Time step, 150
Total amount of heat, 50
Trapezium rule, 59
Travelling waves, 213, 255
Triangular grids, 312
Truncation error, 13, 15, 18, 22
    on non-smooth grid, 16
    on smooth grid, 17
Two level FDS, 152

Unconditionally stable FDS, 254
Unconditionally unstable
    FDS, 254
Uniform grid
    in 1-D, 7
    in 2-D, 27

Variable time step, 150
Vector of normal in corners, 146
Velocity, 196
Viscous pressure, 229
Volume of node, 54, 72

Volume of the cell, 54
Voronoi grids, 312

Weak discontinuity, 261
Work, 197

Zemplen Theorem, 227

Printed and bound by CPI Group (UK) Ltd, Croydon, CR0 4YY

23/10/2024

01778224-0015